LEGALLY POISONED

LEGALLY POISONED

HOW THE LAW PUTS US AT RISK
FROM TOXICANTS

Carl F. Cranor

HARVARD UNIVERSITY PRESS

Cambridge, Massachusetts

London, England

2011

Copyright © 2011 by the President and Fellows of Harvard College
All rights reserved
Printed in the United States of America

Library of Congress Cataloging-in-Publication Data
Cranor, Carl F.
  Legally poisoned : how the law puts us at risk from toxicants / Carl F. Cranor.
    p. cm.
  Includes bibliographical references and index.
  ISBN 978-0-674-04970-3 (alk. paper)
  1. Environmental health—Government policy—United States. 2. Environmental
toxicology. I. Title.
  [DNLM: 1. Environmental Pollutants—adverse effects. 2. Environmental Pollutants—
toxicity. 3. Environmental Exposure—adverse effects. 4. Environmental Exposure—
legislation & jurisprudence. 5. Environmental Pollution—legislation & jurisprudence.
6. Public Health—legislation & jurisprudence. WA 671 C891L 2010]
  RA566.3.C73 2011
  363.73'63—dc22   2010025826

To Crystal, Chris, and Taylor

CONTENTS

ADHD attention deficit hyperactivity disorder
Ah receptor aromatic hydrocarbon receptor
ATSDR Agency for Toxic Substances and Disease Registry
BPA bisphenol A
CAA Clean Air Act
CDC Centers for Disease Control and Prevention
CNS central nervous system
CPSA Consumer Product Safety Act
CPSC Consumer Product Safety Commission
CWA Clean Water Act
DDE dichlorodiphenyldichloroethylene
DDT dichlorodiphenyltrichloroethane
DEHP di(2-ethylhexyl)phthalate
DES diethylstilbestrol
DOHaD Developmental Origins of Health and Disease
ED endocrine disrupter(s)
EPA U.S. Environmental Protection Agency
FDA U.S. Food and Drug Administration
FDCA Food, Drug, and Cosmetic Act
FIFRA Federal Insecticide, Fungicide, and Rodenticide Act
HD Helsinki Declaration
IARC International Agency for Research on Cancer
MTBE methyl tertiary butyl ether
MeHg methylmercury
MPTP methylphenyltetrahydropyridine
NRC National Research Council

NIEHS	National Institute of Environmental Health Sciences
NC	Nuremberg Code
OSH Act	Occupational Safety and Health Act
OSHA	Occupational Safety and Health Administration
PAH	polycyclic aromatic hydrocarbons
PBBs	polybrominated biphenyls
PBDEs	polybrominated diphenyl ethers
PCBs	polychlorinated biphenyls
PD	Parkinson's disease
PFCs	perfluorinated compounds
SDWA	Safe Drinking Water Act
tOP	4-tertiary-octylphenol (a degradation product of BPA)
TCA	trichloroacetic acid, a breakdown product of trichloroethylene
TCE	trichloroethylene
TSCA	Toxic Substances Control Act
TURI	Toxic Use Reduction Institute

LEGALLY POISONED

1

Take a random walk through your life; it is awash in industrial chemicals, many toxic. Of course, we can all be aware that we are exposed to smog in the Los Angeles Air Basin or to a fog of pesticides being sprayed from a crop duster or to probably unhealthy water from a river contaminated with factory wastes. Most exposures are not obvious, however.

Our exposures, understood as substances reaching the boundaries of our bodies, are often much more subtle, invisible, and otherwise undetectable. Drink water or soda from a plastic bottle or eat canned refried beans or soup, and you ingest bisphenol A (BPA). When a dentist uses a plastic dental sealant for your children's teeth to prevent decay or a composite product to fill their cavities, each likely contains BPA. This compound was created in 1891, proposed as a synthetic estrogen in the 1930s (but not pursued), and eventually used to harden many plastics.[1] More than ninety-percent of us have been exposed to BPA. One serving from some cans of food has sufficient BPA to cause adverse health effects in experimental animals.

If you use cosmetics or body lotions, you may absorb lead (in some lipsticks), phthalates, or other products through your skin. Phthalates enter the womb and they have been found in amniotic fluid. Children have higher concentrations in their bodies than adults. Phthalates may contribute to premature breast development, as well as reproductive effects in males, including feminization (retained nipples), infertility, and undescended testes. Lead is a potent neurotoxicant, adversely affecting learning, IQ, and even behavior. It also contributes to cardiovascular disease.[2]

Prepare your dinner in a nonstick frying pan, and you are exposed to perfluorinated compounds (PFCs) from Teflon. Spray your furniture with Scotchgard and produce a similar result. Ninety-eight percent of the U.S. population is exposed to these substances. The PFCs are quite persistent and may stay in your body for most of a decade and much longer in the environment, from which you can be reexposed. Perfluorinated substances were once thought to stable for hundreds of years, but recent research suggests that they may breakdown more rapidly, releasing toxic by-products.[3] They appear to affect neurological development and are a likely human carcinogen.

Many of your couches, chairs, carpet pads, drapes, car seats, televisions, and computers have a brominated flame-retardant (polybrominated diphenyl ethers, or PBDEs) in them. The PBDEs entered the market only in the 1970s, but already millions of pounds are used in the United States. We are all in contact daily with them. Little is understood about their health effects, because legally they could enter commerce without toxicity testing. However, recent studies suggest PBDEs delay neurological development in children and affect the reproductive system by delaying pregnancy. PBDEs' persistence and adverse effects in animals raise many concerns.[4]

Quite ordinary activities bring you into contact with industrial chemicals. When you drink tap water, you likely ingest methyl tertiary butyl ether (MTBE), a gasoline additive. A high percentage of us has detectable body levels.[5] Its toxicity has not been fully determined, but animal studies show it causes several cancers, which should be of concern.[6]

Eat steak for dinner and ingest up to six hormones used to accelerate weight gain in beef. Twelve percent of cattle contain these residues. The European Union (EU) found that there is no safe level of these substances for fetuses. A serving of steak also contains fat-soluble toxicants—long-banned pesticides (DDT (dichlorodiphenyltrichloroethane), chlordane, dieldrin and others), industrial fluids (polychlorinated biphenyls, (PCBs)), and dioxins (contaminants from industrial processes), as well as the more recent PFCs and PBDEs. Concentrations of individual contaminants are low, but their toxic effects can be additive.[7]

We are not merely exposed to most of these compounds, however. Each of us is in reality contaminated by hundreds of substances. That is, industrial chemicals, pesticides, cosmetic ingredients, and other products enter our bodies and then reach and infiltrate our tissues, organs, and blood. They become part of our body burden of chemicals. The Centers for Disease Control (CDC) has developed techniques for detecting them by measuring the amounts in our blood or urine; this process is called biomonitoring.

2

To date, the CDC can reliably identify about 212 substances in our bodies; this number will only increase as protocols for detecting them are developed and refined. The CDC is investigating these compounds because they represent substantial exposures or are known or suspected to be toxic hazards. That is, most have intrinsic toxic properties, or what experts have called "built-in abilit[ies] to cause an adverse effect."[8] (Whether they will pose risks depends upon their concentrations and other factors.)

These substances include, among other compounds, fire retardants (the PBDEs), industrial insulating and cooling compounds (PCBs), gasoline additives (MTBE), discarded rocket fuel components (perchlorate), lead, arsenic, many pesticides, and perfluorinated compounds, as well as plastic-hardening agents and plastic-softening agents. All of these chemicals have penetrated our kidneys, lungs, liver, bones, and in some instances even our brains. The CDC's list includes known or probable human carcinogens; estrogen-mimicking substances; other hormone-disrupting chemicals; thyroid disrupters; developmental toxics; neurotoxicants, which can damage the nervous system and brain; and immunotoxicants, which can damage the immune system.

We cannot escape contamination. Kim Radtke, a pregnant mother-to-be, was determined to do everything right to protect her developing child from worrisome toxicants. She had eaten organic foods for some time, avoided deodorants and scented lotions, and exercised regularly. Yet when she was tested for twenty-three industrial chemicals in a study of ten pregnant women, she was shocked that many of them burdened her body. She was the most contaminated among the small group tested.[9]

Radtke's son, Karson, entered the world already tainted with industrial chemicals. Radtke nursed him because of breast feeding's importance to his development, but in the process she off-loaded some of her hazardous body burden into him.[10] Unfortunately, nursing one's child has become much more of a risk-benefit calculation than it should be. Yet her contamination is not unusual but routine.

Some invaders are naturally occurring elements, for example, lead and mercury, to which humans have been exposed for centuries. Other substances were not present before they were deliberately synthesized for better living through chemistry, such as PCBs, PBDEs, MTBE, DES (diethylstilbestrol), BPA, PFCs, and many pesticides. Sometimes products directly contaminate us; at other times we receive secondary contamination from the environment.

3

Some industrial chemicals are persistent, likely to remain in our bodies for years. The flame retardants can reside in fatty tissues for several years. There they unite with long-banned PCBs. Only about half of the PCB bodily contamination disappears in eight years. In addition, perfluorinated substances have joined the club of long-lasting substances in our bodies with a half-life of up to seven years.[11] This team of three can jointly contribute to similar adverse effects and has considerable time to affect our health adversely.

Bisphenol A and other substances can be quite transient in our bodies. Nonetheless, the CDC found that about 93 percent of people older than six were contaminated by BPA. Children have the highest exposures.[12] This suggests that we have near continuous exposure from numerous sources. Solvents, such as MTBE, or plastic additives such as phthalates, also have a short half-life. Length of residence in our tissues, however, does not determine the risks from these substances.

Both persistent and short-lived substances can pose live or heightened risks to each of us during development. These risks are "live" because of children's contamination by known *toxicants* and their greater vulnerability during development. Most substances in a pregnant woman's body will cross the placenta and defile her developing child. Many compounds will enter her breast milk, be transmitted to her nursing child, and, as Kim Radtke worried, be off-loaded from mother to child. This points to a broader concern. A study of ten newborn babies found more than 200 industrial chemicals in their umbilical cords. Some news reports suggest even greater contamination. Children begin life sullied by industrial compounds and immediately add to them from external exposures.[13]

Moreover, developing children are particularly susceptible to disease, dysfunction, or premature death as a result of prenatal or early postnatal contamination. They also tend to be more highly exposed than adults and have lesser defenses against toxicants. Children are, thus, at live risk for adverse effects and at greater risk than adults.

Lead, mercury, diethylstilbestrol, thalidomide, pesticides, anticonvulsive drugs, sedatives, arsenic, tobacco smoke, alcohol, and radiation are known developmental toxicants. Two hundred known human neurotoxicants are likely toxic to developing children as well. Numerous other products are of considerable concern, based on experimental animal studies on substances such as PBDEs, BPA, phthalates, pesticides, and cosmetic ingredients. There could be as many as one thousand additional neurotoxicants that could pose health problems. No lowest concentration level appears to be safe for many carcinogens, lead, tobacco smoke, and radiation.[14]

Some developmental toxicants—such as thalidomide and anticonvulsive drugs—cause immediate and noticeable effects, such as birth defects. Others have long-delayed, often less visible adverse consequences—lead can lessen children's IQ and contribute to aggressive behavior. In utero exposures to neurological toxicants, such as some pesticides, appear to hasten Parkinson's disease (and perhaps dementia) after a delay of forty to fifty years. Lead appears to have similar effects. Estrogen-mimicking substances may lead to early onset of cancer, increase the chances of its occurring, or contribute to other diseases, but years after exposure. BPA, for instance, can cross the placenta, become more toxic in the fetal environment, change how important genes are expressed, and damage placental cells. Experimental studies indicate that it can contribute to breast, uterine, and prostate cancer as well as obesity and diabetes.[15]

These contributors to disease are newly being understood. Many of the diseases that once ravaged the United States have disappeared. Public health measures eliminated numerous scourges—for example, cleaning up water and sewage reduced cholera and typhoid fever—while vaccinations reduced or eliminated others, such as smallpox, polio, measles, and mumps. Children born today are likely to live up to two decades longer than recent ancestors.[16]

However, there remain noninfectious illnesses of increasing concern, including, as Bruce Lanphear points out, "cardiovascular disease, diabetes, obesity, respiratory ailments, and injuries." And various morbidities affect children—"intellectual impairments" like lowered IQ, poor memory, mental retardation, autism, attention deficit hyperactivity disorder (ADHD), and "behavioral problems, asthma, and preterm birth—[that] are linked with remarkably low-level exposures to . . . noninfectious environmental factors or gene-environment interactions."[17] These must be addressed with better public health protections.

Since diseases typically result from a combination of environmental and genetic influences, and since our genetic sequences do not change rapidly, researchers are looking much more closely at environmental influences, especially molecular contributions. This research focuses on what is called the "developmental origins of health and disease." Scientists believe that some contaminants, while not changing a person's genome, modify how it functions, or "expresses itself," thus leading to disease. These are "epigenetic" effects.[18]

5

Although some activities like smoking, drinking, using illegal drugs, and working in industries with toxicants create obvious developmental exposures, we largely do not knowingly and deliberately invite toxicants into our bodies. Rather, even if we seek to avoid them, industrial chemicals and pesticides still invade us; that is a matter of biology and chemistry. Our laws and institutions permit our contamination by toxic substances and add to our risk of disease.

Many people believe that public health agencies protect them from toxicants, that indeed they have a moral duty to do this. They likely believe that most public health laws require product testing and then review of the tests and products for toxic risks before citizens are exposed, just as the Food and Drug Administration requires testing and review of pharmaceuticals. Unfortunately and sadly, this is mistaken. In fact, the vast majority of substances are subject to "postmarket laws." These laws do not require testing or a public health agency's reviewing of industrial compounds for toxicity before they enter the market. Thus the substances governed by such laws are not necessarily free of risks, especially for the most biologically sensitive among us.

The predictable result is that little is known about large percentages of chemical products in commerce.[19] Only about two percent of 62,000 substances in commerce before 1979 have been reviewed at all for their toxicity by the U.S. Environmental Protection Agency (EPA). Of the approximately 50,000 new substances introduced since 1979, about eighty-five percent had no data concerning health effects; even the chemical identity of a majority of them is veiled by claims of confidential business information. At current rates, testing these products after they are already in commerce could take hundreds of years.[20] At the same time it appears that many industrial chemicals are more toxic than might appear. For instance, for most new substances proposed for manufacturing the EPA has no data on their ability to cause genetic mutations, which can be determined by a relatively simple and reasonably accurate test. Yet a sampling of one hundred chemicals in commercial use and a sampling of forty-six chemicals produced in more than a million pounds revealed that about twenty-percent of each group were mutagenic. When a substance is mutagenic it is also highly likely to be a mammalian carcinogen.[21]

For the overwhelming number of substances, the doors from the factories into commerce are wide open, permitting possibly toxic products to contaminate us. For instance, PBDE flame retardants began to be used in commerce in the 1970s without any review of their toxicity. Before that date Americans essentially had no PBDEs in their bodies. Yet there are now millions of pounds in commerce, and U.S. citizens have the highest bodily concentrations in the world.[22] Moreover, PBDEs are remarkably similar in chemical structure and toxicity to long-banned PCBs. This similarity was not noticed or was ignored, or their use was mistakenly "grandfathered in." Such an outcome is not surprising, since the pertinent laws required no toxicity data about new substances or those in commerce.

As new kinds of technologies are created, there will be even more compounds of unknown toxicity under existing laws. For instance, there is a burgeoning industry in nanotechnology. It seeks to develop products for electronics, medicine, and even pollution control with new chemical structures that are larger than a molecule but smaller than a biological cell. They range in size from one to one hundred nanometers in size (a nanometer is $1x\ 10^{-9}$ meters). A PCB molecule is about 1 nanometer. From 2005 to 2009 the number of such products increased from about fifty-four to more than one thousand. Some are long, rigid, fiber-like tubes similar to asbestos, while others can be tiny particles. Both nanotubes closely resembling asbestos and some small particles have exhibited worrisome toxicity properties. Some nanoparticles have been found to cross the placenta and titanium oxide nanoparticles have been found to cause genetic and chromosomal damage in experimental studies.[23]

Postmarket laws, thus, are reckless and permit companies to be reckless toward adults and children. That is, these policies appear to disregard possible undesirable consequences. And they permit the creation and commercialization of products without sufficient attention to undesirable consequences.

In addition, under postmarket laws a public health agency must carry a considerable legal and scientific burden of proof to secure better health protections once products are suspected of being toxic. The science must be found or generated and must be legally sufficient to support better health protections. Public health officials must have the political will, sufficient staff and adequate resources to act on the science. All can be in short supply. These laws in effect put us at risk because no toxicity data is required about substances, and they then erect substantial barriers to making us safer if there are concerns about the risks of products. And they can be manipulated to forestall health protections. If substances pose risks, any science seeking to reveal them begins well after contaminations have occurred, leaving the public at risk while the science is developed.

The features and the use of scientific studies thus become an essential part of the story that follows. It takes considerable time to detect many diseases considered here. Establishing threats is even more complicated because we obviously cannot test unknown substances on people in labs, but of course we are the guinea pigs right now. Animal studies, early and good indicators of risks, are likely to be denigrated, further delaying health protections. Long lag times to identify risks invite parties responsible for the exposures to exploit these issues to disavow responsibility for them. As a result, legal efforts to protect the public typically become mired in procrastination, obfuscation, and endless disputes about the science and law.

The law permits these problems, but it also holds the answer. Laws governing pharmaceuticals and pesticides do not permit human exposures, even for volunteer subjects, without appropriate testing for the risks of these products. The ethics of medical experiments do not permit people to volunteer before sufficient preparatory research has identified risks and ensured that any risks are reasonable. It is forbidden to experiment on children, unless the data is essential to their well being and there is no other way to obtain it.

In sharp contrast, exposures to the vast majority of chemicals have none of these protections. People, and worse their children, become experimental subjects for industrial chemicals. If substances pose significant or subtle risks, their discovery will likely occur long after exposure has occurred and after some damage has been done.

Moreover, no matter how careful we are about our contamination and our health, the actions and products of others beyond our control can affect us. Atomistic self-help based on individual choices—independent actions on which Americans may pride themselves—cannot solve these problems. It can do little to protect those who try to avoid contamination.

We can, however, design and create a less risky world with sensible legal and regulatory safeguards to protect our health. Those who invent and manufacture industrial chemicals should be legally required to conduct their activities with greater foresight, responsibility, and prudence. They should fund toxicity testing of their products before they are manufactured and commercialized. Such research should focus on developmental and other adverse risks during sensitive life stages. Testing should also aim to identify early the risks of cancers, neurological disorders, lifelong illnesses from impaired immune systems, heart disease, various morbidities due to ill-timed toxicants, and so on. Tests should address the additive effects of substances because chemical soups inhabit our bodies and researchers have identified such patterns.[24] Substances that will have widespread exposures or that will be persistent should receive special attention. The goal should be to provide reasonable certainty that no harm would be done to the public.

In such a world toxic contamination would be reduced, but what else might change? Existing laws have extensive hidden personal, health, social, and economic costs not fully reflected in the price of current products, and these less visible costs would be greatly reduced with better testing. Because companies do not test products, others must engage in a variety of costly and time-consuming activities in order to determine risks and protect themselves. Economists call these hidden costs "externalities."

When companies do not determine the toxicity of their products, taxpayers must pay for public health agencies to determine the toxicity of their bodily contaminants. And, people must educate themselves about the risks of foods, cosmetics, beverage and food containers, furniture, computers, building materials, and neighborhoods. This takes considerable information gathering, effort, and money. Purchasing less risky products is often more expensive. Many rightly worry about their and their children's health. Any resulting adverse health effects are costly: birth defects; retarded or impaired mental development, decreased IQ, poor memory, antisocial behavior, and other neurological deficiencies; early onset of vaginal, breast, prostate or bone cancers; and early parkinsonism or perhaps Alzheimer's disease.

Externalities also reflect moral shortcomings. Companies fail to take full responsibility for their products and their consequences. Currently, businesses have a legal right to contaminate people, who then remain at risk until a burdensome legal case provides better health protections by forcing reduction or elimination of toxicants from products.

As a nation, we should aim for every person to be born healthy insofar as institutions can achieve this goal, as the National Institute of Child Health and Development affirms.[25] This would be part of any reasonable conception of justice. Present institutions fall far short of this moral requirement when people's diseases, dysfunctions, or deaths are the result of poor legal control of toxic substances. Correcting such injustices is also difficult. If a chemical product in fact harms someone, there can easily be too little data to establish a claim to correct the injustice.

While some costs may increase with better testing, these must be properly weighed against existing product costs, risks, and associated externalities. Low estimates of some costs of disease attributable to environmental exposures are about $55 billion per year, lower than the costs of health care due to car accidents but greater than those resulting from strokes.[26] Moreover, the European Union's REACH (Regulation on Registration, Evaluation, Authorisation, and Restriction of Chemicals) legislation suggests a useful comparison. According to an economic analysis of REACH, a high-end estimate for the cost of testing about 30,000 new and existing chemicals is about 3.5 billion euros which is approximately $4.5 to $5 billion. The monetary costs of better testing appear to be much lower than the costs of disease, and perhaps considerably lower than some people might worry. When balanced against monetary and hidden costs of the status quo, increased testing seems quite reasonable, as well as being required by safety and justice.[27]

Even in world with more sensible legal and regulatory safeguards to protect us from toxic products and pollutants, citizens will be contaminated. Molecular contamination per se is not the issue. Contamination by toxicants and reckless contamination by substances of unknown toxicity are the problems. The law permits this; the law can rectify it.

The message of this book can be disconcerting, perhaps depressing, but that may not be the best psychological response. This story with its many dimensions must be understood, whatever our initial reaction. We should be grateful to the scientific community whose research is revealing adverse health effects. We might be incensed at companies for being reckless toward our health and at Congress and public health agencies for failing to protect us. A reasonable reaction might resemble that of Molly Gray, a mother-to-be from Seattle, Washington, "Something is wrong when I, as an educated consumer, am unable to protect my baby from toxic chemicals. I and all other parents should be able to walk into stores and buy what we need without winding up with products that put our families' health at risk. Now that I've learned that companies can put chemicals into products without ever testing for whether they harm our health, I think we need to change our laws."[28]

These issues are developed in the chapters that follow. A deeper understanding of exposure and contamination, as well as of a variety of contaminants, provides background for the scope of the problems of chemical contamination (see Chapter 2). Some substances are well known and have long posed problems—lead, for example—for centuries. Others are of quite recent concern. Some persist for years or decades; others are quite transient. However, whatever their properties, virtually all of these substances will invade our tissues by multiple routes and contaminate us by ways we do not choose. There is no escape. Evidence does suggest, however, that banning even persistent substances over time considerably reduces contamination by them and some short-lived substances will rapidly leave our bodies once exposures cease.

Identifying toxic contributions to diseases and dysfunction is quite unlike causal inquiries into the macroworld of car accidents, airplane crashes, or colliding billiard balls. For toxicants, subtle scientific studies are needed but their strengths and limitations must be understood and particular results scrutinized. Studies reveal hazards but can also be presented in ways that mislead the public (see Chapter 3). Human studies can be so insensitive that adverse effects are missed or underestimated. Advocates whose products may pose risks have commercial temptations to manipulate studies or their reports to falsely reassure the public that products are safe. More seriously, if human data are required before increased health protections are instituted, some people must suffer disease, dysfunction, or death in order to provide sufficient evidence to protect others. This is poor public health policy. Experimental animal studies can provide warnings of harmful effects to people from in utero exposures. Yet this research is often denigrated because animals are not humans and perhaps because conclusions are inconvenient. If the acceptability of animal studies is undermined, our children and we will remain at risk. Citizens should be empowered to understand these issues.

Scientists are quickly coming to understand the extent to which developing children can be susceptible to disease and dysfunction, but this is likely not broadly appreciated. Research into the developmental origins of disease likely will revolutionize some of our ideas of health and disease (see Chapter 4). Most industrial chemicals that contaminate women's bodies can cross the placenta as well as taint breast milk, reaching the tissues of developing fetuses or of newborns. Newborns enter life contaminated with substances of known and unknown toxicity. This puts them at live risk for any toxicants that are present (and typically at greater risk than adults). Substantial harms have already been documented from early life exposures, while experimental studies suggest concerns about a much broader range of substances than those that are currently considered risky.

Ubiquitous contamination and the vulnerability of developing children have profound implications for the law. Postmarket environmental health laws permit products into the market without testing and review of their toxicity (see Chapter 5). We are ignorant of the toxicity of the vast majority of these products. Studies begun after untested products are commercialized are simply too late to protect contaminated children. More fundamentally, postmarket laws appear misconceived to address toxicants that invade the human body.

Such laws so poorly protect our children that they become test subjects for commercial chemicals. This would be completely unacceptable for a medical experiment or for testing a drug or pesticide under existing laws, yet such experimentation is precisely what current law permits. When more protective health standards are needed, unreasonable demands for more certain scientific data create barriers to our safety.

We need a more prudent, sensible legal and regulatory approach to protect us from toxic chemicals (see Chapter 6). Evidence of contamination and children's vulnerability provides substantial support for the testing and review of commercial chemicals before manufacturing begins and these products enter the market. Testing protocols should pay particular attention to susceptible subpopulations and to additive and cumulative effects resulting from exposures to multiple toxicants. Sensitive subpopulations include, among others, developing children, pregnant women, women of childbearing age, those with preexisting diseases and concurrent toxic contamnation, and the elderly.

If better testing and review of chemical products occurs, how might our legal and product world be different? As I will discuss in Chapter 7, successful testing of products and pollutants before exposures would reduce the risks of the developmental origins of disease due to toxicants. Other externalities would be reduced. People would need to spend less time, effort, money, and mental resources trying to reduce or avoid exposures to toxicants. Taxpayers would spend less to support public health agencies doing what companies should do in the first place. Although the costs of some products might increase, these costs must be properly weighed against existing product costs and associated externalities.

The history of toxic contamination should worry us. DES, asbestos, PCBs, PBBs (polybrominated fire retardants), DDT, and radiation caused considerable harm and were eventually significantly restricted or removed from the market. Lead wrecked havoc with children for decades before its exposures were reduced. Some of its risks remain. However, none of these substances was addressed until they had done considerable damage. The

catastrophe of thalidomide was one of the early alerts that prenatal exposures could cause harm.[29] A brief history of that saga is informative.

Thalidomide was developed by Chemie Grünenthal in Germany
about 1953 and promoted by the company as a powerful sedative with
no side effects. Between 1958 and 1960 some twenty countries began to
use the drug.[30] In the early 1960s Grünenthal sought an American li-
censee. A subsidiary of Vicks (as in Vicks VapoRub), Richardson-Merrell
Pharmaceutical, purchased the license. Thalidomide had been poorly
tested in Germany and in the United Kingdom (which required no serious
confirmation of safety or effectiveness). In the United States at the time,
a drug proposed for commercialization, according to Philip J. Hilts, "was
automatically approved unless the FDA, for good reasons, stopped it.
Thus, the legal burden was on the FDA."[31] The law did not specify what
testing was needed.

Richardson-Merrell sought to use the drug as a sedative and "to treat
nausea of early pregnancy." Within nine days of purchasing the license
for thalidomide and before any testing or FDA approval, the company
began delivering about 2.5 million tablets to doctors to distribute to
about twenty thousand patients. The company had no data on its effects
on pregnant women and fetuses. Hilts points out, "Under the 1938 law,
doctors could experiment on patients with new drugs, in any numbers
and with any chemical, so long as they called the work an experiment, . . . [P]atients [did not need to] give their consent to be part of the
experiment."[32]

Thalidomide caused birth defects—shortened limbs and other maladies—in about seven thousand to eight thousand children worldwide,
while another five thousand to seven thousand died before birth. However, birth defects from thalidomide were almost totally avoided in the
United States, thanks to an alert FDA official, Dr. Frances Oldham
Kelsey. As a result, only about forty American children were affected.
Kelsey did not like what she saw in the initial drug application. Claims
were unsupported, and references to foreign studies had little credibility.
She requested evidence that the drug was safe for use during pregnancy;
the company had none. In November 1961 news of birth defects associated with thalidomide use became public in Germany. Grünenthal withdrew the drug from the German market, in the words of Hilts, "blaming
the sensationalism of the press."[33]

The FDA commissioner was frozen in light of this information. He required only that the company send a letter to doctors that thalidomide

should not be given to pregnant or childbearing-aged women, as a precaution.[34] In July 1962 a U.S. news story, heralding Kelsey's suspicions and reluctance to approve the drug, forced the FDA's hand. The FDA sought to retrieve the pills, but poor record keeping hindered that effort.

About the same time, Congress was considering legislation initially aimed at price-fixing in the drug industry, corruption among FDA officials simultaneously paid by pharmaceutical companies, and price gouging. Senator Estes Kefauver, a Democrat from Tennessee, led this effort. He also sought to require the FDA to license drugs, to require scientific tests to determine safety and effectiveness, and to require warnings to doctors and consumers when drugs had hazardous side effects.[35]

Richard Nixon, a California senator, led the Republicans' attack against this legislation. Republicans claimed it would erode, according to Hilts, "individual (read: business) liberties and turn more power over to the centralized national bureaucracy . . . rob doctors of their authority," and be an effort to protect patients against the judgment of their doctors. The industry argued that the bill would increase the cost of drugs, decrease new drugs, and interfere "with the rights of businessmen and doctors to ply their trade."[36] The Kennedy White House initially was no particular friend of Kefauver's bill either, because it was controversial. This opposition led to a watered-down bill for Senate consideration, until the thalidomide story became public.

Senators supporting the industry, fearful of blame for "not caring about deformed babies," dropped their opposition. Both houses of Congress unanimously approved the Kefauver-Harris amendments to the food and drug laws, and the bill was signed into law on October 10, 1962. Following passage of the new law, a major official of the Pharmaceutical Manufacturers Association admitted, as Hilts notes, "that the new rules would pose no serious problems for companies that already had high standards of research."[37]

Hilts, an historian of the FDA, points out, "The lesson of thalidomide is that a good scientific standard, used before market, and based on scientific data rather than impressions or authoritative assurances, is essential for safety."[38]

Several aspects of the thalidomide story are notable. Current postmarket laws provide less protection from commercial chemicals than pre-1960s drug laws did from drugs. New drug applications with "safety evidence" were required for drugs (but the FDA did not always use this authority).[39] For industrial chemicals no toxicity data need be submitted.

14

And most substances contain so little information that the EPA cannot make a decision on their toxicity.

The FDA in the 1960s could easily request more data. Currently the EPA may order a company to submit additional information. However, if the company objects, the agency must take it to court and carry a burden of proof to obtain data or prevent the substance's commercialization.[40] Good scientific standards implemented before products are commercialized are essential for safety. This is currently not required for up to 90 percent of industrial chemicals in commerce.

Thalidomide clearly represents an extreme harm caused by prenatal exposures and a quite dramatic institutional failure. However, passage of legislation protecting adults and developing children from untested or poorly tested pharmaceuticals took a human catastrophe to motivate it. Current laws permit a variety of substances to contribute to more subtle adverse effects. Must we await some other visible misfortunes before taking systemic steps to control toxic contamination that puts our children and us at risk?

In the final analysis, we cannot prevent molecular products from contaminating us. However, laws properly constructed and implemented can greatly reduce contamination by toxicants. If it is wrong for our children to be experimental subjects for commercial products, the law must change. What will we choose to do?

Commercial chemicals of known and unknown toxicity permeate our lives. If you are sitting on your couch as you read this, the cushions likely contain brominated fire retardants: polybrominated diphenyl ethers, or PBDEs. They are not chemically bound in the upholstery, so over time, and perhaps not too much time, they will begin to migrate from your sofa to the floor, the air, and your lungs. Eating red meat and chicken will add to your intake of PBDEs.[1] Walking on or vacuuming your carpet disturbs any PBDEs in the dust. After disturbance, the remnants can settle back into the carpet and be perturbed again and again for inhalation or ingestion. Of course, if your children play on the carpet, they receive greater exposures because they are closer to the dust and have greater metabolic and inhalation rates. If you spend considerable time in front the television or computer, which contain fire retardants, you may be exposed to PBDEs, as there is a chance they migrate from electronics. When you dispose of your furniture and electronic devices, PBDEs enter the wider environment.

Tap water in many areas of the country provides small doses of perchlorate, a rocket fuel and fireworks' component that was poorly discarded and entered the environment. If you eat lettuce from California's Imperial Valley or other regions irrigated with Colorado River water, you probably add to your perchlorate exposure. This compound can interfere with thyroid production, which developing children need for proper neurological growth and function. If mothers have low iodine levels, perchlorate becomes an even greater threat. This is only one toxicant in drinking water; some other toxicants do not yet have health standards.[2]

If you are older than about fifty, you might have long borne and still carry the banned pesticide DDT (dichlorodiphenyltrichloroethane) or its metabolite DDE (dichlorodiphenyldichloroethylene), both chlorinated compounds. If you are a woman who was exposed as a teenager during the height of DDT spraying, you have a fivefold higher risk for breast cancer, and it might appear earlier than in nonexposed women of the same age.[3]

If you have hamburgers for dinner, they will likely contain some remnants from an industrial insulating and lubricating fluid long banned in the United States and many industrialized countries: polychlorinated biphenyls (PCBs). Although barred from commerce, they remain present and in our bodies. Eating seafood does not avoid exposure but could actually increase it, because PCBs have contaminated many fish and marine mammals as well. If you take fish oil supplements for the health benefits of Omega-3 fatty acids, some brands will give you unreasonably high concentrations of PCBs.[4]

If you do not live near a chemical disposal site and are a vegetarian, you might incorrectly think you do not need to worry much about PCBs. Indeed, you can do everything to avoid proximity to many of these products and yet be contaminated. People of the northern latitudes are furthest from the technology and congestion of modern society and live in apparently pristine environments. Yet they are among the most highly exposed to PCBs, DDT, and other chlorinated compounds—substances that disrupt the hormonal system and cause neurological problems—along with mercury, another neurotoxicant. These substances enter the bodies of fish and sea mammals, which in turn enter the human body as part of the Inuit diet.[5] PCBs can also be volatilized and then transported long distances through the atmosphere. Typically, people in northern Canada and the United States receive greater doses of such chemicals than do people in the lower forty-eight states of the United States. People living close to the Arctic also accumulate greater concentrations than those existing in the animals they eat.

All these compounds are known or suspected to be toxicants. They include known or probable human carcinogens, substances that can adversely affect the development of children, reproductive toxicants, and neurological toxicants (see Chapter 4). Other substances are of unknown toxicity. As Philip Landrigan, a leading researcher on children's health, says, "They're in blood, they're in urine, they're in breast milk, they're in the cord blood of newborn infants."[6] They are part of persons' "body burdens."

However, this is not all that might concern us. Can these substances be avoided? Does each person's lifestyle make a difference in contamination?

There are various sources and routes of contamination. Some of us, smokers, for example, deliberately expose ourselves to toxic invaders. Some products may directly contaminate us; others invade more insidiously during routine living, as secondary contaminants from poorly disposed products or wastes. Some naturally occurring chemical substances have been present in human populations for centuries; some newcomers did not exist before they were deliberately created for commercial purposes. Some persist in the environment and in living things; others are quite transient.

The list of chemical substances to which I refer is roughly the same one that the National Research Council (NRC) utilized to review the needs for toxicity testing in 1984, plus additions to the various categories of chemicals since then. The NRC considered the Toxic Substances Control Act Inventory of 48,523 chemical substances in commerce; 3,350 pesticides (active and inactive ingredients) registered for use by the U.S. Environmental Protection Agency; 1,815 prescription and nonprescription drugs approved by the Food and Drug Administration; 8,627 food additives, some approved by the FDA; and 3,410 cosmetic ingredients.[7] These total about 65,725. When considered together with the degradation of products and the growth in the number of substances since 1984, the number probably approaches 80,000 to 100,000 (although there may be some duplication).[8]

People are not exposed to all of these. Many compounds are no longer produced, and some will be industrial intermediates involving little or no human exposure. Nonetheless, there are thousands of substances in commerce to which people are exposed. Probably a few thousand of these chemicals are worrisome because of their production volume or other exposures: about 3,000 high production–volume chemicals plus another 1,000 to 12,000 others. There are about 6,000 chemicals of concern for exposure, according to governmental reports.[9]

What does it mean to say that toxic substances are "in citizens' bodies or tissues, blood or urine," contaminate their bodies, or are part of their "body burdens"? We have some understanding of exposure, but how does that differ from body burden?

Exposure versus Body Burden

"Exposure" is ambiguous. It may merely mean that a toxicant has come into "contact" with a person's body. That is, the toxicant reaches the external surfaces or boundaries of the body. Exposure can also mean that the chemicals at the boundaries of our bodies enter them and penetrate some of our tissues, organs, and blood.[10] The Centers for Disease Control (CDC) determines the concentrations of industrial compounds that are in our bodies, by measuring the amounts in our blood or urine, a process called biomonitoring.

For purposes of this discussion, "mere exposures" or "exposures" to substances are the amounts of substances that reach the external boundaries of persons' bodies. A person's "body burden" of a toxicant is the amount of a substance that can be measured in a person's tissues or fluids by biomonitoring. When people have industrial chemicals in their tissues, organs, or bodily fluids, they are contaminated.

Exposure

People are not ordinarily aware of exposures to suspected toxicants. Of course, on some occasions we can be acutely aware of smoke, pesticide spray, air pollution, or brackish water or see mercury "beads" on surfaces, although we could not detect harmful molecular components of the exposures. We can sometimes readily infer the presence of potential toxicants—smog from the products of internal combustion engines in the Los Angeles Air Basin, or effluent from a factory that we know is upstream in a river—but even this awareness is not a good indication of toxicity. In general our senses—sight, hearing, taste, smell, and touch—are poor guides to both exposure and toxicity. On the one hand, if what we detect smells or tastes bad, we become concerned about it, but even this may not be a reliable guide to a substance's risks. On the other hand, even when air does not smell bad or when foods and fluids do not taste strange, they might well harbor toxicants.

Despite our awareness of some pollutants, it might never have occurred to us that we are contaminated, that these substances are in our bodies—in tissues, organs, blood, and urine. This is one of their most insidious features—we know nothing about the presence of the overwhelming majority of chemical toxicants. For instance, PBDEs are in mattresses, sofas, chairs, carpets, computers, air, sediments, animals, food, dust, and smoke, but we are unaware of them. The former gasoline additive MTBE (methyl tertiary butyl ether) is in the air and water. Many of us are contaminated by it but are completely unaware of either exposure or contamination. Bisphenol A (BPA) is used in the manufacture of plastics and epoxy resins. It can be used for "impact-resistant safety equipment and baby bottles, as protective coatings inside metal food containers, and as composites and sealants in dentistry."[11] It enters our bodies daily from food, drinks, and other sources.[12] Both PBDEs and BPA have found their way from various products or fugitive pollutants into our bodies.

Biomonitoring and Body Burdens

As a result of developments in biomedicine and analytical chemistry, scientists are beginning to detect the concentrations of chemicals in our bodies, by biomonitoring. As the CDC notes, biomonitoring permits determination of people's exposure to toxic substances in the environment "by measuring levels of chemicals that actually are in people's bodies," as detected in blood or urine.[13] This procedure provides systematic information about human contamination—"which chemicals get into Americans and at what concentrations," the percentage of people who might be exposed to toxic concentrations, and a range of body burdens—so that physicians can identify those people with extensive contamination.[14]

Biomonitoring provides a direct measure of a substance in our bodies from, "all routes of exposure—inhalation, absorption through the skin and ingestion, including hand-to-mouth transfer by children." More importantly, biomonitoring reveals the "integrated effect" of different exposures to the same substance. Researchers no longer need to estimate or calculate what concentration of a substance enters human bodies from myriad sources and via different routes. If one inhales tiny amounts of PBDEs from couches or from dust, ingests some with meat, and absorbs some from bedding, biomonitoring identifies the approximate amount that has been concentrated in one's body. Metaphorically, the body acts both as a funnel, collecting substances from several different exposure routes, and as something like a storage unit (for some compounds), holding the collected chemicals within tissues. Biomonitoring "provides unequivocal evidence that both exposure and [biological] uptake have taken place."[15]

Biomonitoring also conveys something very important to all of us—the extensive pervasiveness of chemical substances in our lives and our bodies. Previously, we perhaps knew in some abstract sense that we were exposed to toxicants, that they were at our bodies' boundaries. However, we may have believed that they are merely "out there" and may have given little thought to whether they reach our tissues and organs (although we knew exposures could lead to disease). We now know that we have very intimate contact with hundreds of industrial chemicals; they are in each of us, and in our tissues in ways we probably never have imagined. Scientists also know that many of them can be in our bodies for hours, days, weeks, years, or sometimes decades. The measure of the longevity of toxicants is their "half-life," the period of time it takes for one-half the amount of a substance in our bodies to leave.

Some substances to which we are exposed will enter our bodies and remain; some will enter and quickly be expelled. Of those that remain, some will stay for minutes or hours (like bisphenol A from plastics), some for days up to a few months (like methylmercury), some for years (like some PCBs, PBDEs, or perfluorinated compounds), and some for decades (like lead in our bones).[16]

Each pattern reveals something. Substances with short half-lives in our bodies provide evidence of ongoing exposures, if they are routinely detected.[17] Other substances with longer half-lives indicate exposures at sometime in the past but perhaps not necessarily ongoing exposures (although this could be the case). Exposure to lead in the United States has been reduced with the elimination of leaded gasoline, but it remains in our bones for much of our lifetime. If substances remain in a person's body, internal concentrations are no doubt greater than concentrations of particular individual exposures. And it may be difficult to determine when exposure occurred and especially when exposure ceased.

Biomonitoring also reflects any modifying influences on the external exposures that occur after substances are in the body. According to Sexton, Needham, and Perkle, the "influences of physiology, bioavailability and bioaccumulation . . . can magnify the concentrations of some environmental chemicals enough to raise them above the detection threshold."[18] For other substances, biological mechanisms may detoxify them and expel them from our bodies.

Disturbing Findings about Body Burdens

Our bodies contain varying levels of hundreds of industrial chemicals, many of which are known or suspected toxicants. The CDC *Third National Report on Exposure to Environmental Chemicals* (2005) had reliable methods for identifying 148 chemicals in people's blood and urine. The *Fourth National Report* (2009) developed reliable protocols for identifying approximately 212 substances in people.[19]

The increase in numbers from the third report to the fourth does not mean that there are more substances in citizens' bodies. Rather, more toxicants can now be reliably identified with exposure markers. As markers are identified and measurements are created, standardized, and utilized, more substances will likely be found in our bodies.

Body burdens can be understood in more personal ways by some studies conducted on smaller groups of people. For instance, a Canadian study tested eleven people from widely dispersed areas of Canada for the presence of eighty-eight chemicals in their bodies and found many toxicants at measurable levels.[20] Although this is a small sample, it represents quite a diverse group from many different walks of life with no obvious toxic exposures as part of their lifestyle or employment.

Volunteers ranged in age from twenty-three to seventy-five and came "from the West Coast, the Prairies, central Canada, the North, Quebec and Newfoundland."[21] With their geographical diversity and employment, as well as the absence of any special exposures that might explain their contamination, these Canadians resemble ordinary Americans. Brief personal information about the men and women volunteers and some of their living patterns suggest few obvious sources of toxicants. Those tested appear unremarkable in their pesticide use, in their exposure to chemicals on the job (almost none), and in their smoking status (volunteers included one minimal smoker), all obvious sources of contamination. Moreover, many deliberately chose organic foods as part of their diet. Two made it a substantial portion of their diet. The group included an environmentalist, a family physician, a student, a postal worker, a corporate executive, an artist, a university official, a filmmaker, a small-business owner, a tribal chief and a lawyer. Yet all were substantially contaminated.[22]

The report concludes, "No matter where people live, how old they are or what they do for a living, they are contaminated with measurable levels of chemicals that can cause cancer and respiratory problems, disrupt hormones, and affect reproduction and neurological development." Older people had more PCBs than younger ones. This likely reflects that PCBs some decades earlier were present and entered people's bodies, but because they have been phased out this results in lesser contamination among younger people. One of the volunteers who lives in northern Quebec had the highest levels of PCBs in his body. This is what one would expect, given the northern migration of environmentally persistent substances by evapotranspiration and ascent up the food chain to higher and higher orders of animals.[23]

On average, in each person the study found seven heavy metals, three brominated fire retardants, ten different PCBs, one perfluorinated compound (the only one tested), eight organochlorine pesticides, four organophosphate pesticides, and one volatile and one semivolatile organic compound.[24] Each volunteer was typically contaminated with twenty-eight carcinogens, eighteen hormone disrupters, fifteen respiratory toxicants, and thirty-eight reproductive/developmental toxicants (with, of course, individual variations). Some of these compounds can clearly cause multiple adverse effects.[25]

A similar small study by the Body Burden Work Group and Commonweal Biomonitoring Resource Center reported the biomonitoring of thirty-five ordinary citizens from seven different states within the United States and found somewhat similar results. Of these subjects, thirty-three contained phthalates and BPA. All thirty-five had PBDEs.[26] These three substances are of substantial concern in the scientific community, but none was tested for toxicity before they entered commerce, and only now are they beginning to receive some scientific and regulatory attention.

The NBC program Dateline tested two families to see whether individual contaminants in their bodies were different. The "Green" family included one adult vegan (that is, someone who eats no dairy, no eggs, no other animal products), an adult vegetarian, and two vegetarian children. Family members also tried to purchase "natural" cleaning products and mostly organic foods. They had personal and purchasing habits designed to avoid known toxicants by means of personal choices. In contrast, the "Brown" family was a more typical American family. Family members ate considerable amounts of eggs, cheese, sirloin steak, turkey, and many "convenience" foods because of their schedules. As for other home products, the family mainly worried about how effective they were and their costs, not their toxicity.[27]

Both families were tested for seventy-six industrial chemicals in their bodies. The Greens had forty-two, and the Browns had forty-three. The Browns had three times as many perfluorinated compounds as the Greens did. The Brown children had more phthalates than 76 percent of the people tested in the United States. The Browns had more triclosan, an endocrine disrupter typically in cosmetics, but the Greens "showed barely a trace." Both mothers had contaminants from cosmetics, but the Brown mother had four times as great a concentration as the Green mother.[28]

The Greens seemed to be winning the contamination contest. However, they had moderate levels of bisphenol A, while in the Browns it was barely detectable. The Greens ate many canned foods, especially refried beans. They apparently ingested bisphenol A from the liners of the cans. More detailed results were not available, but the suggestion was that the Greens had somewhat lower levels of the forty plus chemicals than the Browns did, but neither family was living toxicant free. The shocker was that the Greens had such high concentrations of BPA compared with the Browns.[29] The presence of BPA suggests nearly continuous exposure because it is quickly eliminated from the body.

Pregnant women carry compounds of known and unknown toxicity. In one study, ten pregnant women were tested for twenty-three substances of concern; all of the women were contaminated. As noted in the Chapter 1, Kim Radtke for sometime prior to her pregnancy had eaten organic foods and had avoided deodorants, lotions, and cosmetics, as well as exercising regularly. Yet she was one of the most polluted people in the study. She nursed her newborn, Karson, but worried about offloading some of her hazardous body burden into him.[30]

This study does not raise a merely theoretical concern, for a recent study of ten newborn babies revealed that their umbilical cords were contaminated by up 232 synthetic and industrial chemicals.[31] These babies, and others, begin life already sullied by commercial chemicals.

What Do These Contaminations Suggest?

At a minimum we are substantially contaminated by hundreds of industrial compounds. Moreover, the vignettes above suggest there is little way to avoid them. Eating organic foods helps, but perhaps less than one might think. Living in seemingly pristine locations, such as the high northern latitudes or areas near the Arctic Circle, might seem to help. Yet these locations will exacerbate contamination by some persistent compounds.

Systemic approaches such as eliminating the uses of some industrial compounds, or greatly curtailing them, does help over a long period of time. Many developed countries banned PCBs in the 1970s, and the United States greatly reduced lead exposures from gasoline additive in the 1980s. The CDC reports that amounts of both compounds in the environment have decreased substantially since those actions, but both remain in our bodies. For example, in the late 1970s 88 percent of children one to five years of age had unacceptably elevated lead levels, while in the 1990–2004 period that number had dropped to 1.4 percent.[32]

Eliminating the use of toxic contaminants with brief half-lives will likely decontaminate a person's body much quicker. Recent research supports this idea. Twenty-five people stayed in a Buddhist temple for five days eating the vegetarian diet of the priests and following their daily routine. This group had a noticeable reduction in phthalates, antibiotics and an oxidative stresser in their urine after following the modified diet and routine.[33]

On the one hand, we should realize that the mere presence of industrial chemicals in our bodies does not necessarily mean that they pose risks or are harmful. The technology for detecting contaminants is sufficiently accurate that it can detect tiny concentrations. These amounts may or may not pose risks to either adults or children.

Too often almost nothing at all is known about contaminants and even less is known about small concentrations of industrial compounds. For example, with respect to one class of pesticides, a CDC report notes that "the health effects of exposure to organochlorine pesticides on the general population at current . . . [concentrations of the pesticide in tissues and fluids] are unknown." The report contains similar comments about the toxicity of many of these pesticides at low concentrations; for other chemicals it simply does not address this issue.[34] This ignorance is even greater for low concentrations.

All this is worrisome. Too frequently it appears that no one—not the manufacturer, not the responsible regulatory agency, not independent scientists—sufficiently understands the toxicity properties of the vast majority of substances in order to identify risky from nonrisky levels (see Chapter 5). A number of industrial substances are known or likely toxicants—known or probable carcinogens, known or probable reproductive or developmental toxicants, and so on—but whether and to what extent they might be toxic at low concentrations in the body is often not known. Some toxicants do not have a "safe" level below which they do not cause disease; others taken by themselves may have levels below which there are no risks. Thus, the risks of many substances at current concentrations in the body have yet to be determined, if they will ever be.

On the other hand, even if a substance does not pose risks by itself at present concentrations, it may promote the toxicity of other substances already in people's bodies. Or it might function in tandem with natural substances to cause adverse health effects, as some synthetic estrogens do.[35] It might depress immune system responses, leading indirectly to various diseases. There is poor understanding of the additive effects of most toxicants. What is emerging is worrisome (see Chapter 4).

However, it is not only the concentration of substances but also the timing of exposure during development that can pose adverse health risks. For example, MeHg (methylmercury) and DES (diethylstilbestrol) all cause much worse adverse effects from in utero exposures than from analogous exposures to adults (see Chapter 4). The various gaps in scientific knowledge at a minimum reveal how poorly companies that create, produce, and distribute products containing potential toxicants currently understand their own products. And, because companies have not produced such evidence, the wider scientific community may be even less knowledgeable about a substance or class of products.

As if to confirm this point about a much better tested group of substances at least initially, an expert panel of the Food and Drug Administration recommended that the FDA recall cold and flu remedies for young children on grounds that these widely sold medicines did not seem to be effective and that they had not been properly tested for use by children.[36] The fact that the FDA had to recall cold remedies is shocking because pharmaceuticals are among the best-tested products in the United States before they enter commerce. Despite this, there appears to be little understanding about potential adverse effects for children from something as ordinary as cold remedies.

Contaminants

This section considers a few examples of substances found in people's bodies. These have been chosen partly for the varieties of exposure and contamination they illustrate and partly for adverse heath effects that we will consider in Chapter 4. These examples reveal a varied pattern of contamination.

Metals

Lead: Lead is a naturally occurring metal, ordinarily combined with other elements to form compound molecules. It is quite malleable and thus has a wide variety of uses. Lead and its alloys may be found, "in pipes, storage batteries, weights, shot and ammunition, cable covers, and sheets used to shield us from radiation. The largest use for lead is in storage batteries in cars and other vehicles." Lead compounds are also used as pigments in paints, dyes, and ceramic glazes and in caulking compounds. Because lead occurs naturally, one might think that natural sources could lead to human exposure. While there is some small chance this kind of exposure could occur, the same report points out that levels of lead "have increased more than 1,000-fold over the past three centuries as a result of human activity."[37] All significant lead exposures result from commercial uses of this element.

Lead and its health effects have been present in human communities for centuries. The Romans were aware of its toxicity but nonetheless used it for a variety of purposes: in face powders, mascara, rouge, part of the pigment of paints, even as a "spermicide" to help prevent conception. It preserved wine, helped to sweeten it, and disguised "inferior vintages."[38] It was also a major component in a variety of household dishes, pitchers, pots, and other artifacts. It served as piping for an extensive plumbing system.

During the Middle Ages and the Renaissance, lead continued to be extensively utilized. Alchemists tried to make gold from lead. In the late fifteenth century it was used for movable type. With the development of gunpowder, lead became the metal of choice for the projectiles of pistols, rifles, and cannons because of its malleability. When Europeans came to the New World, they quickly began to mine and smelt lead. Its resistance to corrosion was particularly attractive.

In the twentieth century the United States became the world's leading producer and consumer of refined lead, "with the greatest increase in its use occurring," one report notes, "between the years 1950 and 2000 and [reflecting] the increasing worldwide use of leaded gasoline." An important promoter of lead use was the internal combustion engine. Powered by gasoline, early engines had a "knock" that was disconcerting and harmed the engine. General Motors, exploiting its relationship with some chemical companies and fearing loss of the market to Ford's low-compression engine automobiles, patented and marketed tetraethyl lead to eliminate this knock. This innovation led to the development of more powerful and attractive automobiles and ultimately better engines for airplanes. At the time that tetraethyl lead was adopted for this purpose, however, there were several other less toxic alternatives available. By 1980 the United States was consuming about 40 percent of the world's supply of lead.[39]

Lead enters the environment and exposes people in a variety of ways. It can be released from mining and "from factories that make or use lead, lead alloys, or lead compounds." It can be discharged into the air by the burning of coal, oil, or waste products. A public health statement notes, "Before the use of leaded gasoline was banned, most of the lead released into the U.S. environment came from vehicle exhaust." Lead was once used in pesticides and applied to food products such as apples and other orchard plants.[40]

If lead enters the atmosphere, "it may travel long distances if the lead particles are very small." Lead can be removed from the air by precipitation or by the lead particles themselves descending to surface water or land. On land it can attach to the soil and settle in the upper layers, remaining there for many years.[41] When lead enters people's bodies, its half-life in the blood is about twenty-five days and in the bones about twenty-five years.[42] As we will see in Chapter 4, it is typically released from bones during pregnancy and nursing and transferred to developing children.

Although this book is concerned with prenatal and early childhood exposures to toxicants, the lead story represents some progress in protecting the public health. It is no longer used in gasoline in the United States and many other countries, eliminating millions of pounds of exposures per year. It is not routinely utilized in pesticides, reducing lead in our foods. Lead solder is no longer used in newer water pipes and metal cans. It has also been eliminated from new paints, where its use had once been routine. And, as already noted, childhood blood lead levels have dropped dramatically since it was eliminated as a gasoline additive.

Lead has not gone away, however. It remains in dust and soil, as well as in old paints used on buildings bridges, houses, and other structures. It likely is in landfills as part of waste from lead mining, ammunition, or battery production. Worldwide uses are increasing.[43] Poor disposal of lead, of course, can augument amounts in the environment. Higher levels of lead may be found in soil near roadways, an inheritance from the use of lead in gasoline. Old water pipes containing lead solder may contribute to exposures. Another source of lead is in inner cities is the soil around older houses that were covered with lead-based paint. Importantly, lead is also inside older houses, in flaking lead paint on the walls. These various sources in poor areas result in pockets where lead poisoning can occur.[44]

Lead production and use is likely to increase as the need for batteries increases in an electronic world that must reduce greenhouse gases. Proper recycling of lead in batteries or electronics must be closely supervised, or it will increasingly contaminate the environment, animals, and people.

Mercury: Mercury is another metal that occurs naturally, but it can exist in several forms. Metallic mercury is the shiny, silver-white metal, typically in liquid form, probably familiar from childhood in the 1970s and earlier. If it is spilled, it tends to bead up. Metallic mercury is widespread, having been found "at found at 714 hazardous waste sites nationwide."[45] In commerce, it has been utilized in thermometers and some electrical switches. It is also present in fluorescent lightbulbs, which are increasingly used for energy conservation purposes. However, there is less mercury in fluorescent bulbs than in thermometers. The U.S. Environmental Protection Agency estimates that even though mercury in these bulbs will pose a minor disposal problem, overall mercury will be reduced because less will be released from coal-fired power plants.[46] This form of mercury can evaporate, and the high temperatures of power plants release large amounts of mercury vapor, which can reside in the atmosphere, according to researchers, "unchanged for periods of a year or so."[47]

Organic mercury is important for the purposes of this book. This form results when mercury combines with carbon.[48] Although potentially many forms of organic mercury can be formed, a very important one for environmental health purposes is methylmercury, or MeHg. This is the compound that caused health disasters in Japan and Iraq (see Chapter 4). It is also a very common mercury compound in the environment.

Human activities constitute one-third to two-thirds of all mercury releases. Metallic mercury typically enters the air from burning coal in power plants, from incinerating municipal and medical waste, from the production of cement, and from fugitive emissions from plants that use mercury in production processes. Some natural processes also release mercury.[49]

Once elemental mercury is in the environment, it can be transformed into other compounds, often being transported long distances in the atmosphere. Bacteria, phytoplankton in the ocean, and fungi can, as a standard textbook notes, "convert inorganic mercury to methylmercury. Methylmercury released from microorganisms can enter the water or soil and remain there for a long time." Its half-life in human bodies is about 70 days, but once it is lodged in the brain, it can cause prolonged damage.[50]

Methylmercury is especially problematic, since it can enter the food chain and accumulate in the tissues of fish and animals. For instance, small fish might eat foods with MeHg, which then enters and tends to remain in their tissues. If carnivorous fish eat the small fry, the MeHg is transferred as well. Larger, long-lived predatory fish, such as sharks and swordfish, and marine mammals, such as whales, seals, and dolphins, tend to have the highest concentrations of MeHg in their bodies. Goyer and Clarkson report that concentrations in sea mammals can be up to "a millionfold higher than those in the surrounding water."[51] This finding led to cautionary advisories for people to limit their intake of certain kinds of seafood. The content of MeHg in pilot whales and certain fish, major elements of the diet in the Faroe Islands, was so high that health officials had to strongly recommend that women cease eating their traditional diets if they were contemplating pregnancy or were pregnant.[52]

Very little metallic mercury will likely be absorbed in one's body, while mercury vapors and MeHg are much more readily absorbed. Once in the blood stream, they can travel to other parts of the body, especially the brain, which they readily enter. These forms of mercury can remain, as a government report points out, " 'trapped' there for a long time." The same report notes that methylmercury "will easily move into the blood of the developing child and then into the child's brain and other tissues." Some of the methylmercury will enter breast milk and be conveyed to a nursing child.[53]

A quite recent report shows the general ubiquity, persistence, and accumulation of metals in mammalian bodies and the consequences for our world and our health. Sperm whales in remote areas of the world's oceans have high concentrations of mercury, lead, cadmium, chromium, and aluminum in their skin (the bodily area that was sampled). Concentrations in internal organs will likely be much higher. This report encapsulates some great concerns about metals and other persistent chemicals. Roger Payne, leader of the research, as well as founder and president of Ocean Alliance, points out "The entire ocean life is just loaded with a series of contaminants, most of which have been released by human beings." Moreover, this finding suggests how contaminated a very wide environment is and how it is a threat to about a billion people who rely on seafood as a main source of protein.[54]

Persistent Organic Compounds

PCBs: Polychlorinated biphenyls, or PCBs, are strictly a human creation, originally synthesized in 1881, about the time the synthetic chemical industry was developing. Before that date, one report asserts, there were "no known natural sources of PCBs in the environment."[55]

PCBs are part of a related larger family of molecules formed with fluorine, chlorine, or bromine atoms. The "tight" chemical bonds of these molecules tend to make them resistant to metabolic breakdown in mammalian bodies. Because humans and other mammals have not encountered these substances during their evolution, they tend to lack enzymes to reduce the compounds' toxicity.[56] Some are more persistent than others. Perfluorinated compounds, or PFCs, are formed with fluorine atoms and tend to be the most persistent of these compounds because the fluorine atoms are more tightly bound electronically. Brominated compounds are less persistent than chlorinated biphenyls.

PCBs are molecules whose structure resembles a pair of glasses with hexagonal rims that are viewed from the front. Multiple chlorine atoms are attached to two carbon rings joined by a chemical bridge. There are more than 200 variants, or congeners, of PCBs, determined by the number of chlorine atoms on the benzene rings.[57] There are two broad subtypes of PCBs—coplanar and noncoplanar—which have different chemical and toxicity properties. In coplanar molecules the chlorinated rings are roughly in the same geometric plane, like a pair of reading classes; in noncoplanar molecules, the rings resemble the frames of twisted glasses.

Coplanar PCBs resemble dioxin and dibenzofurans, among other compounds. These PCBs are frequently called "dioxin-like" because their mechanism of toxic action is quite similar to that of dioxin. This mechanism is conserved through evolution and is common to many species, including humans.[58] Moreover, because of similar toxic mechanisms, the doses of different dioxin-like substances are additive. That is, the doses of different substances act according to a common biological mechanism to produce cumulative effects.[59]

In the environment, these dioxin-like chemicals are persistent and usually occur as a mixture of substances. The half-life of dioxin-like compounds in people is three to nineteen years. The specific half-life for PCBs appears to be from seven to nine years for both adults and children, although this could be an underestimate.[60] PCBs are fat soluble (lipophilic), persist in fatty tissues, and tend to build up in mammals and fish over time. Even after exposure has ceased or been greatly reduced, because of long half-lives it may take years to reduce a person's body burden.

Noncoplanar PCBs do not resemble the dioxin family of chemicals. As a result, they possess somewhat different toxicity properties than those of dioxin-related PCBs. Noncoplanar PCBs also persist in the environment and in mammalian tissues but pose other adverse health effects, including neurotoxicity (see Chapter 4).[61]

31

PCBs were initially commercially attractive because they replaced "more flammable, less stable and bulkier" substances that had been previously used in products.[62] PCBs were desirable because of their stability, good insulating qualities, and poor burning capacities. Until they were banned in the United States in 1979, PCBs had been used as "coolants and lubricants in transformers, capacitors and other electrical equipment . . . [including older] fluorescent lighting fixtures, electrical devices or appliances . . . , old microscope oil, and hydraulic fluids."[63] Although the first commercial production of PCBs began in 1927, some observers report that by 1914 sufficient amounts of PCBs had escaped into the environment to be detected in measurable amounts in the feathers of stuffed birds in museums today.[64]

Once production of PCBs increased, they entered the environment during manufacture and use as well as from the disposal of waste products. They have contaminated the environment from leaks during transport, accidental spills, or leaks from products containing PCBs. They enter the air as particulates or vapors and can remain there for ten days or more.[65]

PCBs can travel long distances, having been found in water and snow hundreds of miles from where they were released. This results from successive volatilization and condensation (or from the settling of dust particles to which PCBs have been attached). According to the Agency for Toxic Substances and Disease Registry, PCBs "can be found in almost all outdoor air, in indoor air, on soil surfaces, in surface water, and on land plants." PCBs carried in water tend to bioaccumulate and persist in the body fat of fish, birds of prey, and carnivorous sea mammals.[66] They are present in beef and dairy cattle. The concentrations of PCBs increase in fatty tissues as these compounds move up the food chain.

PCBs also tend to become more toxic, or "biomagnify," as they ascend the food chain.[67] This means that "some toxic congeners are preferentially retained," with the result that "bioaccumulated PCBs are more toxic than commercial PCBs."[68] Consequently, PCBs in animals higher up the food chain are both more concentrated and more toxic.

When we eat these fish or mammals, we likely ingest a double whammy of more-concentrated and more-toxic congeners of PCBs than when we eat foods from further down the food chain. As a result of these processes, the breast milk of Arctic mothers in 1987 contained ten times more PCBs than the milk of others in Canada's large cities. Nursing polar bear cubs are more highly contaminated than their mothers, and have concentrations of PCBs billions of times greater than that of the waters of the Arctic Ocean.[69] Interestingly, farmed fish are even more contaminated than wild species.

Because PCBs are lipophilic, a principle source of contamination is, according to the CDC, the ingestion of "high-fat foods, such as dairy products, eggs, and animal fats, and some fish and wildlife. Breast feeding is a substantial source for infants. . . . People have also been exposed as a result of industrial accidents (e.g., after an explosion in a factory in Seveso, Italy), the use of accidentally contaminated cooking oils (e.g., as occurred in Yusho in Japan and Yucheng in Taiwan) . . . and the burning of PCBs contaminated with polychlorinated dibenzofurans, such as in old electrical transformers."[70]

The CDC can reliably identify thirty-five variants of PCBs, both non-coplanar and coplanar, via biomonitoring. In general, however, concentrations of PCBs in U.S. citizens have declined by more than 80 percent since production ceased in 1977.[71] Thus, this is another example of systemic reduction or banning of substances decreasing body burdens over time.

PCBs represent an existing environmental health problem that has been reasonably well studied. Nonetheless, consequences of PCB contamination of the environment are still rippling through the environment as well as through wildlife and human populations. Many of the toxicity properties of PCBs are understood, although subtle effects are still being discovered (see Chapter 4). Even though body burdens are decreasing for many people, PCBs are a special problem for those who live in Arctic regions, because these populations are contaminated with the highest concentrations of the most toxic variants of PCBs. Until PCBs ultimately break down in the environment, their adverse health effects will continue to affect people, animals, and environments around the world.

PBDEs: Polybrominated diphenyl ethers, or PBDEs, are also a subclass of halogenated hydrocarbons. Like PCBs, they did not exist before they were synthesized, with the exception, as one government report notes, of those found in "a few marine organisms that produce forms of PBDEs that contain higher levels of oxygen."[72]

PBDEs closely resemble PCBs. Instead of having from one to ten chlorine atoms attached to two benzene rings connected by a chemical bridge, PBDEs have one to ten bromine atoms attached to two benzene rings connected by a somewhat less rigid chemical bridge.[73] The chemical structure of PBDEs resembles "twisted" frames of glasses, like twisted PCBs.

Like PCBs, polybrominated diphenyl ethers can have different variants, depending upon the number and location of the bromine atoms attached to the rings. PBDEs tend to predominate in three main forms, with the molecules containing ten (deca), eight (octa) or five (penta) bromine atoms respectively. The decaBDEs are larger and heavier and contain more bromine atoms; the octa- and pentaBDEs are somewhat smaller and lighter.

Since the 1960s PBDEs have been used as flame retardants in thermal plastics and foam materials. The EPA notes that PBDEs "react chemically to prevent the spread of a fire . . . by introducing the volatilized [bromines] to react with the product in place of oxygen, slowing combustion." These compounds slow flames but do not prevent combustion.[74]

Smaller PBDEs are used, "in flexible polyurethane foam in furniture, mattresses, carpet padding, and automobile seats." Others have been added to plastics that are in turn are utilized in electronic devices. Larger molecular PBDEs are used mainly in "in high-impact polystyrene plastic that is frequently used to make the back part of television sets, and is also used in certain types of flame-retardant textiles."[75] PBDEs are merely "added" to or "blended" into the products in question, instead of being chemically bonded with the material as part of larger molecular structures. As a consequence, PBDEs "can thus leak into the environment" as the product deteriorates. This leakage is much more likely than if they were chemically bound.[76]

Exposure to the substances can occur during manufacture, from contact with products containing PBDEs, from product disposal, and from disposal of manufacturing wastes. PBDEs are in computers, televisions, mattresses, pillows, sofas, chairs, drapes, carpets, and carpet padding. PBDEs in automobile upholstery may volatilize and "fog up" when temperatures are elevated inside closed cars, entering the internal environment of the vehicle.

Researchers have "found a clear association between consumer products and PBDEs in the indoor environment . . . [and have linked] residential dust concentrations of PBDEs to sources: household furniture and televisions." Possible mechanisms for releasing the PBDEs from televisions are "elevating the temperature of televisions, thereby potentially increasing emission rates, or increasing abrasion."[77] Old, deteriorating sofas or mattresses are also prime sources.

A recent study by some of the same research group suggests that food is a greater source of PBDEs than scientists once believed. Five PBDE variants were significantly associated "with poultry and red meat consumption." Vegetarians who avoided these foods had PBDE contamination on average about 25 percent lower than that of omnivores. No association was found with consumption of dairy or fish. The authors suggest that as products containing PBDEs begin to degrade, more of them will be released into the external environment and contaminate our food supplies.[78]

This is a legitimate concern. When products containing PBDEs reach the end of a useful life, they likely will be disposed of in a manner that results in their further deterioration, tainting the air, sediments, animals, food, dust, and smoke to which we are exposed.[79] In turn they become part of our body burdens. Scientists have measured PBDEs in human body fat, blood, and breast milk, as well as in "fish, birds, marine mammals, sediments, sludge, house dust, indoor and outdoor air, and supermarket foods."[80]

Scientists note that there are approximately 148 million pounds "of PBDEs manufactured annually worldwide," with North America and the United States in particular accounting for much of their use.[81] It is difficult to assimilate such large numbers, but they provide some indication of the enormity of worldwide exposures to these substances and help explain increasing body burdens.

Tasmanian devils, the world's largest carnivorous marsupial, illustrate PBDE's ubiquity. The devils live in remote areas of Tasmania, far from industry. They are clearly not known as users of televisions, carpets, bedding, or furniture. They don't sleep on old sofas. However, they are contaminated by heavier PBDEs.[82] Some scientists worry that these chemicals may be contributing to a virulent, rare, and communicable facial cancer from which these animals suffer. This cancer prevents many Tasmanian devils from eating, and thus they starve to death. The disease appears to threaten the species but needs greater study to understand its cause.[83] Is our industrial society killing them, though they live far from immediate sources of PBDEs? At this point no one can be sure.

Peregrine falcon eggs reveal a more nuanced pattern. Their eggs in large California cities have much greater concentrations of heavy PBDE congeners, while those in more rural areas have lesser concentrations of these congeners but more of the less brominated PBDEs. In contrast PCB concentrations tend to be more or less constant from urban to rural areas.[84]

PBDEs are present in 50 to 97 percent of U.S. citizens. Body burdens in the United States are three to thirty-five times higher than those reported in some small European studies. In 2003 U.S. citizens' serum and breast milk concentrations of PBDEs were the highest in the world. Studies in 2010 report that PBDE levels in California children are "10-to 1000-fold higher than similar aged children across the U.S., and 2- to 10-fold higher than U.S. adults." In general children have much higher levels than adults probably because of exposure to dust and because of breast feeding.[85]

Some PBDEs in people have half-lives estimated at two years.[86] This is a shorter half-life than for PCBs. Nonetheless, with increasing exposures and substantial persistence in human bodies, if they pose health risks, as they seem to, this is quite worrisome (see Chapter 4).

Current body burdens reveal substantial changes from thirty-five years ago. Arnold Schecter and others compared blood samples of citizens from 1973 to 2003 and found that "for all practical purposes, almost no PBDEs were present in the 1973 blood [before widespread use], whereas dioxin, dibenzofurans, and PCB levels were far higher than in 2003." While PCB contaminations have decreased about 80 percent compared with 1973 concentrations, PBDEs have been increasing rapidly.[87] Surprisingly, despite the increase in PBDEs and decrease in PCBs, some congeners of PCBs remain in greater concentrations in citizens' bodies than the rapidly increasing PBDEs.[88] PCBs, thus, have not yet gone away. Moreover, since there is the possibility of additivity of toxic effects between PCBs and PBDEs, this finding increases health concerns (see Chapter 4).

Smaller, lighter congeners of PBDEs tend to be more easily absorbed, more volatile, more toxic, and more persistent. Because of these properties, they can pose substantial problems to humans and wildlife.[89] These variants have been detected in humans, predominating "in U.S. human milk, serum, and whole blood at the present time."[90]

L. G. Costa and G. Giordano report that the "highest serum levels of PBDEs are found in infants and toddlers, as a result of exposure through maternal milk and house dust. PBDEs can also cross the placenta, and similar concentrations are found in maternal and fetal blood."[91] PBDE levels in California children are up to 1000-fold higher than similar aged children across the U.S., and up to 10-fold higher than U.S. adults because California has become a leader in using PBDE fire retardants.[92] This has consequences for our body burdens of these substances. California children have greater concentrations of the lighter, more persistent and more toxic PBDEs than heavier and less toxic variants.[93] Children and women have the highest body burdens. Thus, women could contaminate their children in utero and during nursing, when children are most vulnerable. When children begin to crawl on the floor and move about, they likely add to their body burdens. This poses substantial concerns for developing children, as we will see in Chapter 4.

Since PBDEs chemically resemble PCBs, their disease mechanisms will be likely be similar. PBDEs appear to pose risks of neurological diseases or damage, similar to adverse health effects from noncoplanar PCBs, as well as risks of endocrine disruption and possibly some reproductive effects (see Chapter 4).[94]

Perfluorinated Compounds

Perfluorinated compounds, or PFCs, are another species of persistent organic substances. PFCs are formed with fluorine, creating some of the strongest chemical bonds among the halogenated class of compounds. As a scientific team notes, this "renders them practically nonbiodegradable and persistent in the environment."[95] "Naturally occurring fluorinated organic compounds are rare."[96] The substances of current concern are all humanly created.[97]

Perfluorinated products have been extensively utilized for about sixty years in lubricants, paper, and textile coatings (like Scotchgard), polishes, packaging for foods, and foams as well as fire retardants. They have been used in nonstick cookware and rain-repellant clothes and have well-recognized trade names, such as, Scotchgard, SilverStone, Stainmaster, and Teflon.[98]

PFCs have been detected around the world. They are in the air, soil, sediments, ice caps, and surface waters. They are found downriver from fluorochemical manufacturing plants in Germany and Tennessee, in the Great Lakes, in drinking water, and in coastal waters around the world.[99] PFCs may be distributed by ocean currents or by atmospheric evaporation and deposition.[100]

Major perfluorinated substances are found in more than 98 percent of the U.S. population older than twelve. Moreover, these compounds have been found two years after the 3M Company discontinued production of most of its products made with PFCs. Wildlife and the environment are also widely contaminated.[101] Since 1999, global emissions have dropped by half, although it is less clear how that has affected people, wildlife, and the environment.[102]

It is not surprising that people remain contaminated even after production of perfluorinated substances has ceased. The half-life of several major perfluorinated compounds is 3.5 to 7.3 years, quite comparable to the half-life of PCBs in humans. Because scientists have seen adverse effects in animal studies, these numbers raise concerns.[103] This has led the EPA to actively investigate PCBs because they can cause developmental and systemic toxicity in laboratory animals, they are persistent, and they have a very long half-life in humans.[104]

All the halogenated compounds briefly reviewed here are of considerable concern because of their persistence and ubiquity of distribution, their long half-lives in people's bodies, their bioaccumulation, and in some cases their biomagnification. There are opportunities for multiple exposures over time. If these substances pose risks, as some of them clearly do (see Chapter 4), they will follow and contaminate us for one or more generations.

Synthetic Estrogens

Diethylstilbestrol, or DES, and bisphenol A, or BPA, are two synthetically created estrogens that are similar in some of their biological properties but different in exposures and effects. Neither is persistent nor obviously bioaccumulative. One (DES) has caused substantial toxicity problems; the other (BPA) is coming under considerable scientific suspicion for adverse health effects (see Chapter 4).

Natural Estrogens: Estrogens are one group of human hormones. Hormones in general exercise control over aspects of human biological functioning. As one text notes, hormones, including estrogens, are "secreted into the body fluids by one cell or a group of cells and [exert] a physiological *control* effect on other cells of the body." There are three main natural estrogens: estradiol, estrone, and estriol. Estradiol is the most potent of the group and is considered the main estrogen, but the others have nonnegligible effects.[105]

Estrogen is primarily known as a female hormone, strongly influencing puberty, breast tissue growth, monthly menstruation, and even menopause, but it is present in both sexes.[106] During puberty estrogen increases the size of the fallopian tubes, uterus, and vagina in preparation for reproduction. It enlarges the female breast by causing growth of the duct systems and deposition of fat in the breasts.

Estrogens can contribute to breast cancer later in life (see Chapter 4).

Estrogen substantially influences bone growth during puberty and can contribute to osteoporosis when a woman's body no longer produces the hormone in sufficient amounts. It also affects protein and fat deposition, metabolism, electrolytic balance, and the texture and thickness of the skin.[107]

Researchers are greatly concerned about synthetic estrogens entering the environment and our bodies, because hormones play such an important role in controlling other functions in the body. DES was a pharmaceutical that had limited exposure potential compared with halogenated substances, but it had disastrous consequences for women. BPA, also a synthetic estrogen, is ubiquitous and appears to pose reproductive and developmental problems not only for women but also for men (see Chapter 4).

Diethylstilbestrol (DES): Diethylstilbestrol was first created by Charles Dodds and coworkers in 1938. Shortly after its discovery, several different researchers experimented with high doses of DES as a drug to prevent miscarriage. The FDA approved it for treatment of menopausal symptoms and prostate cancer in 1943. By 1947 the FDA had authorized DES to prevent spontaneous abortions. However, Philip J. Hilts notes that during this period "the companies had no particular scientific or medical guidance from the FDA about what evidence was required to show a drug safe and effective" and that "drugs were often given to humans before they were tested on animals." No one it seemed, knew what constituted "safe." For DES the FDA had animal data in its files showing adverse effects, but seemed to ignore it.[108]

In 1947 the U.S. Department of Agriculture approved the use of a DES tablet under the skin on the heads of male chickens to chemically castrate them in order to create plump birds.[109] While this was successful, when farmers sold the DES-contaminated chicken heads to mink ranchers for food, this sterilized the female minks. Canada subsequently banned the sale of chickens fed sufficient DES to sterilize minks, but the United States did not.[110] This incident shows that there could be some cross-contamination from one animal to another, but this is quite different from the widespread exposures we have seen from some metals and persistent organic pollutants.

If DES fattens chickens, could it not be used for similar purposes in other meat animals? Iowa State researchers soon found that small doses in cattle feed could increase weight gains 35 percent while lowering feed costs by about 20 percent. What's not to like about such a product? The FDA approved this use of DES in 1954. Soon DES was combined with antibiotics in cattle feed to create animals that gained weight more quickly but were freer of disease in feedlots.[111] Industrial meat production received a substantial financial boost.

Even in 1938 there was some evidence that DES caused cancers in experimental animals. Twenty-one years later the evidence had substantially increased. Because of DES's cancer-causing potential and use in animals, chickens dosed with DES were removed from the market about 1959. This did not affect the compound's use in cattle and beef as a fattening agent, with as much as 75 percent of all beef being produced with DES by 1970.[112]

However, beef and sheep producers were sufficiently concerned about DES's cancer-causing potential that they lobbied Congress to enact a DES proviso exempting meat animals from a regulation similar to that for chickens. These producers argued that DES could be administered to meat animals because, "label regulations that were 'reasonably certain to be followed in practice' would preclude residues in the meat." This did not occur. The FDA and USDA did not test for such residues until 1965.[113] Once the Department of Agriculture began testing meat for DES residues, it found some. In some instances DES was present in the livers of beef animals 120 days after DES was last administered. In 1973 the FDA banned the presence of DES in meat.[114]

DES residues in meat led to nearly every U.S. citizen being exposed. This appears to have had adverse consequences for the children of women who ate beef high in DES (see Chapter 4).[115] Other than such residues and transference from chicken heads to minks, there appears to have been little wider environmental contamination from DES. However, women exposed to the drug in utero suffered substantially (see Chapter 4).

Bisphenol A (BPA): Bisphenol A was first synthesized in 1891, within a few years of the creation of PCBs.[116] In the 1930s the same scientists who synthesized DES also discovered that BPA had estrogenic properties.[117] It was considered as a possible pharmaceutical estrogen. However, it was superceded for this purpose by DES's greater potency.

In 1963 the FDA approved the use of BPA in food containers, under provisions of recently enacted amendments to the Food, Drug, and Cosmetic Act. This section of the law required some data about the product before it was approved. It also required the FDA to review new food additives or indirect food additives that could migrate from containers into food, to ensure that they were safe or did not adulterate foods.

The FDA issued no regulations for BPA, in David Biello's words, "because no ill effects from its use had been shown."[118] That is, BPA had been in use for various purposes for some time and, given what little was known about it, no one noticed "ill effects" from its use. (However, it is not clear at this time to what extent BPA was subjected to appropriate testing and review even by the standards of 1963.)

Although Congress passed a law in 1976 authorizing some screening of new chemicals and providing authorization to review existing substances (the Toxic Substances Control Act of 1976, or TSCA), it did not apply to BPA, because this compound was "grandfathered in," being already on the market. Consequently it did not undergo testing because no specific concerns had been raised about it, as they had been about PCBs at that time. No toxicity issues concerning BPA were seen when TSCA was implemented.[119]

BPA, unlike DES, is employed as an industrial substance and incorporated into a number of different compounds and products. It is one of approximately three thousand high-production volume chemicals produced in more than 1 million pounds per year.[120] BPA is a plasticizer utilized to create hard plastic products out of polycarbonate plastic. It is found in baby bottles, large and small water bottles, and soft drink bottles. It is also used in epoxy resins to coat the inside of metal food containers to prevent them from rusting or otherwise deteriorating into food. It has been used to create protective sealants for children's teeth to help prevent them from decaying. BPA is in composite fillings for tooth cavities. The main route of exposure is believed to be from the ingestion of food contaminated with BPA.[121]

A degradation product of BPA, 4-tertiary-octylphenol (tOP), is an intermediate chemical in manufacturing substances, and utilized "in detergents, pesticide formulations, and other applications. Exposure to tOP may occur from contact with personal care products, detergents, water, and food containing tOP." Researchers point out, "BPA and tOP are of concern to environmental public health because of the high potential for exposure of humans to these phenols and their demonstrated animal toxicity."[122]

BPA contaminates children and adults alike. CDC researchers have found that body burdens are in an inverse relation to age: "Children had significantly higher . . . concentrations of BPA (4.5 mg/L) than adolescents (3.0 mg/L) and adults (2.5 mg/L), and adolescents also had significantly higher . . . concentrations than adults."[123] It was once thought that BPA had quite a short half-life in the humans, but recent research by Stahlhut, Welshons, and Swan suggests that it may linger longer, because of "non-food exposure, accumulation in bodily tissues such as fat or both."[124]

BPA's distribution in the environment resembles that of PCBs and PB-DEs. It disperses first from products into people's bodies and then more widely into the external environment, especially as products deteriorate. BPA is dissimilar in that does not persist in mammalian bodies as PCBs and PBDEs do. Thus, the high concentrations currently detected in citizens' bodies show that there is considerable ongoing exposure from their immediate environments.

Pesticides

Pesticides include mixtures or substances, "intended for preventing, destroying, repelling, or mitigating any pest."[125] These substances might be chemical, biological, or physical agents. Pesticides include insecticides (which attack insects), herbicides (which are aimed at undesirable plants), fungicides (which attack fungi), fumigants (which are gaseous pesticides), and rodenticides (which aim to kill rodents). I will consider mainly insecticides.

All insecticides are neurotoxicants that poison the "highly developed" nervous system of insects. A standard toxicology textbook points out that an insect's nervous system is "not unlike that of [a] mammal," although insects' peripheral nervous systems are not as complex as that of mammals. Insecticides have been specifically developed to obtain an optimal substance for a particular "biochemical or physiologic feature of the nervous system."[126]

Because of the properties of insecticides, and because they are not especially selective in target species, these compounds can adversely affect a broader range of animals than those they are designed for. Well-known classes of insecticides are the organochlorines and organophosphates, along with carbamates (the last two of which act by common mechanisms).

Organochlorine Insecticides: Organochlorine insecticides include, among other compounds, DDT, rothane, dicofol, methoxychlor, methiochlor, aldrin, dieldrin, heptachlor, chlordane, endosulfan, and lindane. Many of these substances were the concern of Rachel Carson's *Silent Spring*. Most of these have been phased out in developed countries but continue to be used in developing countries.[127]

As the name organochlorine suggests, all in this group utilize chlorine atoms as part of the compound structure, giving them substantial chemical stability. These substances have a number of attractive properties that made them effective insecticides for sustained periods of time: low volatility, chemical stability, lipid solubility, and slow rates of biotransformation and degradation. As Ecobichon notes, these same features led to removal of these compounds from markets in developed countries because of their "persistence in the environment, bioconcentration and biomagnification in food chains, and the acquisition of biologically active body burdens" in higher species.[128]

Pesticides can enter the environment in many ways: direct application, aerial spraying, runoff from fields, poor disposal practices, and emissions from manufacturing facilities or incinerators. In addition, because of their persistence and bioconcentration, organochlorine insecticides can migrate long distances, much like PCBs.

Organochlorine insecticides are fat-soluble, tending to concentrate in milk, dairy products, meat, fish, and higher animals. Developing fetuses can be exposed in utero, and nursing infants can ingest them via breast milk, which is high in fat. Lesser sources of exposure are through the air or drinking water. People that work with such substances in their manufacture or application, of course, can be exposed.[129]

A recent study of some homes in Arizona illustrates the substantial legacy of chlorinated pesticides. The study found measurable amounts of more than four hundred individual chemical substances.[130] Among those were some organochlorine pesticides that had been long banned in the United States—DDT and its metabolite, DDE, heptachlor, lindane, chlordane, methoxychlor, endosulfan, aldrin, and dieldrin. The researchers noted, for example, "Detection of dieldrin . . . indoors is an indicator of the prevalence and persistence of these pesticides in homes, and suggests that other legacy products may also continue to be exposure hazards." Not surprisingly, several of these compounds are still part of our body burdens, although concentrations of most of them are decreasing, and some are below the limits of detection.[131]

Organophosphates and Carbamates: Organophosphates and carbamates are two other groups of insecticides, and they disrupt insects' biology by a common mechanism. They inhibit the effect of an enzyme, cholinesterase, which stops electrical activity in the nervous system. This inhibition can result, according to Ecobichon, in "continual stimulation of electrical activity," which can disrupt the functioning of various nerves and muscles in both insects and mammals, depending upon where the inhibition occurs (see Chapter 4).[132]

The organophosphates were first created in the 1930s by German chemists. Early organophosphates were quite toxic and the Nazis used some as chemical warfare agents (the nerve gases soman, sarin, and tabun).[133] Current variants are several generations removed from those early formulations and much less toxic. Organophosphate insecticides include chlorpyrifos, diazinon, and malathion. Some carbamate pesticides are aldicarb, carbaryl, and barbofuran. These insecticides typically are effective against many insects and account for more than half of the insecticides sold in the United States. Most residential uses have been phased out. Nonetheless, as the CDC has reported, "about 73 million pounds of organophosphate pesticides were used in the United States in 2001."[134]

People may be exposed to organophosphates by any of three major routes: ingestion, inhalation, or dermal contact. Those who work with the substances typically have higher exposures. High exposures to these substances can cause substantial dysfunctions in the nervous system (see Chapter 4).[135]

The recent CDC biomonitoring study found lower levels of organophosphates compared with levels reported by some smaller and earlier studies and compared with population contaminations in Germany. People who work with these substances could have concentrations fifty times higher than those of the general population. The Arizona home study found several organophosphate insecticides in the houses involved in the study, including chlorpyrifos and diazinon.[136]

Conclusion

Several points emerge from the examples and discussion above. Hundreds of industrial chemicals, ingredients in consumer products, and pesticides find us and penetrate our bodily tissues and fluids. Some chemicals, such as lead and mercury, have been present in human populations for centuries as a result of commercial use. Others, such as PBDEs, PCBs, and BPA, had no significant presence in nature and are newcomers, synthesized for commerce. Some chemicals are persistent;

others are quite transient. Some were designed as analogues of natural hormones; in others hormonal activity is an accidental feature.

Sometimes products directly contaminate us, as DES did, and as BPA and PBDEs do. At other times we receive secondary contamination from the environment through routine living, from PCBs, PBDEs, DDT, mercury, and lead. The Inuits of high northern latitudes need do nothing to interact with modern commercial products in order to be highly contaminated.

For most substances there is nowhere to hide. Families that eat organic food appear to have somewhat lesser concentrations of some toxicants in their bodies but can have high levels of others, for example, BPA. Trying to live freer from toxicants can help somewhat and occasionally a great deal for a few compounds, but perhaps not as much as we might think. Even nonparticipants in human society—such as Tasmanian devils, harbor seals, peregrine falcons, polar bears, and whales—can be invaded, highly contaminated, and perhaps harmed as a result. No matter where on the Earth sperm whales swim, they become highly contaminated with toxic metals.

Industrial contaminants are not picky about the routes they take. PBDEs can be inhaled or ingested. Pesticides and BPA can be inhaled or ingested and pesticides can enter through skin contact. Pharmaceuticals are easier to avoid—we don't have to take the very risky ones, if their toxicity is known and we can avoid them. Yet often discovery of toxic properties occurs too late to prevent risks to some people. DES was poorly tested, and it received no significant evaluation or review when it entered the market (Chapter 4). Pharmaceutical laws have been substantially improved since that time, but even so, the FDA misses some toxic products, leaving some developmental toxicants in commerce. Laws that govern PBDEs, BPA, lead, mercury, and other industrial compounds are post-market laws, requiring no testing or review of substances before the public is exposed. We might expect that they will poorly protect us; indeed, that is the case (see Chapter 5).

Some substances leave an incredible legacy—PCBs, PBBs, DDT, and other organochlorine pesticides, lead, mercury—some newer ones likely will as well—perfluorinated compounds and PBDEs. They remain with most of us in a most intimate sense; they are in food, in the environment, in some homes, and in our tissues. Some, like lead, will reside in our bodies for decades.

Despite recognizing the persistence of these substances, our social institutions seem dense about the past; collectively we seem to going down a path similar to PCBs with their PBDE cousins. As recently as 1973, PBDEs were not detected in people's bodies; now U.S. citizens have the highest concentrations in the world. As we will see in Chapter 4, although perhaps not all the i's are dotted and t's are crossed showing adverse health effects of PBDEs, substantial warning signs are present. Perfluorinated compounds may well have similar consequences going forward because of the persistence resulting from their tightly bonded chemical structure.

3

DISCOVERING DISEASE, DYSFUNCTION,

AND DEATH BY MOLECULES

Not only do industrial molecules invade our bodies; they can also cause harm. Some can kill directly and quickly. High doses of arsenic can kill rapidly. Lower doses might drag out the process but be just as deadly in the end, as presented in *Arsenic and Old Lace*. Animal studies strongly suggest that ill-timed low doses during fetal development can contribute to lung, skin, urinary, and bladder cancers long after arsenic has left a person's body.[1]

During the Middle Ages and the Renaissance the route to a kingship or other positions of high power might lie through long-term lacing of a rival's food with lead, the *"poudre de la succession*—or succession powder."[2] Still other molecular substances can increase our risks of disease, catalyze the effects of other compounds, or accelerate illnesses that we might contract in any case. Early life exposures to diethylstilbestrol (DES) or to DDT (dichlorodiphenyltrichloroethane) increase women's risk of breast cancer. Some neurotoxicants appear to accelerate the onset of dysfunctions of old age, such as Parkinson's or dementia.

However, molecular harms differ from the grosser injuries caused by guns, knives, axes, and the like, some common weapons of violent criminal activity. Gross harms are quickly identified and usually readily understood, the source of harm is comparatively obvious, and adverse consequences typically follow immediately after the violence.

Other than quick-acting poisons, molecular-caused harms are quite different. Molecules are silent, invisible, and typically odorless invaders of our bodies by various routes. They enter and we do not know it. If they trigger bodily harm, we are none the wiser, at least for some time, perhaps forever. Discovery of such harm may not occur for years. Even then it may be difficult to trace the injury to some particular invasion or even to a generic exposure.

Scientific studies are needed to reveal molecular hazards. Consequently, different kinds of studies—their features and use—are an important part of the story that follows.

This requires delving a bit into an unfamiliar and arcane world. Comprehending the outlines of the main kinds of health studies helps explain why risks from some exposures will be difficult to identify and why scientists need to utilize all relevant research, especially experimental animal studies, to uncover hazards. (Readers familiar with these issues could skim or bypass most of the discussion and go to the conclusion of this chapter.)

Research needed to reveal many of the diseases and dysfunctions discussed in this book can take considerable time. For example, it has taken forty to fifty years to identify first vaginal cancer and then breast cancer in women exposed to DES in utero.[3] And the effects of DES may not yet be at an end, depending upon whether it will contribute to more diseases or to diseases in later generations. Of course, not all diseases or dysfunction triggered during development pose such long-term issues, but some do. Difficulties in identifying the sources of disease are also exacerbated when diseases are subtle. It could take years to separate normal variation in mental functioning from the acceleration of dementia in old age caused by pesticides or other neurotoxicants.

Consumers of scientific studies need to appreciate that the use of research tools for revealing risks and harms can also lead to a false sense of security. In part this is because people who are more sophisticated about science but have a commercial stake in the issues can mislead us with carefully phrased assertions about what studies do or do not show. When we are told that "no human studies have shown that substance X poses risks to humans," we may feel there is nothing to worry about, yet this is hardly the whole story. An international group of scientists commenting on an IBM study that was too small to detect cancer risks in electronics plants "predicted that the industry would then cite the predictably negative results, not as inconclusive, but as showing safety, and indeed that is precisely what happened." A later researcher with access to the IBM data from a legal case found elevated risks of cancer among employees.[4]

Such studies may well reveal more about the difficulties of detecting adverse effects or the shortcomings of the studies than about causes of disease. An informed consumer of scientific studies should bring some skepticism toward such assertions, because "no effect" claims are quite difficult to establish. Under special conditions substantiating "no effects" is possible, but it is not easy.[5]

Unfortunately, not all studies are conscientiously conducted; some are designed to minimize, not to discover, or even to hide adverse outcomes. For instance, the chromium industry delayed reporting a study when it was completed, then altered it before publication in order to show that there was little elevated risk of lung cancer from exposure to chromium 6, even among highly exposed employees. Unfortunately this practice is not so rare. There is a history of absent or misleading reports concerning pharmaceuticals, industrial chemicals, cigarette smoking, and various pesticides.[6]

We might also be told that animal studies are irrelevant to humans. When an American Chemistry Council representative said that "tests on lab rats and mice do not prove BPA hurts humans in the same way," this comment obscured scientific understanding of the issues and misled the public.[7] However, a citizenry informed about some scientific tools can better understand public discussion, ask questions about dubious assertions, and better protect itself.

A generic grasp of scientific tools also reveals why postmarket laws are so poor at protecting our children and us from toxicants. Scientific studies take time to conduct and interpret, funding to pay for them, and typically confirmation by other independent investigators. If no public health protections can be implemented until appropriate studies have been conducted, this is poor protection indeed.

Companies whose products may be threatened by scientific findings have commercial incentives to demand unreasonably high degrees of certainty, multiple studies, and "proof" of risks or harm before the public can be protected. If public health agencies acquiesce in unreasonable demands, this can hinder protection of the citizenry (see Chapter 5).

Premarket testing laws would create different incentives. Research would begin earlier on products and would be publicly available to a wider community. This both increases the chances of identifying hazards before exposures and provides other scientists data to follow up and the opportunity to possibly discover more subtle risks (see Chapter 6). Premarket testing also lessens the chances of treating children as experimental subjects.

Finally, some participants in public debates insist that human studies must be conducted before a legal case can be established for reducing risks. Understanding human epidemiological studies, however, reveals the moral bankruptcy of such arguments. A sufficiently large number of people must become sick or die from a toxicant in order to have scientifically significant human evidence that the toxicant poses risks. This is poor public health policy.

Human Studies

Studies conducted on humans seem most attractive from one point of view, since we are usually interested in the adverse effects toxicants can have on adult and developing humans. However, in terms of ethics scientists should not "experiment" on people, exposing them to substances that researchers either know or have good reasons to believe would cause harm. Scientists also should not expose people to substances about which little or nothing is known.

Given ethical constraints on human studies, how can they be conducted to reveal risky exposures without deliberately putting people at risk? In what follows I describe some studies that typically would be utilized to identify risk factors of disease. They are presented more or less in the order they might occur. In each section I utilize an example of MPTP (methylphenyltetrahydropyridine) because research on this and related substances exemplifies a common pattern by which the toxicity of substances can be discovered and illustrates how different kinds of studies can used to gain knowledge about risks. Study of MPTP also provided important research into environmentally caused Parkinson's disease.

Case Reports

In 1982 doctors at a clinic in northern California received one female and four male patients who appeared to be reasonably healthy individuals, except that they had great difficulty moving their muscles. They ranged in age from twenty-six to forty-two years. All had a history of heroin use, and all had recently obtained, according to one report, a "new kind of heroin." The authors of this report note, "All patients became symptomatic within a week after starting to use the new drug." These individuals—"frozen addicts"—appeared to have Parkinson's disease. In the words of J. William Langston and others, each "revealed near total immobility, . . . a complete inability to speak intelligibly, a fixed stare, marked diminution of blinking, facial seborrhea, constant drooling, a positive glabellar tap test [inability to resist a blinking reflex], and cogwheel rigidity [muscle rigidity] in the upper extremities." The patients otherwise had normal brain functions. One day they were reasonably healthy young to middle-aged people; within a week or less they had the physical character-istics of elderly persons with advanced Parkinson's disease.[8] How might the doctors begin to discover the causes of their patients' conditions?

The physicians strongly suspected that the Parkinson's disease in these individuals was caused by methylphenyltetrahydropyridine (MPTP), a contaminant of synthetic heroin. They published their results in a case report, or a "case study." According to the International Agency for Re-search on Cancer (IARC) these reports typically "arise from a suspicion, based on clinical experience, that the concurrence of two events—that is, a particular exposure and occurrence of a [disease]—has happened rather more frequently than would be expected by chance." Case series can pro-vide "evidence that a finding, even though still rare, is repeated."[9]

Case reports can provide early warnings of adverse reactions resulting from exposure to vaccines, drugs, poisons, or environmental and occupa-tional substances. The reports function best to reveal adverse reactions when there is a reasonably short interval between exposure and adverse effects.[10] But they have also been used to identify carcinogens (as dis-cussed below) and other toxicants when there are longer latency periods. Case reports are typically sent to a central location, such as the Centers for Disease Control or the Food and Drug Administration. Centralized data might reveal patterns of adverse effects.[11]

While case studies can provide early scientific evidence, they have some limitations. As single observations, they could be reporting mere coincidences. They also do not comprehensively identify all cases of the same disease in an exposed population and would not estimate the num-ber of cases of the same disease in unexposed groups, thus not permitting comparisons between exposed and unexposed populations.[12]

On occasion, if case reports identify a rare enough disease, they can be used with other data to establish approximate relative risks between an exposed and unexposed population. This occurred, for example, when researchers in Kentucky noticed that three employees in a polyvinyl chlo-ride (PVC) plant contracted a very rare form of liver cancer. Combining data from several other PVC plants where employees had the same liver cancer together with animal data, researchers were able to rule out alter-native causes of the disease. Because of its extreme rareness, they were also able to calculate an approximate and very high relative risk for those who worked in such plants—a risk of disease four hundred times greater among employees than in the general population.[13] Case reports combined with animal or other kinds of evidence can provide quite good data about adverse effects of toxicants.

Some case studies are good, scientifically relevant evidence of causation, and some are less so. Merely descriptive case reports—reports that do not rule out alternative causes of disease, do not assess features of a patient that might have led to an adverse reaction, do not address the biological plausibility of the adverse reaction, or are not at all subtle about the temporal relations involved—are quite poor evidence or may constitute no evidence at all. However, even such data, when combined with additional adverse event reports or other information, may suggest a pattern of causation.[14]

Analytical case reports can be better evidence of an adverse effect. A case study of cardiac effects from the pharmaceutical Parlodel provides an example. Parlodel (generic name, bromocriptine) was a drug approved for new mothers who chose not to nurse. (Subsequently, the FDA found that Parlodel was no more effective in reducing breast milk than letting nature take its course.)[15] A thirty-two-year-old French woman had an uncomplicated pregnancy resulting in a normal birth. She chose not to breast feed her child, and her doctor prescribed Parlodel. Following release from the hospital and soon after taking the drug, the patient suffered visual hallucinations. Ten days after the birth she suffered severe chest pain; she had total blockage of her right coronary artery. Physicians reopened her artery, and she went home, doing well.

One month after the first coronary blockage, her physicians invited her back to the hospital for a "re-challenge" with a deliberately administered dose of bromocriptine to try to determine "the mechanisms of her previous myocardial infarction."[16] She agreed. Two hours after administration of the drug, her arteries began to constrict. Although she did not experience chest pain, she was on the verge of another coronary artery event.

The physicians analyzed the rechallenge. Myocardial infarctions are quite rare in women of childbearing age. The patient had smoked one pack of cigarettes per day, but her lipids and glucose levels were normal, and she had no plaque buildup in her arteries. She had no heart or artery conditions that would ordinarily lead to blockages. The device used to look for clots did not cause the blockage during the rechallenge. Of twenty-four case reports in the literature concerning artery blockages in older childbearing-aged women, eleven were associated with treatments from the same class of drugs as Parlodel, and five of the eleven were associated with Parlodel itself.[17] This suggested that Parlodel might well cause a spasm of the artery.

The patient's physicians concluded that Parlodel was "a possible etiologic agent causing postpartum myocardial infarction."[18] The strength of this case report is that it does not merely describe what happened but analyzes various factors that might have caused the heart blockage when physicians rechallenged the patient with Parlodel and rules out alternative explanations.

Let's return to our Parkinson's patients. Physicians suspected that MPTP caused Parkinson's disease, because of the quite short time period between exposure and effect, the comparative young age of the patients, the unusual onset of the Parkinson's-like symptoms, and the dramatic consequences of the disease. If the disease had been multifactorial with a longer latency period, it would have been more complicated to identify with case reports.

Once an exposure is suspected of causing a disease or dysfunction, what human evidence for a causal connection might be considered better than case reports? Researchers could propose trying to identify the cause of disease by using randomized clinical trials (but ethical restraints would not permit that approach) or by utilizing various kinds of epidemiological studies that permit research on a wider range of exposure possibilities.

Randomized Clinical Trials

An ideal scientific experiment would aim to create, "duplicate sets of circumstances in which only one factor that is relevant to the outcome varies, making it possible to observe the effect of variation in that fact. Achievement of this object requires an ability to control all the relevant conditions that would affect the outcome under study." Unfortunately, in biology, finding duplicative circumstances that vary by only a single factor "is unrealistic." Biological variation greatly complicates this task.[19] This is particularly the case for biologically heterogeneous humans.

In a randomized clinical trial, researchers would randomly divide two groups of people into those exposed to MPTP and those not so exposed, to ensure similarity between the two groups. This would also avoid selection bias in the study.[20] Using comparisons between the two sample populations, the researchers might then be able to determine whether MPTP was associated with parkinsonism or not, and perhaps they could determine how much higher risks from MPTP were for individuals with extensive exposure than for those not so exposed. Of course, a clinical trial on the effects of MPTP would not be conducted, for ethical reasons.

In clinical trials if researchers do not know whether a substance is harmful or beneficial, they must conduct prior research to assess risks from the exposures and to ensure the safety of deliberate exposure. Two major international accords—the Nuremberg Code and the Declaration of Helsinki—plus U.S. regulations place ethical constraints on medical experiments, requiring substantial testing prior to experiments in order to determine the safety of the exposures and any potential risks from them (see Chapter 6 for more on this issue).[21]

Nonetheless, if unexpected adverse effects did appear in a clinical trial despite careful, conscientious prior testing for risks, such an outcome would not be ethically problematic. Prior research would have found no unreasonable risks and the study would have been aimed primarily at ascertaining any beneficial effects; any results showing adverse effects would be adventitious by-products of the primary research.

Epidemiological Studies

Most human studies that investigate potentially toxic substances are non-experimental. These are epidemiological studies. They simply take advantage of any "natural experiments" in circumstances in which people have already been exposed by where they live, where they work, what pharmaceuticals they take, or to what substances they have been environmentally exposed.[22] In the workplace people are exposed to a variety of toxic substances, for example, asbestos, benzene, lead, ethylene oxide, or vinyl chloride. Other individuals are exposed because of voluntary activities, such as smoking, hormone replacement therapy during menopause, or birth control pills. People of the Faroe Islands are exposed to mercury, PCBs, and other toxic hazards merely by their traditional diet.

Nonexperimental studies aim to mimic as closely as they can an experiment by "the selection of subjects." Researchers might utilize case-control or cohort (also called follow-up) studies.[23]

Case-Control Studies: Case-control studies compare a group of individuals who have a disease or dysfunction ("cases") with a very similar, carefully matched group of nondiseased persons ("controls").[24] Researchers then try to determine what differences between the two groups, if any, might have caused or contributed to the disease.

For instance, scientists seeking to identify the causes of Parkinson's disease in young people would typically use such a study. These scientists would compare parkinsonism patients with a carefully matched group of people without the condition, in order to determine if there were exposures or life circumstances common to those with the disease (but not common to the control group) that would likely explain their conditions. Of course, the case reports already strongly pointed to a likely cause— MPTP exposure. However, a case-control study would provide a more systematic means of investigating the likely contributors to disease. A case-control study of those with parkinsonism found that ingestion of MPTP in greater or lesser concentrations was the obvious contributor to the dysfunction.

For another example, researchers conducted an international case-control study to try to determine whether exposures to herbicides and their contaminants caused soft tissue sarcoma and non-Hodgkin's lymphoma. Soft tissue sarcoma is a cancer of tissues such as muscles, tendons, fat, blood vessels, or other supporting tissue of the body. Non-Hodgkin's lymphoma is any of several cancers of the immune system.[25]

Researchers considered eleven patients with sarcoma in an international group and thirty-two with lymphoma and compared them with a large number of controls without the diseases, matched for age, sex, and country of residence. They then reviewed exposures of the diseased people to twenty-one different chemicals or mixtures of these chemicals and their contaminants. Phenoxy herbicide exposures and their contaminants—including, among others, dioxin,—increased patients' risk of sarcoma by about tenfold.[26]

Cohort Studies: A follow-up, or cohort, study is a second kind of epidemiological study that could help identify causal relations between exposures and disease. In such a study, researchers would consider people who were free of disease but who had been exposed to a potential disease-causing substance and compare them with a control group not exposed to the suspect substance. Researchers would seek to determine whether exposure caused adverse effects, at what rate any diseases or dysfunctions might appear, and what range of effects might be associated with exposures. The physicians who first identified MPTP-caused parkinsonism also suggested that such a study would be possible by locating people in industry with known exposure to MPTP and comparing them with similar groups not so exposed.[27]

Cohort studies might be prospective or retrospective. Prospective studies would begin at a certain time and then follow exposed and control groups forward in time to determine if there were differences in adverse effects between the exposed and control groups. A retrospective study would find a group of people who had already been exposed to a substance suspected of causing disease, find a comparable group not so exposed, and then seek to determine if there were a difference in disease rates between the two groups.

Finding appropriate measures of exposure are important in each case, but typically this is more difficult for retrospective than prospective studies because researchers must rely on historical records that may or may not be very accurate. In contrast, while prospective studies can perhaps better monitor exposures, scientists must wait for a substantial period of time to allow any diseases to appear after initial exposures. In the cases of diseases caused by DES, this could require a fifty-year follow-up. Public health protections cannot await such studies.[28]

Herbert Needleman and David Bellinger studied children who had early life exposures to lead, in order to try to determine whether such exposures were associated with lower intelligence quotients, learning disabilities, or other adverse neurological outcomes. They both measured the amount of lead in the placental cord blood and gathered children's milk teeth discarded as their adult teeth came in. They then conducted various IQ and learning ability tests on the children as they grew, to determine if those with greater lead exposures had lower IQs or more learning disabilities than children with no or much less lead burden. Roughly speaking, Bellinger and Needleman found that the greater the concentrations of lead in children's teeth on average, the lower the IQs of these children were and the more learning and other neurological problems they had.[29]

Induction and Latency Periods

Both case-control and cohort studies must allow sufficient time for a disease to appear. They must allow time for a disease process to have been induced as well as sufficient time between initiation of disease and its appearance in a clinically detectable form.[30] Several critical biological events or components may be needed to trigger a full-fledged disease. The time it takes for these events to occur is the "induction period." Even after a disease process has been initiated, it may still take time for it to appear and be reliably identified, for example, for cancer tumors or neurological effects to appear. The latency period of a disease is the time between completion of all events needed to produce the disease (completion of the induction period) and the moment when physicians can clinically detect it.

Diseases might have shorter or longer induction or latency periods. For the addicts who took MPTP, the effect was sufficiently immediate and the period between destruction of some brain cells and manifestation of a Parkinson-like condition sufficiently short, in some cases a few days, that the induction and latency periods were extremely quick. Short induction period, short latency period.[31] Various bacterial or viral infections, such as measles, also have short induction and short latency periods.

Cancers can have a fairly long induction period and a long latency period before the disease is clinically detectable. The general point can be illustrated by DES-caused vaginal cancer. On the one hand, if in utero exposures to the synthetic estrogen DES were sufficient to cause vaginal cancer years later, then one would say that the induction period of this disease was quite short (only in utero exposures were required to complete the disease process). However, the latency period of the disease would then be twenty years.[32] In this case, a very short induction period was followed by a long latency period.

On the other hand, if a natural estrogen surge during puberty were needed to supplement in utero DES exposures before all the antecedents for vaginal or cervical cancer were complete (a second critical step, as it were), then one would say that the disease had a long induction period. However, since the disease process on this possibility would have been completed only during puberty, the cancers would have had a relatively shorter latency period.[33] Here, a longer induction period preceded a shorter latency period.

Consumers of scientific studies must understand these points. If studies do not allow for both induction and latency periods, they will not detect the diseases for which they search. The public, courts, or regulatory agencies can be presented with studies that are inadequate in this regard and be told that exposure does not cause disease. In reality, adverse effects would have escaped detection because of a poorly or misleadingly designed study.

This issue could be quite acute for some diseases triggered during development, such as some neurological diseases or dysfunctions. Based in part on discoveries that MPTP, as D. B. Calne and colleagues note, "can enter the central nervous system and . . . induce neuronal damage" in an area of the brain called the "substantia nigra," some scientists have proposed a similar mechanism for other substances.[34] Thus, exposures to certain environmental toxicants early in a person's life might diminish neurons needed for normal brain functioning. However, the remaining healthy neurons would have sufficient adaptability to compensate for missing or poorly functioning neurons for much of a person's life. As people age, however, there is some natural loss of neuron functioning. This loss, combined with neurons that were damaged years earlier, would then lead to diminished neurological functioning earlier than would have occurred without the toxic insults. This has been illustrated in experimental animal studies.[35]

There is an additional social problem related to the scientific problem of the follow-up period. Not all researchers are seriously interested in detecting diseases or dysfunctions; their employers or funding sources may not welcome such results. Consequently, there have been occasions in which studies have been conducted that are too short, given the induction and latency periods of disease, to have detected a disease even if the exposure causes it. Yet those who reported the studies claimed that the exposure caused no adverse effect.[36]

This is one of the most basic mistakes of epidemiology. "No evidence of an effect" from a study does not imply that the study shows "evidence of no effect." An analogy sharpens the mistake. Just because we cannot see germs with our unaided eyes, it does not mean there are no germs on a desk. We must have instruments of sufficient power and use them properly to detect the germs. The process is similar with epidemiological studies.

The public and other consumers of scientific information must be alert to such errors or deliberate strategies. Was a study conducted for a sufficient period of time to detect any disease or dysfunction of interest? Were a sufficiently large number of people included in it (was it powerful enough) to reveal the adverse effect? Consensus scientific committees have identified conditions under which "no effect" studies correctly show that there is evidence of no adverse effect between exposures and disease; it requires multiple studies conducted carefully, followed by careful interpretation.[37]

Strengths and Limitations of Each Kind of Study

Case-control and cohort studies have different strengths and can provide different information about exposures and disease. Cohort studies permit scientists to evaluate a range of possible adverse effects related to a single kind of exposure. For example, do organophosphate pesticide exposures contribute to cancers, neurological dysfunctions, or other adverse effects? If children had in utero exposures to methylmercury, what range of adverse effects, if any, would be manifested?[38]

The National Children's Study in the United States has begun to examine the effects of environmental influences on children's health and development. As a prospective cohort study, it will follow children from birth to age twenty-one. Using good exposure data about various "environmental" factors (broadly construed), it seeks to determine which factors, if any—including toxicants, lifestyle factors, and geographic or cultural influences—might be correlated with adverse outcomes. Discarded umbilical cords preserved over time would provide in utero exposure data. Researchers would then analyze the blood from these cords to determine any toxicants or other anomalies and their concentrations. Researchers would seek to identify what, if any, adverse effects appeared and try to determine environmental factors that might have caused them.[39] Since this prospective study was begun in 2009, it could be some time before definitive results will be known. Some results might appear earlier, if they are evident at birth or in early childhood, but others might not be seen until much later, if diseases or dysfunctions are delayed many years or are quite subtle.

In general terms, a cohort study might reveal several different diseases or dysfunctions resulting from a particular exposure. For instance, lead not only is a neurotoxicant but also can elevate blood pressure, cause strokes and kidney damage, and injure other organ systems. In contrast, a case-control study can investigate a wider range of potential causes or contributors of disease, because it begins with diseased patients and then seeks to determine what contributed to the disease.

Each study has limitations. Cohort studies must have large enough samples to detect rare diseases. For example, the National Children's Study will study one hundred thousand children. However, if a study is not large enough, it might miss rare adverse effects, even if toxicants cause them. Large studies are expensive.[40] Smaller studies cost less but might not detect risks of concern.

Case-control studies can much better study the causes of rare adverse effects because scientists consider only those individuals with a dysfunction or disease and seek to infer what antecedent conditions might have contributed to it. Early research on DES used such a study, when early case reports were soon followed by a case-control study of seven women with the cancer.[41] Each woman was compared with four matched controls. Arthur L. Herbst and colleagues found that the mothers of women with vaginal cancer had been treated with diethylstilbestrol during the first trimester of their pregnancy, but none of the mothers of the controls had been so treated. This revealed a very high risk of cancer from DES exposure in utero.

Cohort studies can have difficulties tracking exposed and control subjects over the study period. The longer the interval of study, the harder this tracking will be. But case-control studies that rely on subjects remembering what drugs they took or what exposures they might have had can, according to Rothman, suffer from "recall bias in classifying exposure" because subjects may misremember exposure conditions. Subjects might also recall exposures greater or lesser than they really were. However, this is a less serious problem with cohort studies.[42] When scientists have objective measures of exposure, such as lead in teeth or toxicants in placental cord blood, this problem of faulty recall is avoided. However, there can be problems estimating exposures with cohort studies, when they have to be inferred from workplace records or, even worse, from what might have been environmental exposures.

Other Limitations of Epidemiological Studies

Consumers of scientific information must understand other limitations of epidemiological studies in order to comprehend what they can and cannot show. Epidemiological studies can be insensitive in other ways that create misleading results or frustrate the detection of diseases.

The natural history of a disease might be much longer than estimated. If an exposure is comparatively recent, there may not yet be "sufficient human experience with the agent to determine its full toxicological potential."[43] This might be a problem, as noted earlier in this chapter, if some pesticides accelerate or worsen neurological diseases typical of old age, which some experimental studies have shown and recent epidemiological studies are suggesting. A research design that is too short for the natural development of the disease would not yet reveal any adverse effect, even if pesticide exposure causes the disease. This would not be blameworthy science but an error from insufficient understanding of the disease.

Exposure estimates can be what some experts call "crude and retrospective."[44] Crude estimates can lessen the impact of the research but still might point to clues. Obtaining data for prenatal exposures could be especially difficult. How would researchers know, several years after the fact, to what concentrations a developing fetus might have been exposed? The lead studies by Needleman and others solved this problem, as we have seen, by utilizing preserved placental cord blood and discarded baby teeth. This imaginative solution may not be available to all researchers, however.

For rare diseases, cohort studies may, as Vincent James Cogliano and colleagues note about cancer, be insufficiently "sensitive to identify a carcinogenic hazard except when the risk is high or involves an unusual form of cancer."[45] When the frequency of disease is rare in an exposed population it likely would not be detected by most studies.

Adverse effects can be so subtle that early research might not detect them. For example, studies of lead in the early twentieth century focused on quite visible and dramatic adverse effects, especially in workers. It was only much later that Herbert Needleman, Richard Canfield, Bruce Lanphear, and others developed methods to detect subtle neurological effects of lead in children, including diminished IQs compared with their peers or particularly aggressive social behavior.[46] Similarly, major birth defects, such as shortened limbs from thalidomide exposure, were obvious and easily seen. The more subtle neurological deficits from thalidomide were identified later.[47] Thus, one should not expect that early studies will reveal all adverse effects or all possible long-term effects. Yet critics might be tempted to assert on the basis of early studies that the substances in question have been exonerated.

Researchers can overcome this kind of insensitivity, but it takes time and understanding of the disease process to identify the more subtle of its manifestations. More-subtle consequences must be recognized and more-refined measures developed to detect adverse effects; greater care must used in designing experiments; and probably increased time and efforts at interpretations are needed to be sure the effects are real as opposed to being artifacts of studies.

Epidemiological studies are limited also in their "ability to identify and adjust for confounding exposures or genetic susceptibility," as M.

Carbone and colleagues note.[48] Consider the genetic susceptibility of sub-populations in the United States. Many adults do not appear to be susceptible to adverse effects from organophosphate pesticides at lower doses, simply because their bodies can sufficiently detoxify the agents before they cause harm. However, most infants and a subset of adults do not detoxify these substances well. They lack a particular genetic allele that makes a major contribution to this function.[49]

A study that lumped people with normal genetic abilities to detoxify organophosphate pesticides in with a more susceptible subpopulation might fail to identify those who are genetically susceptible to organophosphate exposures. Moreover, sensitive subpopulations are unlikely to be detected in early studies because scientists will need to refine their knowledge over time, identify possible genetic contributions to the diseases of subpopulations, and then design studies to determine whether there are adverse effects in subpopulations.

A quite serious long-term issue is implicit in the data concerning human contamination discussed in Chapter 2. Human populations might become so contaminated by industrial chemicals that over time this will undermine the efficacy of epidemiological studies to identify dysfunctions and diseases. At their best epidemiological studies are not always particularly sensitive in identifying real effects, and the rarer the disease, the greater the difficulties (up to a point). However, with more and more contamination of the general population, there will be fewer (if any) uncontaminated control populations against which to compare exposed adults and discover adverse effects.

Finally, an important social point: because of the latency periods of cancers and some developmental effects, if the public must rely upon human studies, fellow citizens must become sick or die before epidemiology would uncover diseases and dysfunction. Thus, years of preventable contamination and diseases would likely occur before studies could be completed.[50] Socially and for public health protections, there is a premium on identifying contributors to cancer and developmental diseases in children and adults. Identifying toxicants is even more important for protecting children's health. There might well be many annual cohorts of adversely affected children, whose diseases would go undetected and unprevented while slow, expensive, insensitive human studies were conducted.

Some Considerations in Interpreting Epidemiological Studies

In order to interpret epidemiological studies, scientists must consider several kinds of potential errors that can creep in. These researchers must be alert to sources of bias or systematic errors in their studies. These are, the IARC points out, "factors in study design or execution that [suggest] a stronger or weaker association than in fact exists between disease and an agent, mixture or exposure circumstance." For example, volunteers in a colon cancer study might have lower disease rates simply because they are more health conscious and adopt a diet that is less likely than others to contribute to colon cancer. For another example, some have proposed that birth order in a family leads to Down's syndrome. However, it is not the order of the birth but the age of the mother at the time a child is born that offers the better explanation of the likelihood of a child having Down's syndrome.[51]

There are other kinds of systematic bias that could influence a study, for example, recall bias or misclassification.[52] One kind of misclassification would consider some who were exposed as being unexposed. This error would dilute the exposed population and put more diseased people in the "unexposed" group, thus increasing any disease rate in the unexposed group, which would suggest a lesser difference between exposed and unexposed groups. A related but opposite misclassification might occur when people who are not exposed (or who are less exposed) to a toxicant are classified as being exposed (or more highly exposed). This would tend to reduce any disease rate in the highly exposed group, and it would tend to produce a study showing a reduced relative risk between exposed and control populations. This bias can be a deliberate strategy.

For instance, the chromium industry authorized a four-plant study specifically designed to assess lung cancer risks from chromium 6 across the plants. However, as unfavorable data came in, researchers changed the protocol according to which the study was conducted. They combined intermediate and low exposure subjects (who would have lesser adverse effects) and then compared high exposure employees with this combined data. Thus, researchers reported that the most highly exposed employees had some elevated risks from chromium 6 but much less risk than a proper comparison of data would show.[53]

What epidemiologists call the "healthy worker effect" is a more subtle selection bias. Workers are healthier than the general population, and they ordinarily have better health care.[54] If these two groups are compared without adjustment, the study will underestimate any disease from work-place exposures. This bias is important because many epidemiological studies are conducted on available populations of workers exposed to in-dustrial chemicals. Such studies can easily underestimate risks to the public.

Another concern in assessing studies is random statistical error. The best studies must be sufficiently large to minimize outcomes that could occur by chance alone—statistically misleading results. Since epidemiol-ogy can consider only representative samples of larger groups of exposed and unexposed populations, there is some probability that by random chance alone a research project would study a sample that was not prop-erly representative of the larger groups in question.[55] For instance, by sta-tistical chance an exposed group might have more people with a tendency to disease than the larger group they represent. Or a "control" or "unex-posed" group might accidentally have more individuals in it with a ten-dency to the disease under study.

Studies that are too small might misleadingly suggest a relation between an exposure and an adverse effect that appeared by chance alone, a "false positive." In contrast, small studies might not detect an adverse effect, even if it were present. The outcome would result in a misleading "no effect" or "false-negative" result. Both kinds of errors would be statistical flukes.

Statistical methods provide a means of assigning probabilities to out-comes so as to estimate the odds that they are true positives or false pos-itives. So-called statistical significance is one measure of the chances of a positive study being mistakenly positive because of random error. (Lead-ing epidemiologists have long opposed the idea of statistical significance, arguing for more informed methods to determine what the data show.)[56] If a study were to show that there was no association between an exposed group and an adverse effect, this would be a "no effect" or "negative" study. Such a study result might be correct, that is, a true negative, or in-correct, a false negative.

Scientists tend to design and analyze studies to avoid false positives, in order not to introduce mistakes into the body of scientific knowledge. However, for public health purposes it is particularly important to avoid studies with mistaken "no effect" outcomes, because people's health and lives are at stake. Larger studies can assist in avoiding false "no effect" outcomes. However, increasing the size of studies tends to increase their costs. Since I have addressed this issue in some detail elsewhere, I do not pursue it further here.[57]

Once scientists have identified an association between exposure and a disease or dysfunction, they would assess whether the observed relationship is causal. I do not consider these assessments here, as I have discussed them more extensively elsewhere.[58]

Although human epidemiological studies can be quite useful for identifying risks and harms to humans, they do have limitations, which we have just reviewed. When faced with such limitations in human studies, what do scientists do? If they are concerned that an exposure contributes to adverse effects, they must turn to other kinds of evidence.

Experimental Animal Studies

For several reasons scientists regard experimental animal studies as especially important for inferring that substances can cause adverse effects in humans. To identify substances that can cause diseases, scientists need not wait for humans to become sick or die. Any latency periods are shorter in animals that have shorter life spans than humans do; thus researchers may be able to identify adverse effects more quickly by studying animals.

Animal studies are controlled experiments in a way that human epidemiological studies are not. Animals are randomly assigned to control and exposed groups. Exposures can be specified quite accurately, and there are few confounders. Conditions of the experiment can be designed to test the plausibility of a particular hypothesis or to investigate a very specific issue. Experimental animals can be sacrificed and autopsies conducted on their tissues in order to see what effects, if any, exposures had on particular tissues or organs. This is a great strength of animal studies, but it is also a source of controversy among animal rights advocates.

In addition, because animal studies are experiments, they better establish causality between exposures and disease, and experiments can be replicated across labs. And, again, the short lifetimes of animals permit quicker insights into causes of disease.[59]

Scientists utilize animal experiments not only to understand the biology of particular animals but also to understand the biology and toxicology of humans. Some scientists regard epidemiological studies as "merely" statistical associations between exposures and adverse conditions. That is, there are scientists who are reluctant to infer causal relationships between exposures and human disease on the basis of human studies alone. Researchers who seek to understand better the causal relationship between a substance and disease typically use animal models for the purpose.

Experiments are usually conducted on rodents, such as rats or mice, but may also involve other species, such as hamsters, monkeys, dogs, rabbits, or sheep, as the circumstances and similarities to humans for the effects under study recommend. Developmental researchers aim to find an appropriate species to serve as a model for humans when studying effects of substances that would cause diseases in humans. A good model increases researchers' confidence that adverse effects in the chosen species will provide biological insights to risks in humans and to the doses at which risks are likely to occur. For instance, spider monkeys turned out to be a good model for MPTP-induced Parkinson's disease. Experiments on them helped to confirm that MPTP caused damage in a certain region of the brain that exercises voluntary control over muscles.[60]

In addition, because animal studies are controlled experiments, they can be utilized to shed light on the mechanisms that contribute to toxic effects. In fact, experimental studies on animals yield virtually all understanding of the basic mechanisms of toxicity, whether these result from carcinogenic, reproductive, or developmental toxicants.

Bernard Goldstein, former dean of the University of Pittsburgh Graduate School of Public Health, points out that a good deal of toxicology rests on experimental animal studies: "There is a commonality of biological function across species. . . . The similarity in cellular and organ function is particularly strong among mammals such that extrapolation of effects from one species to another is accepted by the scientific community as a means of evaluating the toxicity of external agents. . . . Although extrapolation can be complex, there is sufficient information to permit reliable extrapolation in many situations."[61] The reason for this similarity is that "nature is inherently conservative," according to Linda Birnbaum, director of the National Institute of Environmental Health Sciences. Consequently, if "chemicals . . . cause a multiplicity of effects in several species, why would we think that some people are not going to be susceptible to the effects?"[62]

How might experimental studies on animals be conducted? In seeking to identify whether pesticides could cause Parkinson's disease, researchers assigned mice to control and experimental groups and then exposed pregnant mice in the experimental group to one pesticide (paraquat) or one fungicide (maneb, each for a week total).[63] Maneb is used to prevent crop damage and to prevent deterioration of harvested crops. It is moderately toxic to humans and of moderate to low toxicity in animals, but it is unstable in the environment, with a short lifetime. At typical exposure levels in food, it is not considered to have adverse effects on humans. Paraquat is an herbicide used to control weeds. It is highly toxic to animals and can cause human toxicity. In contrast with maneb, paraquat is persistent in the environment.[64] Later, researchers exposed the offspring of controls and experimental animals to one pesticide or the other.

The result: mice exposed to paraquat and maneb in utero and then dosed with either paraquat or maneb in adulthood came down with parkinsonism. Another experiment by Barlow, Cory-Slechta, Richfield, and Thiruchelvam dosed the mice only with one maneb or paraquat dose, not both, then dosed the offspring just once in adulthood with one or the other. Males suffered a dramatic (95 percent) decrease in locomotor activity. And the pertinent portions of the mouse brains had been damaged.[65] For ethical and legal reasons such studies would be beyond the pale for humans.

Moreover, since even genetically very similar animals have some variation between them, some control animals may naturally develop some tumors or other adverse effects. Scientists would then compare the disease rates in the control group with the disease rates in the experimental groups to determine whether there were statistically significant differences between them.

Humans Are Animals

As many people have pointed out, humans are animals.[66] Even so, experimental studies may appear problematic because there are such obvious visible differences between laboratory animals and humans. Animals are smaller, are furry, have tails, and, if they are rodents, have pointed noses. They may smell bad to humans and also eat different foods. Biologically, laboratory animals have faster metabolic rates, process substances somewhat differently, and are more genetically homogeneous.

In contrast, humans are larger and taller than rodents. They weigh more; lack fur; have no tails; have short, stubby noses; and usually smell better. Humans also have, as James Huff and David P. Rall point out, longer life spans, have slower metabolism, suffer concomitant diseases, are exposed to various exogenous chemicals, and likely "process (i.e., metabolize, store, excrete) the agent in question in a different manner and . . . are much more genetically heterogeneous."[67]

The differences between animals and humans tend to be highlighted by different groups for their own purposes: by some scientists, by the public, and by companies whose products may be legally challenged in the regulatory or the tort law. Moreover, the differences often receive greater attention than the similarities do, but the scientists are agreed: the similarities are more important for understanding biology and toxicology for human health purposes than are the differences.

The public may be persuaded about the differences because of obvious visible characteristics and size differences between humans and animals. The public may also be reluctant to identify human biology and toxicology with that of animals that have a history of unpleasant associations (those animals that are often called "vermin"). Some may even reject the idea that humans evolved from lower animals and are consequently biologically similar to other mammals. These individuals might even dismiss animals as having any biological relevance to humans. (During the last twenty-six years, 38 to 48 percent of the American public has rejected the idea of evolution.)[68]

Physicians who normally conduct research on and diagnose diseases in humans may downplay experimental studies that do not use humans as their research subjects. It is not clear how wide a problem this is, but it could be seen as part of larger problem. As if to illustrate this issue, the California Proposition 65 Developmental and Reproductive Toxicity Science Advisory Panel, composed of seven physicians, recently rejected evidence that bisphenol A is a reproductive or developmental toxicant. Almost all the evidence about bisphenol A is based on animal studies. Board members asserted that "none of the studies they reviewed offered *clear* evidence of the chemical's toll on human health."[69] Did implicit bias against animal studies influence their assessment?

There can be a tendency by experts in a given discipline to value their own research field more than other areas of inquiry. I have observed this dynamic on a science advisory panel concerning possible adverse effects from electric and magnetic fields. Epidemiological studies in a variety of circumstances conducted by different researchers kept reporting similar low-level adverse effects from exposure to electromagnetic fields, or EMFs, thus satisfying important considerations pointing to causation.[70] Epidemiologists gave those studies considerable credence and believed that EMFs could have some adverse effects, although they were probably weak. Biophysicists, who had long studied ionizing radiation from atomic reactions (radiation that can break chemical bonds in humans and other mammals), expressed the view that the EMF fields were simply too weak to cause cellular damage in humans and mammals. Thus, they argued, epidemiological studies must be mistaken. In short, each group had some tendency to more strongly credit their own fields than to accept research results from other fields. Consequently, it may not be surprising that physicians, who typically conduct research on humans, might undervalue experimental animals studies for bisphenol A.

Goldstein notes that mistaken publication bias may also contribute to the idea that "human responses most often differ from other mammals." Differences between humans and animals are scientifically interesting because these differences can assist our understanding of human physiology and responses. Consequently, the results of studies focused on such dissimilarities are "readily publishable in a good scientific journal. This means that review of the overall literature comparing animals and humans will be [misleadingly] biased toward differences rather than similarities."[71]

Companies subject to legal proceedings concerning the health effects of their products might devote considerable effort to discrediting experimental studies or trying to show that they are inapplicable to the issue. Firms have incentives to delay the scientific and legal reckoning about their products as long as possible and to insist on the very best and most certain evidence before their products are threatened by regulatory action or by a tort suit. Since human studies are difficult and expensive to conduct and are often too insensitive to detect an adverse effect of interest, companies are motivated to insist on this kind of evidence. They also tend to denigrate experimental animal evidence for reasons known only to themselves, since many of these companies utilize such studies in product development and assessment. Of course, there are some biological differences between animals and humans, but companies whose products are threatened by scientific findings tend to exaggerate the differences, for legal audiences as well as for the public at large. Is this because this stance may have some traction with an uninformed public?

Despite such reservations about and such denigration of experimental animal studies, there are very good reasons for scientists to use them to identify and estimate human risks from exposures to toxicants. We considered some of these reasons above. There are others. For one thing, major scientists have pointed out that "experimental evidence to date certainly suggests that there are more physiologic, biochemical, and metabolic similarities between laboratory animals and humans than there are differences." As we have seen, Goldstein echoes this idea.[72]

Other experts discuss these similarities in greater specificity: "Biological processes of molecular, cellular, tissue, and organ functions that control life are strikingly similar from one mammalian species to another. Such processes as sodium and potassium transport and ion regulation, energy metabolism, and DNA replication vary little in the aggregate as one moves along the phylogenetic ladder." This similarity is manifested at the level of DNA in the close relationship between mice and humans. At the gene and protein levels, scientists have now confirmed this relationship in studies of the human and mouse genomes.[73]

James Huff, a leading scientist at the National Institute for Environmental Health Sciences has pointed out that the more "we know about the similarities of structure and function of higher organisms at the molecular level, the more we are convinced that mechanisms of chemical toxicity are, to a large extent, identical in animals and man." The U.S. Environmental Protection Agency and scientists explaining the science for federal judges make the same points.[74]

In an area of considerable concern to the public and often of substantial interest in legal venues, scientists have devoted considerable effort to understanding and interpreting studies that assess carcinogens. Thus this effort has led researchers to believe that, based on current information, there is great similarity in the carcinogenic processes between animals and humans. Others find the relationship to be even closer.[75]

Scientists, thus, are quite comfortable using animals as models for biological responses in humans. This is the point of animal studies—good animal models are utilized to show either beneficial or harmful effects in humans. By being able to carry out such studies, researchers can modify experimental circumstances carefully over time and better understand the effects of a substance and establish causality.[76]

Generic conclusions from major consensus scientific committees concur about the importance of experimental animal studies. For instance, the IARC points out that "animal studies generally provide the best means of assessing particular risks to humans," noting that mechanistic and other data can be utilized to address the relevance of animal studies to human cancer.[77] And the IARC adds that "it is biologically plausible that agents for which there is *sufficient evidence of carcinogenicity* in experimental animals . . . also present a carcinogenic hazard to humans. Accordingly, in the absence of additional scientific information, these agents are considered to pose a carcinogenic hazard to humans."[78]

Well-known researchers from the National Institute of Environmental Health Sciences conclude, "Chemicals that are carcinogenic in laboratory animals are likely to be carcinogenic in human populations and . . . , if appropriate studies can be performed, there is qualitative predictability. Also, there . . . can be a quantitative relationship." Other distinguished scientists and scientific committees agree, including the National Academy of Sciences.[79]

A joint committee of the Institute of Medicine (IOM) and the National Research Council (NRC) strongly supports the use of experimental studies, echoing many earlier reports. Animal studies are especially powerful because they can be controlled scientific studies that can reveal effects not likely to be revealed by human use of a substance until harms actually occur. And animal studies can detect risks when similar studies would be ethically barred for human subjects. The report continues, *"Unless there is scientific evidence that raises significant doubt regarding the relevance of specific toxicity findings to humans, it is prudent and scientifically appropriate to consider animal studies relevant in evaluating potential human toxicity, especially . . . where sufficient human data are not available."*[80] The IOM-NAS conclusions apply even more strongly to developmental issues.

When it comes to developmental issues, Kenneth Korach, a leading reproductive biologist, points out that various factors—pharmacological, environmental, and natural—can disrupt development with devastating results. Because many factors can contribute to such consequences, it is difficult to identify one as a critical factor. Thus it is important to use the latest science to understand the phenomena, including observations from experimental animal studies. These observations, Korach notes, "will have direct relevance toward human exposure and its resultant consequences for fertility and normal development."[81]

Developmental toxicologists in a standard textbook reinforce this general view of the relevance of animal studies and add details about quantifiable similarities for developmental effects. In general, experimental studies predict the potential for adverse effects in humans. Concordance between results in animals and humans "is strongest when there are positive data from more than one test species, although even in this case the results are not applicable to extrapolating specific types of effects across species. . . . *In a quantitative sense, the few comparisons that have been made suggest that humans tend to be more sensitive to developmental*

71

toxicants than is the most sensitive test species." Kavlock and Rogers then cite a variety of studies finding concordance between adverse effects for specific animal species and those for humans.[82]

J. L. Schardein and K. A. Keller compare the concordance between fifty-one human developmental toxicants and animal data for the same substances. They found that "overall, the match to the human, regardless of the nature of the developmental response, was rat, 98 percent; mouse, 91 percent; hamster, 85 percent; monkey, 82 percent; and rabbit, 77 percent." F. R. Jelovsek and others, reviewing about 165 substances, found surprisingly good predictive value in animal studies for human developmental toxicity.[83]

When it comes to quantitative measures of doses that cause developmental diseases or dysfunctions, Schardein and Keller found that for nineteen of twenty-one chemicals humans had a lower threshold of sensitivity than animals did. That is, a lower dose of a substance (up to ten times lower than the dose for experimental animals) would cause adverse developmental effects in humans.[84]

There are good reasons for this concordance. Studies are conducted on genetically quite similar strains of animals that have been well characterized and studied over time. However, their very similarity typically underrepresents the much wider genetic and biological diversity of large human populations. For example, when pharmaceuticals or other substances that have been tested in animals and on small groups of people are released into commerce, exposing a wider and more varied human population, there are a greater variety of adverse effects. That humans are more sensitive than animals is worrisome because one would hope that experimental studies could reliably detect adverse effects to humans and the doses at which they were likely to occur. Humans' greater sensitivity makes this reliability harder to achieve. This point is reinforced by recent results from transgenerational studies in animals. Michael Skinner has found that inbred animals are less susceptible to some diseases than outbred or wild types. He believes that the very process of inbreeding makes animals resistant to adverse effects that more genetically diverse animal would exhibit.[85] To the extent that inbred animals are less sensitive to toxic insults results from such research may well underestimate risks to humans.

Of course, there is no mathematical certainty between adverse effects in animals and those in humans. But to insist on such a high degree of certainty would be to demand too much for biology to show. Nonetheless, there are very strong reasons for taking seriously adverse effects in animals as a guide to probable adverse effects in humans.[86]

Experimental studies should be reviewed in light of the best scientific evidence for their relevance to humans. For example, in carcinogenicity studies scientists often need to use high doses and to evaluate a large numbers of tissues for tumors in order to identify carcinogens. There is a concern about whether high doses might change the metabolism of a toxicant in animals compared with lower dose effects in humans. Researchers also sometimes test strains of laboratory animals that are prone to high numbers of spontaneous tumors. These testing protocols can sometimes produce results that may not be likely to occur in humans under normal exposure conditions (although we have just seen that humans may be more sensitive to developmental toxicants).[87] However, even studies such as these may be quite pertinent to several groups of people: to workers exposed to higher concentrations, to especially susceptible individuals, to those subject to accidental exposures, and to those exposed to similarly acting substances. Finally, sometimes chemical agents may induce cancers in animals as a result of mechanisms that do not operate in humans, but this should be established by evidence, not merely asserted as a speculative hypothesis, before rejecting well-established studies.[88]

A National Academy of Sciences committee has just addressed an issue that bears on animal studies as a guide to adverse human health effects: thresholds for adverse biological effects. Some biological processes at the organism level, or perhaps processes below the organism level when dose-response relationships are identified, can be subject to doses of a toxicant below which no adverse effect is seen. This has been a long-standing view—when there are threshold or nonlinear mechanisms involved—that there is a "safe" exposure level below which individuals will not be harmed or be at risk for adverse effects. A threshold constitutes what the NAS calls a "bright line" of safety. However, the Academy committee argues that this view should be modified.[89]

Thresholds apply to individuals, not necessarily to populations. Background exposures, underlying disease processes, substantial biological variability, and other factors may turn a threshold, or nonlinear, mechanism into something much more like a linear dose-response mechanism. (A linear response would be one for which lower and lower doses may result in lesser risks but for which there is no concentration that results in no risk; there is no "safe" dose.) Thus, even if thresholds are appropriate for individuals within a population, they do not necessarily apply to heterogeneous populations with wide genetic variability, sensitive subpopulations, and exposures to other toxicants that might have additive effects.[90] Consequently, often there may be no human population threshold. And even if there is one, it might be quite difficult to estimate. If it could be identified, it would be the threshold for the most sensitive person or group of people within a larger, more biologically varied population.

The significance of this? Animals often exhibit threshold responses when exposed to toxicants. However, because they typically constitute a comparatively homogeneous population, they have less biological variability than humans do. Animals are not exposed to other toxicants, as are humans, nor do animals generally have personal habits that add toxicants to their bodies—they do not drink, smoke, use illicit drugs, or have other environmental or occupational exposures through ordinary living. Consequently, they are likely to exhibit similar thresholds as a group, or at least a group threshold could be identified. When such results are extrapolated to human populations, threshold effects from animal experiments may not be particularly representative of human biological responses, because of differences between experimental animals' and humans' variability, confounding exposures, personal habits, work experiences, and so on. Consequently, threshold doses of substances that cause adverse effects in experimental animals may be underestimated for humans, and the related adverse effects in human may also be substantially underestimated, as Kavlock and Rogers have noted.[91]

Concluding Remarks on Experimental Animal Studies

The upshot of this brief review of a body of evidence concerning experimental animal studies is that when such experiments show adverse developmental or carcinogenic effects, this should be regarded as *strongly presumptive evidence for developmental or carcinogenic toxicity in humans, unless there is some well-substantiated scientific reason for not so regarding it.* A mere hypothesis that animal results do not apply to humans, or a mere speculative reason that they do not, as has sometimes been asserted, should not contravene the well-established use of animal studies for identifying and estimating toxic responses in humans.[92] Similarly, when companies demand "proof" with high degrees of certainty approaching that of math or logic (e.g., that experimental studies prove adverse effects in humans), this suggestion misconstrues the law and science of sensible public health safeguards—perhaps deliberately.

Finally, a cautionary note: animal studies, despite a number of ethical and scientific advantages, are not costless. At a minimum, cancer studies in animals take five years (perhaps more) to conduct and interpret and many person-hours of effort. (Studies of developmental adverse effects may be similarly time consuming.) These studies are also expensive. These time and monetary challenges are difficult enough for basic research, because it does not provide quick and easy tests for determining toxic effects. Moreover, research on animals is routinely used to assess the toxicity of pesticides, pharmaceuticals, and new food additives. However, for purposes of testing a wider range of substances that enter commerce over time, scientists need to find quicker, but still accurate surrogates for toxicity or other short-term tests that can be utilized for these purposes, as I argue in Chapter 6.[93]

Other Evidence Relevant to Toxicity Assessments

Evidence of other kinds can assist researchers in inferring a causal relationship between exposure to a substance and a disease: chemical structure–biological activity relationships, genetic and chromosomal damage data, mechanistic data, and some other tests.

A chemical's molecular structure can affect its biological consequences for mammalian or other systems. (We will see this concerning PCBs in Chapter 4.) A standard textbook points out, "An agent's structure, solubility, stability, pH sensitivity, electrophilicity, volatility and chemical reactivity can be important [in identifying hazards caused by substances]." It adds, "historically certain key molecular structures have provided regulators with some of the most readily available information on the basis of which to assess hazard potential."[94] An Institute of Medicine and National Research Council Committee concurs: "The biological effects of chemicals, including toxic effects, are implicit in their molecular structures (referred to as toxicophores when they are associated with toxic effects)."[95]

Certain classes of structure-activity relationships have been important in providing clues to toxicity or in revealing developmental toxicants or chemical groups that are known to interact with proteins or mammalian DNA. These relationships have provided insights about some carcinogens as well as structural alerts for aromatic amine groups and identified certain dyes as potential carcinogens. Molecular structures similar to valproic acid, retinoic acid, and glycol ethers—all well-established developmental toxicants—

75

provide clues about other developmental toxicants. Such relationships provide strong, but not infallible, reasons for thinking that substances with similar chemical structures have similar biological activity. Industry utilizes such information to identify toxicants during product development.[96]

In addition, several consensus scientific committees as well as federal and state agencies have classified a large number of dioxins and related compounds as carcinogens, based not on human or whole animal studies but on their propensity to bind to a certain molecular receptor (the aromatic hydrocarbon, or Ah, receptor). This receptor tends to increase the toxicity of some substances (and can sometimes decrease other toxicants). These molecules fit into particular receptors as a key into a lock. Moreover, this receptor is, according to the International Agency for Research on Cancer, "conserved in an evolutionary sense and functions the same way in humans as in experimental animals."[97] Thus, if an individual chemical binds to the Ah receptor, this is quite important data for toxicity judgments.

Synthetic substances that mimic estrogens tend to attach to estrogen receptors in mammalian bodies and can contribute to various adverse effects triggered by too many estrogens, especially in women, including breast cancer, among other disorders.[98] This point is quite important in understanding the significance of synthetic estrogens (see Chapter 4).

Other kinds of molecular evidence and evidence of biological mechanisms can be quite valuable. When scientists understand a good deal about the toxicity of a substance, much of this data is at the molecular level. For example, some substances bind to molecules by means of electron transfer, so-called covalent binding. It is irreversible, and if a chemical covalently binds to DNA in a way that interferes with DNA replication or other functioning, this may well trigger a process resulting in cancer or other serious conditions. Other compounds may activate or inactivate some molecular pathways that lead to adverse effects.[99] For example, if researchers know that a substance has a chemical structure that will bind to an oncogene (a gene that tends to cause cancer) or, alternatively, that the substance will bind to a tumor-suppressor gene (one that helps to suppress tumors if they develop) so as to interfere with the gene's functions, this data is very helpful for making toxicity assessments.

Another illustration of molecular-level data contributing to researchers' understanding of disease comes from our MPTP example. Physicians conducted a positron scan of the MPTP patients' brains and were able to see reduced brain activity in some regions compared with the brain activity of those who did not have parkinsonism.[100] In addition, researchers identified squirrel monkeys as an appropriate animal model and were able create standardized doses that would produce predictable effects in the monkeys. They dosed the monkeys with MPTP and, after euthanizing them, found "marked depletion of the neurons in the zona compacta of the substantia nigra in four consecutive animals" that had been systematically dosed.[101] This study further confirmed the toxicity of MPTP, which had first been revealed by case reports, followed by case control studies and the positron scans of people suffering from parkinsonism.

A further point from the MPTP research shows how molecular data can provide clues to other toxicants, thus broadening our understanding of toxicity. MPTP is structurally quite similar to the pesticide paraquat. If MPTP can cross the blood brain barrier and damage neurons, leading to Parkinson's disease, can paraquat do so as well? Despite a difference in electric charge between the two, researchers found that in fact paraquat can have similar effects in animals.[102] Moreover, a very recent human epidemiological study shows that paraquat and maneb increase the risks of Parkinson's disease among people with certain genetic susceptibilities, echoing results from animal studies.[103]

Molecular-level data plus information about how toxicants are transformed within mammalian bodies and other mechanistic information can assist the discovery of other toxic effects. Thus, the International Agency for Research on Cancer utilizes a variety of data from short-term tests, data about gene mutation and chromosomal damage, cell transformation, and even short-terms tests in simpler biologic systems to assist in scientific judgments about human carcinogens. Distinguished cancer researchers recently called for increased research to identify potential carcinogens earlier and less expensively, research based *"primarily [on] mechanistic information,* even in the absence of epidemiological or experimental animal studies in which the tumors are observed."[104]

This sketch of some other kinds of data about toxic potential of substances is far from complete; it merely points to additional categories of data than can assist scientific inferences. Consensus scientific bodies, such as the International Agency for Research on Cancer and the U.S. National Toxicology Program, routinely utilize all such information as clues to toxicity and as part of the body of evidence that support identification of substances that have adverse effects in humans.

Conclusion

We are now in a position to appreciate better strengths and limitations of scientific tools to identify and estimate risks and harms to humans. Science is needed to discover the risk factors of disease or dysfunction caused by molecules. Scientific inferences about risks are much more subtle than identifying the causes of grosser kinds of harms. Substantial induction or latency periods for disease exacerbate these challenges. Researchers must also tease out toxic contributors of disease from natural processes. Such research can require especially refined approaches in human studies. While experimental animal studies are particularly useful in assisting such inferences, they can face unwarranted skepticism. Both epidemiological and animals studies can take considerable time. Such research can be arcane, yet we need to understand its outlines for empowerment to judge scientific claims and for public health protections.

Failure to recognize the limitations of human epidemiological studies may induce a false sense of security. For example, human studies can mistakenly yield "no effect" results. Specifically, studies can be too small to reliably detect the adverse effect of interest, too short for the induction and latency periods or the natural history of disease, or insufficiently subtle to detect an adverse effect from the exposure of interest. When a study has one (or more) of these shortcomings, a researcher could announce that there was "no effect" from the exposure but thereby mislead the public, as both IBM and the chromium industry may have intended with reports of their studies. If such claims are taken as exoneration of the substance, this is misleading and conveys false reassurance.

An interested party might assert (even correctly) that no study had found that exposure to the substances caused harm to humans. At the same time, if animal studies revealed adverse effects, providing reasons for concern for humans, these studies should be taken very seriously. However, if those studies were deliberately attacked and denigrated as not relevant to humans, this might fit comfortably into an incorrect view of the unimportance of animal studies, and people would tend to make the mistake of disregarding the outcome.

This dynamic may be on display concerning bisphenol A. If a company says that there is no human evidence that bisphenol A has caused diseases or dysfunction in humans, this could be technically true but misleading. BPA is a synthetic estrogen (see Chapter 2). BPA is present in the bodies of men and women at levels that cause adverse effects in experimental studies, revealing extensive ongoing exposures. However, it will be difficult for an epidemiological study to find that such concentrations increase women's chances of breast cancer, because estrogenic diseases are common and BPA's contributions would be small. Yet synthetic estrogens added to a woman's natural estrogens could contribute to her risks of breast cancer (see Chapter 4). How much a contribution this is might be difficult to determine at present, but women's exposure to too much estrogen over a lifetime can increase their risks of this disease. Moreover, several recent consensus scientific committees have expressed concern about low-level human exposures to BPA.

Experimental studies should be taken very seriously as early "canary in the coalmine" responses for humans. This is especially important for studies in which exposures are "environmentally relevant" (when adverse effects in animals occur at concentrations of toxicants similar to human exposure levels). Even at higher levels, these studies can provide evidence of causation, models of investigation, and evidence of harm for more highly exposed people. Denigration and dismissal of such studies undermine legitimate early warnings and lessen public support for increased public health protections from the products in question. This may be the purpose of such criticisms.

Postmarket laws, I will argue in Chapter 5, greatly slow the already deliberate pace of science needed for determining risks and protecting the public. Roughly speaking, if no scientific studies are conducted until after products are in the market, discovering hazards from exposures to products will be enormously delayed. If products contaminate people and cause harm, citizens will suffer diseases, dysfunction and perhaps premature death because studies would reveal adverse effects too late, if at all. Diseases with long induction or latency periods only exacerbate the problems.

In contrast, recognizing the time-consuming nature of studies as well as some of their limitations strengthens the argument for beginning research much earlier in the lives of products and before they enter commerce. This approach provides a much better chance of revealing some of the more significant risks before the workforce and broader public are contaminated and put at risk. Moreover, a public research record on health effects might inspire independent researchers to follow leads earlier and to conduct independent studies of the substances (see Chapter 6).

Finally, for firms whose products might present health risks, there are incentives and temptations to urge that human studies must support legal actions. However, it should be clear why demands for this kind of study before public health protections are provided are morally offensive. If this policy were followed, a sufficiently large number of people must become sick or die from a toxicant in order to have scientifically significant human evidence that the toxicant poses risks and harms. This is simply unacceptable public health policy.

Scientific ignorance of health effects is not the best way to protect the public's health. In our modern technological and science-based society, we can no longer afford scientific illiteracy on issues of such importance. Public awareness of the generic strengths and weaknesses of scientific studies will assist the public in its own health protection.

4

Until about forty years ago the scientific community viewed a woman's womb as a sheltered, capsulelike environment, safe from the intrusions and dangers of the outside world. Not only did scientists subscribe to the idea, but they also conveyed it to their students as the received view.[1]

The public understanding of the womb was likely not different from that of the scientific community. If a woman were pregnant, she could continue her eating and personal habits—a predinner cocktail, one or two glasses of wine with dinner—because her developing child was tucked safely inside her. There was no need to change simply because she was pregnant. Indeed, there was a time when a little alcohol was considered good for a pregnant woman, for reducing some of her anxiety. This recommendation might have been given as late as the early 1980s.[2]

If she smoked, why should she stop? If mosquitoes plagued her home, she could spray them with DDT. And a pregnant woman likely gave no thought to the possibility of exposures to industrial chemicals or to the possibility that she might be contaminated and be transmitting that contamination to her developing child. She might well have assumed what she ate, drank, breathed, touched, or otherwise came into contact with would affect only her. Alcohol might relieve anxiety, give her a buzz, or make her dizzy. Cigarettes at timely intervals could be relaxing and satisfying. Handling some substances with her bare hands might cause them to tingle but be of no further consequence.

None of this did she and her physicians believe was pertinent to the fetus developing in her womb. Her body provided it a safe harbor and sustenance via the umbilical cord. Of course, pregnant women's behavior should not reflect poorly on their judgment; they were responding to the best, but mistaken, medical advice of the time. Probably few or no researchers were considering the possibility of prenatal contamination of developing children. If a mother's personal behavior was unlikely to affect a fetus in utero, how could lower concentrations of industrial chemicals, pharmaceuticals, pesticides, and contaminants in drinking water, food or the air pose problems?

The placenta suggests a similar picture. This is, according to the American Pregnancy Association, a "pancake-shaped organ that attaches to the inside of the uterus and is connected to the fetus by the umbilical cord." It provides pregnancy-related hormones and is something of a "trading post" between the baby's and mother's blood supply. Nutrition and oxygen are transferred from the mother's blood to the fetal blood, with wastes being transferred from fetal blood to the mother's blood, but with no blood mixing.[3] Perhaps because the two blood supplies do not merge, scientists once considered the placenta as a "barrier" that afforded "great protection to the embryo and fetus." However, as one expert notes, scientists "now know that the degree of protection is often modest at best, and that instead of being a barrier, the placental membrane acts more as an ultrafilter."[4]

Shattering the Picture of The Protective Womb

The presixties view of the womb and of the effect of chemical body burdens on development were called into question by two major catastrophes in the 1960s–1970s. Children born to women exposed to methylmercury in fish and to the pharmaceutical thalidomide raised the early alarms.[5] These incidents were followed by a more silent problem that was revealed over the course of two decades.

Methylmercury

A Japanese petrochemical company disposed of about twenty-seven tons of methylmercury (MeHg) in Minimata Bay, Japan, from 1953 to 1968. Humans have mined, refined, and used mercury for a variety of purposes for centuries. As we have seen (in Chapter 2), mercury released into the environment ultimately changes into an organic version, methylmercury. Once in the environment, MeHg enters the food chain and, according to the EPA, "is easily absorbed into the living tissue of aquatic organisms and is not easily eliminated."[6] It thus accumulates in fish and birds of prey, as well as other meat-eating predators.

Methylmercury is also used in industrial processes, such as in the manufacture of acetaldehyde, an organic chemical used as an intermediate in the production of other industrial chemicals.[7] This was its origin at Minimata Bay, where it was used as an industrial catalyst for some time, with the excess dumped into the bay and ultimately absorbed by fish. Since fish were an important source of food for local residents, they consumed the methylmercury-contaminated fish, and MeHg accumulated in their bodies.

Because MeHg has a half-life of seventy to eighty days in human bodies and is particularly toxic, adults who ate the fish developed neurological problems. Some had numbness and loss of feeling, some suffered ataxia—clumsy motion because of failure of muscle coordination—some had tunnel vision, and some went blind. Some were permanently disabled. In more extreme cases people experienced paralysis, seizures, convulsions, and sometimes death. Animals who ate fish were also affected. Cats imported into the area developed disease symptoms within two months. They "danced" strangely; some fell into the sea. Birds fell from the sky.[8]

Children contaminated in utero by MeHg were at greater risk. They exhibited symptoms similar to cerebral palsy at a rate ten times greater than that of unexposed children. As the National Institute of Neurological Disorders and Stroke describes the disorder, cerebral palsy is a dysfunction of muscle coordination and movement that affects voluntary control of muscles, causes stiff or tight muscles, and produces "muscle tone that is either too stiff or too floppy." They also suffered mental retardation, inhibited bodily growth, primitive reflexes, and death. More subtle effects appeared over time: sensory disturbance, constriction of visual field, muscle weakness, and tremors, *inter alia*.[9]

Not only were developing children more susceptible to the effects of MeHg; they were contaminated by greater concentrations because of the way MeHg behaves biologically. Researchers have found that methylmercury concentrates in fetal blood at greater levels than in the mother's blood "because of the active transport of MeHg to the fetus through the placenta." MeHg is further concentrated by active transport into the brain.[10] Consequently, concentrations of MeHg could be at least five times greater in the fetal brain than in the mother's blood.

Scientists extended their knowledge about the prenatal effects of MeHg from another contamination disaster, this time in Iraq in the 1970s. The Iraqi government had distributed about eighty thousand tons of wheat grain seed for planting. The grain had been dosed with a MeHg fungicide to preserve it during storage. During this time a severe drought had so depleted the food stocks of wheat that many families began using some of the fungicide-treated grain for making bread.

Adverse effects began to show up within a few months. At the lowest exposure levels, adults experienced paresthesia, a tingling, pricking, or burning sensation on the skin. Importantly, children's in utero exposures caused adverse effects not seen in their mothers when they were similarly exposed. Children born to mothers with only temporary paresthesia had severe cerebral palsy at a much higher rate than unexposed children. Lesser fetal exposures led, as Philip J. Landrigan and colleagues have noted, to "psychomotor retardation, blindness, deafness and seizures." Many mothers of these children were without effects.[11]

Ultimately about five thousand people died from consuming the bread made with MeHg-contaminated wheat. The MeHg exposures from the Minimata and Iraqi catastrophes and the resulting disease aftermaths that caused such serious neurological consequences have led scientists to refer to "the exquisite vulnerability of the developing brain to MeHg."[12]

Later research not related to these incidents revealed that exposure to MeHg from eating contaminated fish could cause acute myocardial infarction and even cardiovascular death. The risk of death was two to three times higher in the exposed group than in comparison populations.[13]

Thalidomide

A second major catastrophe revealing the effect of chemical body burdens on development came from a minimally tested sedative and sleeping pill, thalidomide, which we have already met in Chapter 1. Chemie Grünenthal first synthesized thalidomide in 1953. It entered the market about 1958. The company trumpeted this pharmaceutical as "a strong sedative that was also remarkably safe . . . a drug that was almost as powerful as a barbiturate but with no noticeable side effects."[14]

Potential pharmaceutical distributors who sought to license thalidomide found that it was more like a mild sedative than a barbiturate. Still later, others found that it had neurological side effects: peripheral neuropathy (poisoning of the nerves), which created a "tingling sensation and a feeling of numbness or cold" that could progress to "cramps, weakness and loss of strength." These symptoms were largely reversible, but not always. German physicians who were critical of the drug estimated that there were about forty thousand cases of peripheral neuropathy in adults. Chemie Grünenthal claimed there were only about four thousand.[15]

When Grünenthal teamed with Richardson-Merrell, as Philip J. Hilts, an historian of the FDA, has written, "it met a kindred spirit—financially aggressive, inexperienced in drug work, and not too concerned with medical issues. Richardson-Merrell was willing to bring the drug to market in the United States without any testing." Pharmaceutical laws at the time did not specify what testing was needed. Immediately after obtaining a license from Grünenthal, Richardson-Merrell began distributing thalidomide to about twenty thousand "patients, including pregnant women." It had no data on how thalidomide might affect developing fetuses and did not know whether it would cross the placental "barrier," as it was then called. When a few tests were conducted on animals, many of them died.[16]

More serious problems arose from prenatal exposures to thalidomide. Children born to mothers who took the drug during pregnancy developed "phocomelia—meaning 'seal limbs' . . . [or lacked the] long bones in the arms and legs, which meant that the hands and feet or just the fingers and toes of the infants sprang directly from the trunk. . . . It was also common for the baby to be born with no bowel opening, no ear openings, and segmented intestines. Of course, many of those with the disorder died and were never counted as thalidomide babies." James L. Schardein and Orest T. Macina note other malformations in children exposed to thalidomide in utero: eye problems; central nervous system difficulties ranging from facial nerve paralysis to deafness and respiratory abnormalities; congenital heart disease; and some abdominal malformations.[17]

Moreover, there appears to be a connection between effects on the nervous system of adults and birth defects, as Hilts points out: "Drugs that attack nerves, as [thalidomide] did, often cause birth defects." (Anticonvulsive drugs taken during pregnancy also have neurotoxic and morphologic effects.) Consequently, thalidomide was also found to cause a variety of functional deficits ranging from autism (at thirty times the rate in the normal population), mental retardation, to epilepsy and severe learning disorders. Some children did not survive the fetal period because they were spontaneously aborted; others died within the first year of life (about 40 percent). It appears that exposure during development—twenty to thirty-six days following conception—was the period when thalidomide induced birth defects.[18] This is about the time during pregnancy when women begin feeling tired and experience morning sickness and might seek sedatives. Indeed thalidomide was found to cause human limb malformations as a result of a single exposure. The disease or dysfunction rate among children born to mothers who took the drug was two hundred times the background rate of similar birth defects in nonexposed children.[19] Conservative estimates place the number of thalidomide babies at seven thousand to eight thousand, with another five thousand to seven thousand dying before birth.

In the United States, however, because of the alert actions by Frances Oldham Kelsey, an FDA scientist, the drug was never approved for marketing. Unfortunately, about forty thalidomide babies total were born in the United States, as a result of Richardson-Merrell distributing more than 2 million tablets.[20]

Methylmercury and thalidomide were dramatic and catastrophic instances of in utero exposures causing injuries, because in each case the children were so obviously harmed. Thalidomide caused shortened limbs that were obvious at birth. Some dysfunctions from methylmercury may have taken slightly longer to be recognized because they involved neurological deficits, but they were ultimately identified. The older view of the womb as an impermeable capsule within which development occurs protected from the external world and the mother's exposures was on its way out.

Diethylstilbestrol (DES)

About the same time another in utero drug-caused disease was silently developing. As discussed in Chapter 2, the synthetic estrogen diethylstilbestrol (DES) was created in 1938, introduced to clinical medicine as an orally active estrogen in 1946, and approved for use in pregnancy in 1947 with minimal or poor testing. The drug was prescribed to prevent spontaneous abortions and premature deliveries. As early as 1953, however, researchers showed that it was not even efficacious for this purpose.[21]

About 3 million pregnant women (1 to 2 percent of the total) took the drug from 1938 to 1971. The year of peak use was likely 1953. Approximately nineteen years after mothers took DES during the first trimester of their pregnancies, their daughters were at greatly increased risk of vaginal and cervical cancers. Arthur L. Herbst and associates first identified this relationship in 1971. This kind of cancer had never been seen before in women younger than about the age of thirty.[22] Based on initial calculations DES daughters had an extremely high risk of these cancers compared with women not so exposed (theoretically infinite because no unexposed controls had the cancers compared with seven diseased women exposed to DES *in utero*). Recent research puts the relative risk at 40:1

compared with the general population, much higher than the ten-fold risk of lung cancer from smoking a pack of cigarettes per day.[23] It is difficult to determine an actual number, since this form of cancer is, according to a CDC report, "virtually non-existent among premenopausal women not exposed to DES." The DES cancer cases led to one of the largest series of product liability cases in U.S. history.[24]

Cancers were not the only effects. The exposed women also had cervical or vaginal abnormalities, more spontaneous abortions and preterm deliveries, and perhaps a higher incidence of pregnancy losses. There is also a less frequently told story of hidden adverse effects. A group of physicians conducted autopsies of 281 female stillborns and neonates, some of whose mothers took DES and some of whose did not. The stillborn females born to mothers who took DES had a prevalence of vaginal adenosis of about 70 percent approximately eighteen times greater than among 159 offspring who were also autopsied but who had not been exposed to DES. Vaginal adenosis is an abnormality of the vaginal wall resulting from incomplete development. It can occur in female babies not exposed to DES, but is a moderately rare condition that usually improves with time. In some cases (one in a thousand to one in ten thousand) it can develop into a form of cancer (adenocarcinoma of the vagina).[25] And in other cases of adenosis there are risks of other reproductive tract abnormalities.

In fact, DES appears to have caused reproductive dysfunction in 85–90 percent of the DES daughters.[26] Was this how the hidden adenosis in stillborn DES children manifested itself among survivors, even if they would not have developed cancer later in life? This may not be known or, at this late date, may not be capable of being studied.

Major adverse effects in developing children from DES exposures remained hidden for many years, unlike those caused by methylmercury and thalidomide. Cancers typically have substantial latency periods, ranging from a few months up to forty years. The cancers caused by in utero DES exposures took nearly two decades to appear. Male babies were also affected, revealing a "host" of genital abnormalities, including abnormalities of the location of the urethra, malformed genitals, difficulty passing urine, and one reported case of testicular cancer.[27]

Methylmercury, thalidomide, and DES may be the most visible examples showing that in utero exposures can cause death, serious disease, or various forms of human dysfunction. However, subsequent to their discovery, numerous other chemical agents or environmental exposures have been implicated in causing adverse effects from in utero exposure. Some prominent ones include lead, tobacco smoke, ethanol from ingestion of alcoholic beverages, polychlorinated biphenyls, vitamin A, anticonvulsive drugs, and pesticides, as well as other drugs and chemicals.[28]

Adverse effects from in utero exposures are typically categorized as "death [of the fetus], malformation, growth retardation, and functional deficit."[29] Fetal deaths or malformations tend to be obvious, although the cause may not be clear, but growth retardation and various kinds of functional deficits are subtler. Functional mental deficiencies are more complicated to detect, and their causal antecedents are more difficult to identify in human studies. Some textbooks on developmental toxicology seem mainly to focus on death or malformation, perhaps because these are obvious adverse outcomes.

The Placenta Is No Barrier to Toxic Contamination

Scientists revised their views of the risks from in utero exposures as a result of these early shocks to the older paradigm. A mother's body and the womb, as well as the placenta, create a less protective environment than scientists once believed. They now know that toxic contaminants in pregnant women's bodies can also contaminate their developing children prenatally.

"It is clearly evident that there really is no placental barrier per se: The vast majority of chemicals given the pregnant animal (or woman) reach the fetus in significant concentrations soon after administration," notes James Schardein, a major developmental biologist. Others echo this point. Kavlock and Rogers, for instance, indicate that "virtually any substance present in the maternal plasma [blood] will be transported to some extent by the placenta." The womb is no longer considered a protected, shielded capsule within which a child's genetic program plays out over a period of nine months. Rather, as Ana Soto has noted, it is perhaps better to see the womb as an internal environment within a woman's body, supplied by light, sound, food, and fluids. However, just like environments outside the body, this one can expose a child to toxicants by the same routes that provide nourishment and fluids.[30] Moreover, as we will see later in this chapter, a person's genetic sequence is quite stable, but exposures to toxicants and other factors in utero can affect how the genome functions. In this way, some exposures can lead to disease or dysfunction.

Scientists not only understand how porous the placenta can be but also now have a more fine-grained understanding of its permeability. A variety

88

of factors influence whether a given chemical agent, pesticide, drug, or industrial chemical will pass through the placenta: among other things, molecular size of the substance (smaller sizes are more likely to cross the placenta), its fat solubility (the more soluble in fat a substance is, the more likely it will cross the placenta), its degree of ionization (less highly charged molecules are more likely to cross), and its molecular complexity (the simpler and less complex a molecule is, the more likely it will be to cross the placenta). Most drugs, for instance, are small enough to cross the placenta.[31]

Many substances will also reach nursing newborns via breast milk. Factors of substances similar to those that affect the diffusion of toxic hazards across the placenta also affect diffusion from a mother's blood into the milk. Compounds that have low molecular weight, greater lipid affinity, less acidic properties, and lesser binding to blood plasma, among others, will tend to enter breast milk. Fat-soluble and bioaccumulating substances are especially worrisome, as are many endocrine-disrupting compounds.[32] For instance, PCBs are concentrated up to 100 times greater in a mother's breast milk than in her tissues.[33]

Developing Children Have Critical Periods of Susceptibility

When industrial chemicals cross the placenta, they can have adverse consequences. A human's susceptibility to a wide range of adverse effects is increased during development, from the period before conception through adolescence, depending on the organ system. One developmental toxicology textbook makes the general point: "Organisms tend to be significantly more sensitive to many adverse environmental influences during early developmental stages. . . . Many tissues are undergoing rapid cell division, and the embryo, and to a considerable extent, the fetus, has much less capacity to metabolize xenobiotics than does the adult." Others agree. Luz Claudio and colleagues point out, "The toxicology and pharmacology literature documents that children often react quantitatively and/or qualitatively differently to many toxins and drugs as compared with adults. In most organ systems, these differences amount to an increased susceptibility to many hazardous environmental chemicals."[34]

Various substances can change or disrupt the developmental process, with devastating or subtle consequences. Pharmacological, environmental, and industrial and natural chemicals with diverse chemical structures, notes Kenneth S. Korach, can pose problems by means of "a single action, or in combination, and can influence either individual or multiple cellular signaling pathways in a tissue or organ system."[35]

Moreover, exposures to toxicants during critical developmental periods "have been suggested as possible causal factors responsible for some of the increases in [childhood] leukemia, lymphoma, brain cancers and testicular cancer," as well as other diseases and dysfunctions.[36] Recently, researchers at the University of California, Berkeley, and elsewhere have been able to "backtrack" chromosomal damage from newborns or young children to chromosomal translocations in the womb. According to Martyn Smith, "Overall, these findings show that most childhood leukemias begin before birth and that maternal and perinatal exposures such as chemical and infectious agents are likely to be critical." During the earliest stages of development, more of the body's cells are comparatively undifferentiated stem cells, capable of developing into a number of different tissues in the body. When the DNA in these cells is damaged, it is capable of conveying permanent changes in DNA to other cells and tissues in future cell divisions.[37]

Stages of Development

Developing children, and mammals more generally, go through several major stages in utero, at any one of which they can be susceptible to external insults. Development is typically divided into gametogenesis, fertilization, the blastocyst stage, gastrulation, organogenesis, and finally the fetal stage. Various organ systems continue to develop after birth, some until the person reaches adulthood.

During the first stage, gametogenesis, eggs in females and sperm in males begin to develop. Young girls are born with their lifetime supply of eggs, one of which is released during each ovulating period after puberty. Adult men produce 150 to 200 million new sperm per day and continue to do so throughout their lifetime. The formation of each kind of gamete creates one period of vulnerability. Exposures to some toxicants can affect the quality of the sperm or eggs.[38]

Several known toxicants can affect the sperm. The fungicide 1,2,-dibromo-3-chloropropane (DBCP) greatly decreased sperm counts of and even caused sterility in men working in a DBCP plant. Ethylene dibromide can cause a decrease in sperm viability and sperm count, as well as structural abnormalities. Various solvents have adversely affected sperm in animal studies.[39]

Even more seriously, during a critical period when the male germ line that produces sperm is undergoing development in utero, exposure to any one of several antiandrogenic substances, which block the activity of the male hormone androgen, can cause permanent changes that are transmitted across several generations. These are transgenerational effects on the sperm. They include sterility plus other unexpected adverse effects, including, according to Matthew D. Anway and colleagues, "breast tumors, prostate disease, kidney disease, and immune abnormalities." These occur in great-great-grandsons from exposure to the pesticide vinclozolin during one period of the development of their great-great-grandmother. Other substances cause similar problems: bisphenol A (BPA), phthalates, and the pesticide methoxychlor.[40] We will return to these issues later in the chapter.

Women's exposures to toxicants "may cause infertility, menstrual disorders, illness during pregnancy, chromosomal aberrations, breast milk alteration, early onset of menopause, and suppressed libido," Grace K Lemaster and Sherry G. Selevan have noted. For another example, a recent study reports that when pregnant mice were given low-dose exposures to bisphenol A, chromosomal damage occurred in the eggs of the pregnant mouse, her female offspring, and her granddaughters. Thus bisphenol A appears to create adverse effects via the eggs of the exposed pregnant female through several generations.[41] Are the manufacturers of bisphenol A contributing to transgenerational chromosomal damage in humans?

The second stage, fertilization, occurs when a man's sperm fertilizes a woman's egg. During even early periods of fertilization (the first six hours) scientists have found that exposures to toxicants can result in malformed fetuses. Ethylene oxide, a sterilizing agent, is just one substance that can cause these adverse effects.[42]

During the blastocyst, or embryonic, stage the fertilized embryo "moves down the fallopian tube and implants in the wall of the uterus." The embryo also undergoes rapid cell divisions. Exposure to toxicants during preimplantation may slow or have other adverse effects on growth or cause the death of the blastocyst. DDT, nicotine, or methylmethane sulfonate exposures during this period can cause body or brain weight deficits and embryo death; some other toxicants can lead to fetal malformations.[43]

After the embryo is implanted, as Rogers and Kavlock note, it "undergoes *gastrulation*." Gastrulation is the process of forming "three primary

91

germ layers—*the ectoderm, mesoderm, and endoderm,*" the tissues from which organs develop. During this time cells migrate to form basic templates leading to the development of organ systems. This stage is a prelude to organ development, or organogenesis, and is a particularly sensitive period. A number of toxicants can cause malformations of the eyes, brain, and face.[44]

During organogenesis most bodily organ structures begin to develop. The embryo goes through "rapid and dramatic" changes from weeks three to eight of human pregnancy, at the end of which time the embryo has become a fetus and can now be recognized as a human. This is an especially vulnerable period because cells multiply rapidly, migrate to specific locations, and begin to communicate with one another and because tissue structures undergo changes. During the development of different organ systems, there are different windows of sensitivity to toxicants and numerous opportunities for exogenous substances to disrupt the developmental process and to damage organ systems. There are different periods of vulnerability for different parts of the developing body, coinciding with key events in the changing organ structure: eyes can be affected quite early, long bones of the body somewhat later, and the palate during two different periods of sensitivity. Moreover, a particular toxicant may affect more than one developing system.[45]

At the end of organogenesis is the beginning of the fetal period of development, about eight weeks into a human pregnancy. This stage of development is "characterized primarily by tissue differentiation, growth and physiologic maturation." Bodily organs may not be totally complete by this time, but, as one textbook puts it, "almost all organs are present and grossly recognizable." During the fetal period there can be "functional anomalies of the central nervous system and reproductive organs—including behavioral, mental, and motor deficits as well as decreases in fertility."[46]

A summary of recent research results concerning toxicological effects on development is the Endocrine Disruption Exchange, located at www.endocrinedisruption.com/home.php. This summary was created by Theo Colborn, a well-known researcher with long-standing interests in identifying the adverse effects resulting from endocrine disruption.

The Sensitivity of Developing Organ Systems

A brief review of some critical systems undergoing development suggests their susceptibility to the adverse effects of toxicants.

Brain and Nervous System: The brain not only is central to who we are as persons but is also biologically critical to other organ systems. The developing human brain is much more vulnerable to toxic agents than the adult brain is. During a short nine-month period, the brain must grow from a few cells into an extremely complex organ of billions of cells. These must be interconnected, very specialized, and properly located in order to communicate with one another and to function properly. Thus there are numerous opportunities for toxicants to interfere with neurological function during development that do not exist for mature brains or for other organ systems. The brain may be uniquely sensitive among the body's systems.[47]

From about the third trimester of pregnancy to about the age of two, humans undergo what Claudio and colleagues call a "brain growth spurt." Brain weight "continues to increase after puberty, until a person is about the age of 20," undergoing "several fundamental developmental changes." As others note, brain cells proliferate; that is, they divide and greatly multiply during development. When such changes occur, there are more opportunities to damage DNA and, thus, more chances that it will be improperly repaired or that it will actually be altered.[48]

These possibilities of damage during cell proliferation are especially critical in the case of the brain. Unlike some tissues and organ systems, for which there might be ongoing opportunities to repair or remove damaged DNA or cells, "neurons proliferate primarily during development, and each specific cell type only during a limited period, reducing the ability to repair lost cells and function."[49] Philippe Grandjean and Philip Landrigan agree, pointing out that if brain development is halted or inhibited, "there is little potential for later repair, and the consequences can therefore be permanent.[50]

In addition, not only is the brain and neurological system more vulnerable in utero, but this vulnerability continues, in the words of P. M. Rodier, "well into the first year of life." A large variety of different neuron types are generated during this period, and agents that "interfere with all cell production seem to injure the CNS [central nervous system] more than other organs." Toxic effects on the nervous system might include a number of different end points, contributing to various sensory deficits, memory and learning dysfunctions, and poor motor coordination, as well as altered states of arousal.[51]

Immune System: The immune system is similar to the brain in the precise, sequential nature of development and potential sensitivity to exogenous insults. As one team of researchers has said, "The development of the immune system results from a series of carefully timed and coordinated events during embryonic, fetal, and early postnatal life." As Rodney R. Dietert and Michael S. Piepenbrink point out, the immune system goes through several "discrete functional changes representing critical windows of differential vulnerability to toxicants." Mark D. Miller and colleagues note that exposing pregnant animals to a number of toxic substances at concentrations that cause "only transient effects in adults produces long-lasting or permanent immune deficits in their offspring."[52]

In general, early life stages are more sensitive to most toxicants. For example, the developing immune system has end points that are sensitive to lead concentrations that are three to twelve times lower than concentrations that affect adult immune systems. Dioxin and mercury also affect developing immune systems at concentrations lower than those that affect adults.[53]

In addition, some toxicants adversely affect the developing immune system in ways that produce, "a different and unpredictable array of alterations when the exposure occurs *in utero* or in the early neonate versus the adult." Some aspects of the immune system are simply more sensitive as a result of in utero exposures compared with adult exposures. Thus the assessment of risks in adults does not necessarily predict how sensitive developing children will be to concentrations of particular toxicants or what the range of long-term consequences might be.[54] Other toxicants produce different ranges of effects, depending upon the age of the mammal at exposure. Lead, the insecticide methoxychlor, and ethanol, the active ingredient in alcohol, cause effects in developing children that would be unpredictable from studies on adult animals or humans.[55] More broadly, in a review of five immunotoxic chemicals a research team concluded that until there are data to the contrary, one should assume that "development and maturation of the immune system constitutes a period of greater sensitivity to xenobiotic exposure."[56] Moreover, the immune system is particularly important because it plays the major role in protecting other organ systems from toxic effects. If it does not function well, is immature, or is partially disabled because of any adverse toxicity effects, it will be limited in carrying out this task.

Finally, time bombs—permanent changes resulting from in utero chemical exposures—can cause problems long into adulthood. For example, Robert W. Luebke and colleagues note that DES produces outcomes that are "among the most persistent reported for any chemical," suppressing two types of cells that are important in combating disease. A wider range of substances now appears to cause similar long-term immunological effects. Dioxin-like compounds, including some PCBs, that block a particular cellular receptor (the Ah receptor) cause long-standing "functional alterations in the developing immune system, whereas the impact on the mature immune system is transient." In utero exposures to lead cause adverse postnatal effects when the immune system is subjected to a viral infection. Moreover, experimental studies "place the developing immune system on par with the developing neurological system for lead sensitivity."[57]

During windows of vulnerability, the immune system can suffer injury that can lead to an array of diseases and dysfunction over a lifetime. Exposure to a toxicant that does not cause the death or obvious dysfunction of a developing fetus, Luebke and others note, "may produce an . . . immunotoxic alteration [that is unrecognizable] until the postnatal immune system is placed under subsequent stress." Others suggest there are certain "patterns of disease" related to immune system deficiencies. For example, "entry-level," or early appearing, allergic reactions such as asthma may be followed over time by ear infections, respiratory infections, and even lung cancer, in an identifiable pattern. Autoimmune reactions can also result from developmental insults to the immune system. A delay in some comorbidities can be latent for thirty to forty years. Substances must be tested before exposure to eliminate such effects.[58]

Most of these kinds of effects likely would not result from exposures to adults. However, a person with early exposure to one of these substances or a combination of them might have a greater number of diseases, some serious, over a lifetime. Someone might think he or she was just more susceptible to viral infections, which would be true, but the explanation might well be early exposure to toxicants that reduced the robustness of the immune system for a lifetime.

There is one qualifier to these concerns about the immune system. When the mother and the fetus or neonate are exposed to an active bacterial or viral infection, the child receives substantial protections from the immune system of the mother directly (in utero) or indirectly (from breast milk after birth). However, even though a mother's immune system protects a developing child, it is well understood that its effects are reduced because it must protect two bodies, but does not have the capacity to do so fully.[59]

Consequently, the developing child is not as quite as susceptible to bacterial or viral infections as the immaturity of its own immune system would suggest. However, it receives less protection than the mother could provide for herself, if she were not pregnant. Nevertheless, none of this detracts from the point that the developing child's immune system itself can be damaged during this period of time so that it does not function well early or even later in life.

Lungs: Lung development goes through a long process that is not completed until children reach adulthood. Somewhat similar to the brain, the complex organization of the respiratory system has numerous critical developmental steps, including, as Radhika Kajekar notes, "branching morphogenesis, cellular differentiation and proliferation, alveolarization, and maturation of the pulmonary immune, vasculature, and neural systems." During this period, "the developing lung is highly susceptible to damage from exposure to environmental toxicants [because of] the protracted maturation of the respiratory system, extending from the embryonic phase of development in utero through to adolescence." These exposures may adversely affect the lung's structure, function or maturation.[60]

A child's respiratory system is immature at birth. According to Kajekar, children's "unique differences in . . . physiology and behavioral characteristics compared to adults . . . [increase] the vulnerability of their developing lungs to perturbations by environmental toxins. Furthermore, an interaction between genetic predisposition and increased opportunity for exposure to chemical and infectious disease increase the hazards and risks for infants and children."[61]

Ozone, sidestream tobacco smoke, by-products of combustion (polyaromatic hydrocarbons), and particulate matter pose special risks to the developing lung. For example, ground-level ozone can aggravate asthma, emphysema, and bronchitis, as well as inflame lung cells, which, when repeatedly injured, may cause permanent long-term health effects. Experimental studies indicate that ozone "may reduce the immune system's ability to fight off bacterial infections in the respiratory system."[62]

Reproductive System: The reproductive system develops not only in utero but also throughout childhood and puberty. During this long period there are numerous opportunities for a variety of toxicants—pharmaceutical, environmental, or natural—to interfere with development, with potentially devastating results. Recent toxicological studies show "that exposure during specific periods of development results in long-term effects that occur following sexual maturity and adulthood." Potential targets for toxic insult include the developing fetus, infants, children, and adults. According to one text, "Exposure during these sensitive periods either alters normal development, resulting in immediate or acute effects, or may subsequently compromise normal physiology and function later in life."[63]

DES is only one example of a toxicant causing effects on the reproductive system as a result of in utero exposure. Barbara J. Davis and Jerrold J. Heindel note that "literally hundreds of chemicals cause reproductive dysfunction in animal models." Other researchers note that BPA, methyoxychlor (a pesticide), and nonphenol (related to BPA) at doses equivalent to DES doses produce very similar effects in experimental studies.[64] These studies, as we have seen, serve as important warnings for similar effects in humans.

Developing Children Have Lesser Defenses Than Adults

The sensitivity of developing organ systems might not be of concern if humans had adequate defenses to prevent toxicants from reaching their organs, to eliminate toxicants before they could pose risks, or to repair any DNA or other biological damage that might occur. However, this happy possibility does not exist, because fetuses and children have reduced defenses compared with adults. We have already seen that the immune system is not fully developed in utero or at birth. Consequently, during these periods it offers less protection against toxicants than an adult immune system would.

The blood-brain barrier does not develop in children until about six months. This "barrier" normally protects adults by keeping many exogenous toxicants out of the brain and other parts of the neurological system, such as the spinal cord, retina, and peripheral nervous system, but is not fully functional at birth. The barrier is created by the tight relationship between cells just outside the brain, in contrast to somewhat larger "gaps" between cells in other parts of the body. As Douglas C. Anthony and colleagues describe it, "To gain entry to the [nervous system,] molecules must pass *through the cell membranes* of the endothelial cells of the brain rather than *between the endothelial cells,* as they do in other tissues."[65] It is more difficult for molecules to pass through cell walls than

97

it is for them to pass between cells. Thus this barrier can keep out of the brain some toxicants and beneficial drugs that would enter tissues in other areas of the body via the circulatory system.

This is an important protection for sensitive brain and nervous system tissues against exogenous agents. Before the blood-brain barrier develops, Rodier points out, "some toxic agents that never enter the mature brain enter the developing brain freely. Examples include cadmium and monosodium glutamate." Consequently, any toxicants in the mother's body that cross the placenta will expose the fetus and may not be prevented from entering the brain, because the blood-brain barrier is not yet effectively functional. We should recall, as well, that mercury actually increases its concentrations in a fetal brain compared with amounts in the mother's blood. In addition, one study notes, the "brain continues to change throughout life . . . with possible age-specific periods of susceptibility to neurotoxicants."[66]

Enzymes, a defense mechanism not normally considered a part of the immune system proper, can also be undeveloped or underdeveloped. Enzymes, according to the dictionary, are "biochemical catalysts in living organisms." These catalysts increase the rates of biochemical reactions. Some enzymes increase biochemical reactions that reduce the toxicity of substances, while others enhance the toxicity of exogenous chemicals. Consequently, when detoxifying enzymes have not developed at all or have not developed to full strength in young children, these individuals will be more vulnerable to toxic substances. In contrast, when undeveloped enzymes would increase the toxicity of other substances, children will be better protected, simply because they cannot transform less toxic molecules into more toxic variants. Recent research has found that some metabolic routes that increase the toxicity of substances are more active in embryos than previously believed. Moreover, this vulnerability now appears to last until the age of seven, much longer than researchers had previously believed.[67]

Children's skin also offers less protection than that of adults, because children have a greater skin surface area to body weight ratio and because their skin is more permeable than adults'.[68] Both features provide more opportunities for toxicants to enter through the skin into children's bodies. How serious a vulnerability this is depends upon toxicants with which children would have contact via their skins. If they play on the floor and are exposed to toxic molecules such as polybrominated fire retardants or the pesticide chlorpyrifos, these compounds would more easily penetrate their skin than that of adults.

The ability of the body's blood proteins to bind to toxicants and assist in their elimination is lower in newborns than in adults. Trichloroacetic acid (TCA) is a breakdown product of trichloroethylene (TCE), a common solvent and degreaser, probable human carcinogen, and suspected neurotoxicant. TCA seems to be less able to bind to blood proteins and to be eliminated from the bodies of newborns than from adults. Thus, Miller and others report, TCA would have "greater potential for toxicity" because it remains in the body longer.[69] If TCE, a common contaminant, is in the drinking water that infants and young children ingest, in principle they would seem to be at greater risk than would adults.

The kidneys have an important function in eliminating toxicants and other waste from the body. The more slowly risky substances are eliminated from the body, the more time they have to contribute to disease or dysfunction. Studies conducted on a large body of drugs found that a majority of them are eliminated more slowly by neonates and newborn infants than by adults. This research also found that methylmercury has a longer half-life in neonates than in adults.[70]

Not all aspects of children's defenses or sensitivity to toxicants are more worrisome than those of adults, however. Children can tolerate some pharmaceuticals better. Children also have some repair mechanisms that are more effective than adults' (and these would provide better defenses to toxicants than adults' would). Also, as noted above, since children's enzyme system may not be as well developed as adults' for substances that require enzymes to increase their toxicity, children are better protected.

Genetic Variation

Up to this point we have considered general or typical biological tendencies in developing mammals in utero, as neonates, or as youngsters. However, individual people can be more or less susceptible to toxic effects as a result of genetic variability and diversity.

For example, Frederica Perera and others found that environmental exposures to polycyclic aromatic hydrocarbons (PAH), including those from sidestream and secondhand tobacco smoke, cross the placenta and create adducts on DNA. When a substance forms adducts on DNA, it is bound to the DNA, altering its function and often causing mutations or incorrect repair, which lead to cancers or other diseases. Polycyclic aromatic hydrocarbons are formed during incomplete combustion of organic compounds, such as coal, gas and oil, garbage, tobacco, and charbroiled meat. Some PAHs are manufactured for use in dyes; pesticides; medicine; and various tars, oils, creosote, and roofing products. Miller and others report that in urban areas or households with smokers, a common source would be sidestream and secondhand smoke. Subpopulations of fetuses that have more PAH-DNA adducts will show increased sensitivity to genetic damage compared with the mothers and compared with other fetuses. PAH exposure can compromise fetal development and presage smaller head circumference, which is associated with other adverse effects, as well as genetic damage in the newborn.[71] The Perera team now has data from two countries finding that above average prenatal exposures to PAHs reduce a child's IQ 5 years later by several IQ points.[72]

Similarly, vulnerability to organophosphate pesticides can, as Brenda Eskenazi and colleagues report, "vary by age and genotype." Specifically, these researchers found that children and adults with a variant of a particular gene had lower levels of an enzyme that assists in metabolizing organophosphate pesticides. As a result, these individuals "may be at higher risk of health effects from organophosphate exposure." In addition, given the variation in enzymes that detoxify organophosphate pesticides, Clement E. Furlong and others estimate that "most, if not all, newborns, as well as a subpopulation of adults, will exhibit significantly increased sensitivity to organophosphate exposure." Some adults can also have low levels of this enzyme and thus can be more susceptible to organophosphates, simply because of genetic variation. Potential effects include neurotoxic consequences as well as some cardiovascular end points.[73]

Consequently, even when the average or typical child might not be as susceptible to a particular contaminant compared with adults, human genetic variability can increase or decrease the extent of sensitivity. Thus, public policies that aim to prevent or reduce risks from toxicants must take into account the range of genetic variation that can increase sensitivity to toxicants.

Children Typically Have Greater Exposures to Toxicants Than Mothers Do

Developing children are typically subject to greater exposures than adults are. According to the consensus scientific statement from the Faroe Islands Conference, "the mother's chemical body burden will be *shared with her foetus or neonate,* and the child may, in some instances, be exposed to *larger doses relative to the body weight.*"[74]

As we have discussed, methylmercury concentrations in the fetal brain can be at least five times higher than concentrations in the mother's blood. Since breast milk contains considerable fat, when newborns are breast fed, any toxicants that are lipophilic (fat soluble) may have greater concentrations in breast milk than in the nursing mother's body. A breast-fed infant's daily dose of PCBs, according to Grandjean and colleagues, "may be 100-fold higher than the mother's," leading to greater toxic concentrations in a developing child than in the mother.[75]

Lead in a pregnant woman's bones can be mobilized by the same hormonal and other biological processes that release calcium from her bones for her child's skeletal development in utero and during nursing (lead follows the "calcium stream"). Lead can also easily cross the placenta with concentrations in the fetus nearly identical to those in the mother. Thus substantial concentrations of lead can enter her developing children. Consequently, a mother's prior lifetime contamination, not just her immediate exposures, can affect her children during pregnancy or nursing by the same calcium stream that provides critical nutrients.[76]

Newborns enter life already substantially contaminated by more than 200 substances. If a mother nurses, she off-loads some of her body burden of chemicals, adding to the accumulated industrial chemicals of the newborn. The choice to nurse should not be a cost-benefit calculation but it is. Nonetheless, researchers urge on balance that women nurse their newborns.[77]

Once children begin life as more or less independent organisms, they have a number of features that typically increase their exposures to toxicants. Specifically, they have higher metabolisms and inhale much more air and any toxicants in it than adults do. Over the first twelve years of life, a child on average has a breathing rate that is double that of an adult. Children also drink more than twice as many fluids and eat more food on a per body weight basis than do adults. Children eat five or ten times as much fruit. In sum, according to Grandjean and Landrigan, children have "augmented absorption rates, and diminished ability to detoxify many exogenous compounds, relative to that of adults."[78]

Also, as any parent knows, young children also tend to play close to the floor and test objects in their environment by "mouthing" everything within their grasp: toys, cooking utensils, electric cords, pet food, and objects in the grass, in the dirt, or on occasion even in mud. Children are also much more active than adults.[79] All of these behaviors increase children's contaminations. For instance, the pesticide chlorpyrifos can be found in carpeting when people have sprayed the inside of their houses for pests. Other toxicants may well accumulate in carpets, fuzzy toys, blankets, or furniture, increasing exposures.

Recall that PBDE flame-retardants are released from plastics and furniture, becoming part of house dust. Children likely ingest, inhale, and assimilate more of these substances than do adults. Recall also from Chapter 2 that there is an inverse relationship between PBDE concentrations and age with the highest concentrations of PBDEs in the youngest children.[80] This is clearly a worrisome relationship.

Consequently, children have at least four major routes of exposures to contaminants: (1) Before they are even conceived, the biological material that will go into their formation will have been suffused with contaminants in their parents' bodies. (2) Once conceived, children will be contaminated in the womb. (3) During nursing, they will receive contaminants from their mothers' body burdens. (4) Once born, they will also be exposed from the general environment. That is, a newborn will be subject to most of the same exposures as adults, except, for example, for occupational exposures. Moreover, children's exposures in general will be higher compared with adults' for physiological reasons. Thus, during development a child is quadruply exposed—by three sources from the mother's body burden and then after birth by the same sources as most adults but at increased rates on average.

Children Have More Years of Future Life

Children have more years of future life ahead of them than do adults. This provides more time for the development of diseases or dysfunctions that are initiated by early exposures to toxicants. A disease process might require one, two, or three critical steps to occur before the disease is fully initiated. If one or two steps occur in utero, as DES exposure likely did, then fewer steps would need to occur later in life for full-fledged disease to materialize.[81]

For example, Miller and others report, "cancer is a multistage process and the occurrence of the first stages in childhood increases the chance that the entire process will be completed, and a cancer produced, within an individual's lifetime." Even though cancers can have long latency periods, if the disease process is completed during the late teenage years or early adulthood, a cancer is likely to appear much earlier in life. We should remember also that childhood cancers are initiated in utero. In addition, other diseases can be conceived as one-, two-, or three-hit processes.[82]

It's important to keep in mind also some of the mechanisms for Parkinson's disease discussed in Chapter 3. D. Rice and S. Barone Jr. note, "If this process were accelerated by chronic or historic exposure to a neurotoxicant, the effect as the individual aged would be a further decrease in functional capacity from that typically observed during aging" This "decrease in functional capacity" would likely lead to earlier onset of dysfunction. Concerns about this process may well increase with the aging of the large baby boomer cohort, especially among those who have been exposed to neurotoxicants, such as industrial employees, mechanics, farmers, farm workers, and so on.[83]

Children Tend to Be More Susceptible
Than Adults Are at Similar Exposures

Children tend to be more susceptible to many toxicants than adults are because of contaminants entering the womb and the children's more sensitive organ systems, lesser defenses, and greater exposures. In short, recall that Claudio and others explain, "that children often react quantitatively and/or qualitatively differently to many toxins and drugs as compared to adults . . . [and as a result have] an increased susceptibility to many hazardous environmental chemicals."[84] Moreover, a number of diseases or dysfunctions caused during development can lead to lifelong health or functioning problems.

For very good general biological reasons, developing children are more susceptible than adults at the same concentration of toxicants, and there is some quantitative evidence on this point, but it is somewhat limited because of the difficulty of conducting human studies. Nonetheless, a variety of human and experimental animal studies have substantiated the increased susceptibility of young mammals to external toxicants. Consider a few of these.

Although DES causes vaginal cancer and increases breast cancer in DES daughters as they reach middle age, the DES mothers did not suffer from the same problems. However, DES mothers now are at a 20 to 30 percent elevated risk of breast cancer.[85] Thalidomide presumably had beneficial effects in adults (along with some peripheral neuropathy), but it caused terrible effects in newborns.[86]

The anticonvulsive drugs diphenylhydantoin and valproic acid have beneficial effects (reducing convulsions) for the mothers, but their children are at risk for what Landrigan and others call a "broad spectrum of abnormalities" from diphenhydantoin and "neural tube defects and heart, craniofacial and limb anomalies" from valproic acid. These same researchers point out that isotretinoin, a pharmaceutical similar to vitamin A that is a treatment for cystic acne, "is associated with a characteristic pattern of malformations." It also caused profound mental retardation in children.[87]

Children with in utero exposure to Coumadin, an anticoagulant blood thinner, can have underdeveloped cartilage of the nose and body, as well as atrophy of the optical nerves.[88] Yet the mother experiences none of these conditions. Similarly, children exposed to radiation had higher rates of leukemia and thyroid cancer than did adults similarly exposed. Women exposed to radiation during puberty also had higher rates of breast cancer than those exposed to similar levels as older adults.[89]

Experimental studies of immunological toxicants illustrate that the developing immune system is, in the words of Luebke and colleagues, "at greater risk than that of the adult, either because lower doses produced immunotoxicity, were more persistent, or both." Until there are data to the contrary, these researchers note, scientists should assume that "development and maturation of the immune system constitutes a period of greater sensitivity to xenobiotic exposure." These findings were further confirmed for dioxin when studies revealed that very low prenatal or nursing exposures to dioxin caused animals when they reached adulthood to be more susceptible to influenza virus than their mothers were.[90]

Young women under age fourteen who had high DDT exposure had rates of breast cancer that were five times higher than those for women who were older when they were exposed.[91] Barbara A. Cohn and colleagues point out that the human breast appears to have critical periods when it is susceptible to carcinogenic insults: "during fetal life, adolescence, and early reproductive life, particularly before the first full-term pregnancy." These researchers also state, "Radiation . . . increases breast cancer risk most strongly when exposures occur early in life." Moreover, these results are consistent with experimental studies showing critical periods of mammary development.[92]

104

Landrigan and others, citing a number of sources and a report by the National Academy of Sciences (NAS), note that "in utero and early life exposures to lead, polychlorinated biphenyls (PCBs), certain pesticides, and other environmental neurotoxicants are known or thought to contribute to the burden of [neurodevelopmental] disorders," including learning disabilities, dyslexia, mental retardation, attention deficit disorder, and autism. These dysfunctions are of particular concern, according to the NAS committee, because "3% of developmental disabilities are the direct consequence of neurotoxic environmental exposures, and . . . another 25% arise out of the interplay of environmental factors and individual genetic susceptibility." Boyle and others, in an older article, estimate that about 17 percent of children have exhibited some kind of "developmental disability." However, this study has not been revisited and may overestimate some effects.[93]

Since humans have wide genetic diversity, the estimate that up to 25 percent of such disabilities are partially a function of environmental exposure is particularly worrisome. Public health policies must account to a large extent for genetic variation. Consequently, if about one-fourth of developmental disabilities result from environmental-genetic interactions, citizens have a strong interest in scientists' finding the antecedent factors contributing to disease.

Some Representative Toxicants

In order to have a better picture of some issues of developmental toxicity, consider some well-studied substances and some about which less is known. These examples illustrate a range of adverse effects and add to the picture that developing children are more susceptible to adverse effects than adults. Some of them also provide some useful comparisons with related but less well-studied chemicals. For coming chapters, these examples help reveal some of the shortcomings of the law.

Lead

Lead is ubiquitous. According to Robert A. Goyer and Thomas W. Clarkson, it is "detectable in practically all phases of the inert environment and in all biological systems." It has been utilized for many purposes from the time of the ancient Greeks to the present. It maims and causes disease, dysfunction, and death. It is toxic to most living things. Adult, prenatal, and neonatal exposures to lead can cause adverse effects in humans, with

105

long-term consequences. Researchers have not found a dose below which there are no adverse effects.[94]

The Romans knew that lead was toxic and even caused death. The patron saint of metal smiths, Vulcan, who worked closely with this metal, showed "advanced lead poisoning: lameness, pallor and wizened expression." Mentally deficient people were considered as "crazy as a painter." Even though lead was utilized in some cooking utensils and water pipes, the Romans recognized that earthenware pipes and containers were safer for food and water. Lead mining was reserved for slaves.[95] During the Middle Ages and Renaissance, people continued to be aware of the toxic effects of lead—as a silent, long-term poison, it was difficult to trace, facilitating assents to positions of power.[96]

In 1848 a French researcher noted that children who mouthed lead painted toys developed "colic." In 1887 an American researcher noted that nine members of a family developed lead poisoning from eating bread that had been colored yellow with lead chromate, a common coloring ingredient at that time. In 1904 an Australian physician, J. Lockhart Gibson, began what David Rosner, Gerald Markowitz, and Bruce Lanphear call "the worldwide uncovering of the role of paint pigments in creating an epidemic of lead poisoning that has damaged literally millions of children in the United States." Lead dust from deteriorating wall paint damaged "the optic nerve and the motor nerves to the eye [in a way] that interfered with the children's sight and eye motion." In 1914 physicians at Johns Hopkins documented the death of a two-year-old who gnawed lead paint from the railing of his crib. Prior to his death, he exhibited typical lead poisoning symptoms—pain in his face and head, vomiting, and ultimately convulsions and a coma.[97]

Major public health problems with lead increased when General Motors and the lead industry strongly lobbied for the use of tetraethyl lead in gasoline to improve the function of internal combustion engines in 1921. Within three years employees working in plants that produced the additive, "fell sick and died." Journalists writing about these issues called fuel with tetraethyl lead "looney gas."[98] Even when people subject to high exposures did not die, they could still experience severe health effects, including damage to the brain and kidneys. In the 1940s, "19 of 20 survivors of acute poisoning were noted to have severe learning and behavioural problems."[99] High exposures in men can also damage sperm production.

The phasing out of lead in gasoline began a modest public health success story. Levels of lead in the environment and in people's bodies have declined substantially. However, lead has not disappeared from our bodies or the environment, because too many avoidable sources remain—including lead paint in older buildings—for this public health action to be a greater success. Lead poisoning in children from paint still occurs in "poor urban pockets" in older areas of cities.[100] In addition, worldwide use of lead is increasing because of demand for its use in batteries for microprocessors and fuel efficient vehicles.[101]

The neurological system appears to be the most sensitive to lead, but other targets include the gastrointestinal tract and the reproductive, cardiovascular, and skeletal systems. Low lead exposure in adults can decrease performance in tests that measure nerve function and can cause weakness in fingers, wrists, or ankles. Ellen K. Silbergeld and Virginia M. Weaver report that even much lower lead levels in adults (less than five micrograms per liter in blood) can cause chronic kidney disease, while mortality from cardiovascular disease occurred at blood levels even lower. Blood lead levels in the same range can affect adult blood pressure, with no lowest threshold for adverse effects. These levels can also contribute to heart attacks and strokes.[102] All these effects occur below legally recommended "safe levels" of lead. Workers exposed to lead are at an increasing risk because of greater worldwide production.

In addition, because lead accumulates in bones and tissues, there is a further need, the same researchers point out, "to protect against cumulative effects on target organs, such as the heart, brain and kidney, during the life span." The EPA reports that once lead enters people's bodies, it takes twenty-five days for half of it to leave the blood, and more than twenty-five years for one half of the lead in bones to be eliminated from the body.[103] This finding is even more important in terms of women's exposures to lead, since lead stored in bones can in turn be mobilized and contaminate children in utero and during nursing.

Children are even more susceptible. According to the Agency for Toxic Substances and Disease Registry, higher concentrations of ingested lead can cause "kidney damage, colic (severe 'stomach ache'), muscle weakness, and brain damage, which ultimately can kill the child." Lesser quantities can affect the blood, mental development, and even behavior. A Collegium Ramazzini report notes that in general lead increases "the risk for diminished intelligence, shortened attention span, reading problems,

107

attention deficit hyperactivity disorder (ADHD), school failure, delinquency and criminal behavior." Fetal exposures when mothers' bodies contain considerable lead can lead to premature birth and lower birth weights. Even quite low exposures in the womb might slow mental development and decrease IQ. And, such effects may well "persist beyond childhood."[104]

Quite low prenatal exposures to lead can cause delays in early cognitive development that remain persistent even when blood lead levels are above what the U.S. Environmental Protection Agency considers a "safe" level of exposure. Both fetal exposures and postnatal exposures can be harmful, although postnatal exposures appear to pose greater problems. Even minor prenatal increases in blood lead levels to fetuses tend to increase antisocial behavior, as indicated by arrests for violent behavior. Kim M. Cecil and colleagues along with David C. Bellinger have found that effects on the executive function of the brain (the decision centers) contribute to antisocial behavior, violent crimes, attention deficit hyperactivity disorder, and teen pregnancy and, in animals, to certain forms of substance abuse.[105] Other recent studies raise substantial concerns about lifelong effects from the body burdens of lead. Since lead persists in the human body for decades and because people can continue to be contaminated from products or workplaces, these facts raise concern about long-term adverse effects on cognitive functioning.[106]

Recent research indicates that there likely is no threshold for adverse effects from childhood lead poisoning. That is, researchers have found no lowest "safe level" for children with postnatal exposures. And the slightest increments in lead exposure above zero are now seen to cause a greater per unit effect on cognitive functioning than increments at higher levels.[107] It has taken substantial time for epidemiological studies to identify such subtle effects, because of the need to develop sensitive measures to reveal them.

Finally, a cautionary note to baby boomers. As people's bones demineralize with age, Edwin van Wijngaarden and colleagues point out, "lead can be released back into the blood stream and gain access to critical target organs, including the brain," which lead can enter quite readily. The effects on the brain are likely to be persistent. Thus there is a significant concern that lead can contribute "to cognitive dysfunction and decline in older adults. . . . on a par with decrements in cognitive function explained by aging; lead may thus be associated with accelerated aging in the brain." Consequently, some "proportion of what has been termed 'normal' age-related cognitive decline may, in fact, be due to exposure to neurotoxicants such as lead."[108] Some cases of Alzheimer's or other neurogenerative diseases may be due to lead or other neurotoxicants stored in aging citizens' bodies decades ago. This concern has recently been heightened as researchers have been able to identify areas of the brain in children with known lead exposures that functioned less well than those areas in children without such concentrations.[109]

Scientific studies of the adverse effects of lead suggest several larger points. As epidemiological studies have been refined over time, researchers have been able to identify more subtle effects, especially those resulting from in utero exposures. However, the crude tool of epidemiology had to be substantially refined in order to "see" these effects, thus illustrating (from Chapter 3) one of the ways in which epidemiological studies can be insensitive. These lead studies also illustrate that, when the human studies are sufficiently refined, humans are more sensitive than animals. It does not follow that every substance will be more toxic in humans than in animals, but these epidemiological studies add to more general studies suggesting the same point.[110]

Developing children appear to be exquisitely sensitive to lead because no "safe" level of exposure has been identified. For how many other substances might this be true?

PCBs and PBDEs

Recall from Chapter 2 that PCBs and the PBDE fire retardants are part of a generic family of halogenated persistent molecules. These two classes of substances are quite similar to each other chemically and have some similar adverse health effects.

PCBs: There are two types of polychlorinated biphenyls (PCBs): dioxin-like and non-dioxin-like. Dioxins and some other halogenated chemicals act by means of similar biological mechanisms in human and other mammalian bodies; these mechanisms have been conserved through evolution and activate enzymes that increase the toxicity of some substances.[111]

Dioxin-like compounds tend to be associated, "with reproductive, immunologic, teratogenic and carcinogenic effects." Moreover, since numerous substances have these properties, there can be additive toxic effects across seemingly quite disparate chemicals. Thus the amount of dioxin-like PCBs in a person's body does not accurately predict the disease effects to which they might contribute. A proper toxicity assessment should take their additivity to related substances into account.[112] Since these chemicals are especially persistent in humans, this additivity is a further concern.

Exposures to dioxins at comparatively high doses alter pigmentation, are associated with developmental delays and lower IQs, can cause cognitive delays, and probably affect sex-related behaviors, with none of the effects being reversible. PCB exposures can also cause male deficits in spatial reasoning, lack of endurance, clumsy movement, and IQs approximately six points lower at age eleven. Children's susceptibility to PCBs is much greater prenatally than during the immediate postnatal period. Background levels of dioxins can also adversely affect the human immune system, reducing its efficiency.[113]

Recent studies add to the neurotoxicity concerns about PCBs related to dioxins (although both kinds of PCBs are likely implicated). Researchers have found that prenatal exposures are associated with attention-deficit/hyperactivity disorder (ADHD), which manifests itself in inattention and impulsive behavior in children.[114] Considered together, human studies, according to Linda Birnbaum, Director of the National Institute of Environmental Health Sciences, "suggest that levels present in the general population may be associated with subtle signs of neurological dysfunction, delays in psychomotor development, alterations in thyroid hormone status, and changes in immunological functions."[115]

Recent news reports suggest that PCB exposures also may skew sex ratios in newborns among the Inuits and other groups inhabiting the Northern Hemisphere, but further research is needed to substantiate this point.[116] A 2008 study of San Francisco women reported related results. Women with higher concentrations of PCBs in their blood had a lower male-to-female birth ratio than those with lower exposures. The researchers concluded either that women with the highest concentrations of PCBs favored fertilization by female sperm or that male embryos were more susceptible to PCB exposures.[117]

Non-dioxin-like PCBs possess somewhat different toxicity properties than dioxins and affect "various enzymes in a different manner from dioxins." Non-dioxin-like PCBs also are persistent in the environment and accumulate in mammalian tissues. This subclass poses other adverse health effects, including especially neurotoxic risks. In the environment, of course, both kinds of PCBs are mixed, producing a variety of adverse health effects. Perhaps the better-known toxicity effects of PCBs are traceable to the non-dioxin-like variants that largely affect the nervous system and the brain.[118]

The first evidence of PCBs' neurotoxicity came from major poisoning accidents involving fairly high exposures, one in Japan in 1968 (the Yusho incident) and the other in Taiwan (the Yucheng incident) ten years later. Both involved PCB poisoning of rice oil, widely used in cooking, which was then ingested by adults and children.[119] Children born to mothers who ate the contaminated rice oil exhibited both hyperactivity and hypoactivity compared with controls. The Taiwanese children had substantially lower full-scale and verbal IQs and some hearing problems. Subsequently, epidemiological studies have found learning and memory deficits in children whose mothers consumed fish that were probably contaminated with PCBs.[120]

More subtle effects from PCB exposures have also been suggested. Prenatal exposures have been associated with poorer performances on various psychomotor tests for children. Kodavanti reports that in utero exposures to " 'background' PCB concentrations [have been] associated with poorer cognitive function in preschool children." In the Faroe Islands, researchers found that PCB concentrations in cord blood were associated with deficits in learning, but these children were also exposed to mercury, making it difficult to identify with confidence which neurotoxicants made the contribution to learning problems.[121]

PCBs have been studied in monkeys, rats, and mice. Prenatally and neonatally exposed animals manifest behavior typical of attention deficit hyperactivity disorder. This disorder includes hyperactive motor activity, impaired learning, and decreased cognitive function in several species and sensory deficits, especially hearing.[122] These results support the human findings, although it is difficult to determine subtler adverse effects in animals that might be manifested in humans.

PBDEs: The PBDEsPolybrominated diphenyl ethers (PBDEs) are chemically and biologically similar to non-dioxin-like PCBs, providing reason to believe that once PBDEs are well studied, they will be found to pose neurological and other adverse effects in humans. Neurotoxic effects are of particular concern, but others are beginning to appear.

Some human studies point to adverse reproductive effects. Recall that ninety-seven percent of adults are contaminated with PBDEs, with concentrations doubling every four to six years (Chapter 2). Women with greater concentrations of PBDEs in their blood take significantly longer to become pregnant.[123] The higher the concentrations, the longer it takes. Compounds similar to the PBDEs appear to affect male sperm via thyroid pathways.[124]

Any neurotoxicity from PBDEs is likely to be slow to be revealed, but researchers have seen lower mental and physical development among children with higher PBDE exposures.[125] For several reasons toxicity problems in humans have been difficult to attribute to PBDEs. Because there are multiple neurotoxicants, teasing out the contributions of one versus others is difficult. Neurotoxic end points may not have been studied sufficiently to separate different causal contributions with the crude tools of epidemiology. In addition, PBDEs may not have had a sufficient period to manifest their adverse effects in people. Nonetheless, there are good chemical, biological, and toxicological reasons to believe that PBDEs will pose adverse health effects in humans.

One such reason is that PCBs cause similar neurotoxic effects in animals and humans. And there are known similarities between the adverse effects of PCBs and PBDEs in animals. Together these results suggest likely similar adverse effects of PBDEs in humans.[126] Substantial research shows that PCBs cause neurological developmental effects in both animals and humans. PCBs decrease cognitive function in rats, primates, and mice; impaired visual discrimination; altered spatial perception function; and deficits in learning and memory. As we have seen above and Kodavanti reports, PCBs can also "cause neurotoxicity in humans . . . and it is believed that in utero exposure is more important than lactational exposure in causing the neurotoxic effects." These effects include behavioral changes and learning deficits resulting from both adult and perinatal exposure.[127]

PBDEs cause neurotoxic effects in experimental animal studies similar to those caused by PCBs. PBDEs appear to alter motor activity and cognitive behavior. They affect the thyroid hormones that in turn can cause neurological effects. More broadly, PBDEs are of "concern as a result of their association with endocrine disruption, reproductive and developmental toxicity, including neurotoxicity, and cancer," reports Arnold Schecter and colleagues.[128] In experimental studies the adverse effects of PBDEs are seen, "at exposure levels relevant to humans, at least in North America."

Relatively high concentrations of PBDEs are found in breast milk in the United States, with dust and food also constituting substantial sources of PBDEs. In people, the concentrations of PBDEs on a per body weight basis seem to be highest in infants and lowest (but still of concern) in adults. PBDEs are one to two orders of magnitude higher in the United States than in Europe and the highest in the world.[129]

According to L. G. Costa and G. Giordano, the half-life of PBDEs in rodents "is in the order of several days or months, [while] the terminal total body half-lives in humans have been estimated to be much longer, in the order of years for lower brominated congeners" and months for the heavier, more brominated ones.[130] This is quite worrisome because, as we saw in Chapter 2, the lighter PBDEs are more easily absorbed, more toxic and more persistent. Thus the longer human half-life suggests that PBDEs will have more time in humans to inflict damage that can lead to tissue or organ damage. Based on this body of evidence, Prasada Rao S. Kodavanti concludes, "Considering the structural similarities of PBDEs and PCBs and the known health effects of PCBs in humans, these two groups of chemicals . . . could conceivably work through the same mechanism(s), to cause developmental neurotoxicity."[131]

Very recently, major researchers working on neurological effects have suggested that exposure limits based on seemingly sensitive data from animal studies may not protect the public. Despite the concordance of human and animal responses, they argue, "intake limits for these compounds established exclusively on the basis of rodent studies have not been sufficiently protective of human health compared with epidemiological studies."[132] How widespread this result is remains to be seen, but it is a warning that standard assessments of toxicity and risks based on experimental studies may not be adequate to protect humans.

In addition, it is important to recall that individuals are neither merely exposed to nor contaminated by one substance at a time. Researchers have found that some low doses of PCBs alone or PBDEs alone do not show adverse effects in rodents, but "co-exposure [of] the same low doses of BDE-99 and PCB-52 produced significant behavioral alterations. These were equal, if not greater, than those caused by a high dose of each compound alone. This finding is suggestive of a strong synergistic effect." Adding to this, we have seen that people are also contaminated with other developmental neurotoxicants—lead, mercury, perchlorate, dioxins, and so on—which could have additive or synergistic effects, enhancing the chances of disease or dysfunction.[133] Since many of these compounds have long half-lives, they jointly remain in people for long periods of time, potentially interacting to produce toxic effects.

This additive function has now been more conclusively demonstrated in experimental studies for PBDEs and perfluorinated compounds (PFCs). Researchers have found that some heavier PBDEs combined with certain PFCs at low doses can cause "deranged spontaneous behavior, decreased learning and memory abilities, reduced cognitive function," and changes in levels of important proteins needed for neuronal growth.[134] Moreover, because PBDEs are increasing, and because PFCs and PCBs also contaminate us, the potential is substantial for adverse effects to be induced by triply cumulative exposures.

There are gaps in the PBDE-neurotoxicity picture, but these are largely because of the absence of human studies. The body of animal evidence on PBDEs, as well as their similarity to non-dioxin-like PCBs is quite worrisome, however. PBDEs' potentially additive effects with PCBs and other neurotoxicants add to the concerns. Some human evidence is beginning to appear however. As this book goes to press a recent study found children one to six years of age with higher concentrations of less brominated PBDEs scored lower on mental and physical developmental tests than children with lesser contamination. Significant results were seen for general mental development, full-scale and verbal IQ, and performance IQ. Although more research will be needed to confirm these findings, they demonstrate a relation between cord blood concentrations of PBDEs and adverse neurodevelopmental effects. The authors also call attention to potential interactive effects between PBDEs and other compounds as suggested in this book.[135] Moreover, researchers have now found that higher levels of PBDEs in pregnant women suppress their thyroid hormones. This can affect neurological development in utero and may help explain the adverse neurological effects referenced in the preceding paragraph.

A major social point is that legislatures and public health agencies appear not to have learned from the chemical and toxicological similarity between PBDEs and PCBs; are we going to repeat the PCB "mistake," with devastating long-terms consequences for people and the environment?

Some Endocrine Disruptors

In recent years researchers have become increasingly concerned about industrial chemicals and pesticides that can mimic our bodies' hormones. In what follows I illustrate this issue by focusing largely on the synthetic estrogens DES and BPA, but also mention briefly some problems with phthalates.

DES and Bisphenol A: In the United States and many western European countries, breast cancer rates among women have been high and in some cases are still rising. A U.S. woman has a one in eight or a one in nine chance of developing breast cancer during her lifetime. A woman in the United Kingdom has about a one in nine chance of developing breast cancer. Although breast cancer rates have leveled off or actually begun to decline in the United States, they remain at high levels. (The incidence of breast cancer in the United States is about five times the rate in Japan.) Moreover, breast cancer incidence rates are not due to improved diagnostic screening.[136]

A large percentage of breast cancers appear to arise from environmental influences. Only about 5 percent of, or one in twenty, breast cancer cases are traceable to inherited genetic predispositions. Some women carry a defective breast cancer tumor–suppressor gene. These genes create proteins that prevent or reduce the division of cells, an especially important function when cancer cells have been initiated and are dividing. If tumor-suppressor genes cannot inhibit cell division, the result is uncontrolled cell division and proliferation, an important step in causing cancer.[137]

Other contributors to the disease are "environmental influences." This terminology is ambiguous. Environmental factors could be any non-genetic factors that contribute to breast cancer, but this definition is too broad. A better conception of environmental factors would be, according to Andreas Kortenkamp, "avoidable contributions" to breast cancer, including "work place exposures, food contaminants, pharmaceuticals, chemicals in consumer products, air, water and soil, and physical factors such as radiation." What percentage of breast cancer cases result from avoidable factors? Studies of identical twins indicate that about 33 percent of breast cancers in these pairs are attributable to genetic factors and that 67 percent are the result of avoidable environmental factors beyond the genes.[138]

The female hormone estrogen is one important contributor to breast cancer. Too much estrogen from natural or synthetic sources over a lifetime increases risks of breast cancer. Bernardino Ramazzini, an eighteenth-century Italian occupational physician, observed that childless women or those in nunneries had greater rates of breast cancer than women who had children. He conjectured that their celibacy and childlessness contributed to their breast cancer.[139] His observations were correct, but he did not understand the role of hormones, including one, namely, that

115

childless women produce more estrogen over a lifetime than those who have children. For example, the earlier women give birth, the lower their rates of breast cancer, while the later they give birth, the greater their odds of contracting the disease. This protective effect of early childbirth appears to result from more complete growth of the breast tissues early in life, thus reducing the ultimate number of cells in the breast that are susceptible to cancer.[140] Researchers can now explain Ramazinni's observations.

The longer women are exposed to their own estrogens, the greater their risks of breast cancer. For example, early puberty or late menopause each increases the risks of breast cancer. In contrast, delayed puberty or early menopause both tend to lower breast cancer rates, because either of these occurrences somewhat shortens exposure to estrogens.[141] Higher levels of exposures to mothers' estrogens in utero also increase breast cancer rates. Women who are twins, identical or fraternal, are exposed to higher concentrations of estrogens in utero and have 1.6 to 2 times the rate of breast cancer compared with women who are the product of single childbirths.[142] Female offspring who are nonidentical twins of a brother have a somewhat higher rate of breast cancer than those who are in utero with an identical female twin. Similarly, exposure to synthetic estrogens in utero increases breast cancer rates. For instance, as DES daughters reach middle age, their breast cancer rate doubles compared with the general population of women. We have also seen that mothers who took DES had an increased breast cancer rate.

Research from hormone replacement therapy adds to this view. In a large British study, women who received some form of hormone replacement—estrogen only or a combined estrogen-progestin treatment—had "increased breast cancer risks."[143] Use of this therapy may have increased breast cancer cases in the United Kingdom by twenty thousand (versus the number of cases among women with no use of hormone replacements). U.S. researchers identified a similar risk from hormone replacement therapy and recommended against using it, resulting in a downturn in breast cancer rates.[144]

This total body of evidence strongly supports the claim that a woman's increased exposure to estrogens, natural or synthetic, can increase her risks of breast cancer.[145] In utero and early exposures seem even more important, as the DES and twin studies illustrate. What does this have to do with body burdens of industrial chemicals, pesticides, and adverse health effects? Put another way, are the only women who have to worry about breast cancer those who are twins or who have bad genetic luck? The quick answer is no.

DES is no longer utilized as a pharmaceutical, so it is one less thing for women to worry about, except for women who took it, their children, and their grandchildren. However, since synthetic estrogens, like DES, can contribute to breast cancer, and since DES has been well studied, it provides some evidence and a model for the effects of other estrogenic compounds in animals. Bisphenol A (BPA) is another synthetic estrogen, which was synthesized about the same time as DES. BPA is produced and used at the rate of about 2.3 billion pounds annually in the United States, and virtually all citizens are contaminated with it. Moreover, most citizens are continually exposed from various sources. BPA seems less potent than the most potent natural estrogen and than DES; should it be of any concern? Should citizens have concerns about its adverse effects by itself? The answer to both questions appears to be yes. Based on animal results, a consensus scientific panel on BPA led by Frederick vom Saal concluded that BPA likely can contribute to breast cancer, and also have other adverse effects. The same panel points out that current body burdens of BPA in adults are "within the range that is predicted to be biologically active in over 95% of people sampled."[146]

Experimental studies reveal a wide range of adverse effects at low doses during development as well as during adulthood. This research suggests possible similar adverse outcomes in humans. These outcomes would include increases "in prostate and breast cancer, uro-genital abnormalities in male babies, a decline in semen quality in men, early onset of puberty in girls, metabolic disorder including insulin resistant (type 2) diabetes and obesity, and neurobehavioral problems such as attention deficit hyperactivity disorder (ADHD)." Many of these effects are irreversible and can result from even brief "exposure during sensitive periods in development." For example, adding to the number of cells that are vulnerable to breast cancer increases the possibility of breast cancer, and this phenomenon is not reversible.[147] At concentrations typical of human blood levels BPA also kills human placental cells, raising concerns that it could affect pregnancy outcomes.

Very recent research adds to these concerns. At environmentally relevant doses PBA can freely cross the human placenta (in an experimental model) in a manner that is quite similar to another substance with this property, anipyrine. Moreover, BPA is not conjugated, which would protect the fetus from toxic effects, but remains in "free" form, which is much more toxic. The authors conclude their paper, "The underdeveloped fetal liver cannot protect the fetus against this constant exposure. Hence, the fetus is potentially quite vulnerable to the adverse effects of BPA, and it is critical to evaluate the harmful effects on the health and well-being of the fetus from such an exposure." Similar results are also seen in experimental studies on rats. In addition, BPA can change how important genes are expressed, and can damage placental cells.

BPA may not be detected in a person's body when the disease or dysfunction appears—it might even be long gone by that time, because it has such a short half-life in mammalian bodies. If so, there could be a long latency period between exposure and potential disease, which increases the difficulty of tracing the causal contributions. However, because most people are contaminated with BPA, this suggests, as the consensus BPA committee points out, that "there is chronic, low level exposure of virtually everyone in developed countries to BPA."[148]

A National Toxicology Program (NTP) committee largely echoed the consensus expert group. The committee reiterated unease about "neural and behavioral effects in fetuses, infants, and children at current human exposures." It also had some concerns "about bisphenol A exposure in these populations based on effects in the prostate gland, mammary gland, and an earlier age for puberty in females." The reason for this concern is that animal studies show that " 'low' level exposure to bisphenol A during development can cause changes in behavior and the brain, prostate gland, mammary gland, and the age at which females attain puberty." An earlier National Research Council report expressed similar concern about the additive effects of estrogens in animals: "No threshold exists for exogenous estrogen, because endogenous estrogens are at a sufficiently high concentration to exceed the threshold for sex determination. . . . If the threshold for response to estrogen has already been exceeded before exposure to an environmental estrogen, the additional load of an environmental estrogen might cause a significant increase in the occupied receptors required for the [sex determination] response."[149]

As the above results show breast cancer in humans is not the only concern from synthetic estrogens. Moreover, there may be much wider consequences because of the cumulative effects of different estrogens in people's bodies. For one thing, there is a consensus that substances that attach to certain similar receptors or sites in mammalian cells can make additive contributions to disease or dysfunction. Such effects have been identified for dioxin-like substances (as noted above) as well as for two groups of pesticides, organophosphates and carbamates (see Chapter 5). In addition, natural or synthetic estrogens similarly attach to particular receptors. Thus, Kortenkamp argues, "human risk assessment could work on the basis of the rebuttable hypothesis that dose addition [of estrogens] is applicable." More significantly, "joint effects occur *even when all mixture components are present at levels below doses that cause observable effects* . . . . [Consequently] the ground is prepared to seriously consider group-wise regulation of EDs [endocrine disrupters]."[150]

Yet some scientists or some companies affected by a legal action will argue that there is no evidence that at very low exposure levels substances in fact cause adverse effects. There are many problems with such assertions. As already noted, epidemiological studies can easily be too small or too short to discover risks of harm (see Chapter 3). Moreover, human studies can also be too crude to identify subtle effects, at least in the early stages of research. Diseases may have multiple causes, some of which mask the slight effects caused by low-dose exposures. And it is difficult for studies to identify new causes of adverse effects from variation of disease patterns in existing diseases, especially if they are common.

A biological problem is that the effect of low concentrations of synthetic substances added to background levels of similar substances together can cause adverse outcomes. This is the case with BPA in experimental studies. Thus concentrations of individual substances that might seem quite low can add to natural levels of estrogens or other sources of estrogenic compounds and cause diseases or dysfunction. Moreover, if there are several natural or synthetic estrogens that attach to similar receptors, the problems increase. It is even possible that no one synthetic estrogen taken by itself at tiny concentrations may pose problems, even when added to natural background rates. However, when all sources are added together, the total estrogens may be more than sufficient to produce adverse outcomes. Consequently, just because a substance is present in persons' bodies at levels below which adverse effects would be seen in animal or human studies, it does not follow that there is no effect when the effects of substances are additive. There is not necessarily a "no effect" level, according to Andreas Kortenkamp unless the concentration of the substance is at zero.[151]

Phthalates: BPA is not the only synthetic hormone-like substance that can disrupt endocrine systems. The Endocrine Society of American is sufficiently concerned about these effects that it has called for an alert on endocrine disrupters and noted several for specific concern: the pesticides vinclozolin, DDT and its metabolite DDE, along with PCBs, dioxins, and phthalates.[152]

Phthalate esters are another class of endocrine disruptors. Adults and children alike have widespread exposures from toys, food packaging, pharmaceuticals, medical devices, cosmetics, and personal-care products, as well as cleaning and building materials. They have been found in the amniotic fluid of animals and humans. Children have higher concentrations in their bodies than adults.

Males seem especially susceptible to diseases or dysfunctions caused by phthalates. Phthalates cause reproductive tract disorders in animals, including "undescended testes, a cleft phallus (hypospadias), and indices of feminisation such as retained nipples." Others include infertility, decreased sperm count, and "other reproductive tract malformations." Recent human data suggest different effects; children exposed to phthalates *in utero* are more likely to exhibit hyperactivity attention deficit disorder.

As the above shows men are not home free. Endocrine disrupters also have the potential to affect their health adversely. Human studies have found associations between prenatal contamination with phthalates and the early stages of penile dysgenesis syndrome, echoing experimental studies. Moreover, we have already discussed some transgenerational effects on male animals from their great-great-grandmothers' in utero exposure to several pesticides, BPA, and phthalates. U.S. citizens are extensively contaminated by both BPAs and phthalates. And we have considered some synergistic effects of antiandrogens in experimental studies. Since many men and women (of childbearing age) are contaminated by many of these substances, experimental studies raise warning flags about developmental effects from multiple endocrine-disrupting chemicals.[153]

Generalized Additive Effects

We have seen how some substances, such as estrogens or dioxin-like substances, can attach to particular cellular receptors, like a key to a lock, and have additive toxicity effects. However, substances do not need to attach to similar cellular receptors to add to the toxicity of other substances. Tracey Woodruff and others have reported several compounds that interfere with or perturb different "upstream pathways" and that cause adverse effects without necessarily attaching to the same cellular receptor (as dioxin-like compounds do).[154] These compounds produce similar adverse health consequences by different mechanisms. This is a broader pattern of additivity about which we should be concerned.

Substances that reduce thyroid hormones illustrate this kind of adverse effect. When thyroid hormones are decreased in pregnant women, this can effect developing fetuses that need these hormones for normal neurological development. Similar effects can occur in newborns. In turn, this increases the risks of adverse nervous system effects. Dioxins, dibenzofurans, and dioxin-like PCBs activate one set of liver enzymes, with adverse effects, while non-dioxin-like PCBs activate another set of liver enzymes, with similar results.[155] According to Woodruff and others, these chemical classes appear to have "a dose-additive effect on [thyroid hormones] at *environmentally relevant* doses . . . demonstrating exposures to chemicals acting on different [biological] pathways can have cumulative effects."[156] Environmentally relevant doses are those close to the concentrations humans actually receive. Given what is known about the properties of PBDEs, they may also contribute to such adverse effects.

In addition, researchers from Andreas Kortemkamp's laboratory have found synergistic outcomes by different mechanisms caused by chemicals that have antiandrogen effects on the male reproductive system. Four compounds, each at doses below which they have caused adverse effects, when administered together can cause adverse effects. Jointly acting by different mechanisms, they produce greater adverse effects than would be seen from adding the doses of single substances.[157]

The National Research Council has also recently cautioned against focusing too narrowly on cumulative effects resulting from toxicants attaching to particular cellular receptors. It strongly supports the idea of more generalized additive effects: "the focus in cumulative risk assessment should be on the *health outcomes* and not on the pathways that lead to them, whether defined as mechanisms of action or as modes of action."[158]

Toward a Paradigm of the Developmental Origins of Disease

Scientists have formulated a new paradigm for some sources of disease— the developmental origins of health and disease—as they have accumulated a body of evidence for it and as they have also begun to understand some of the biological mechanisms by which disease and dysfunction arise in utero and in early childhood. That is, they have gone beyond merely identifying perinatally induced adverse effects in humans and in experimental studies to a broader explanation of what is occurring.

This paradigm has at least two components. The first, more important one is that many diseases, in the words of Heindel, "may have their origin during development and not during adult life when the disease becomes apparent." The second component, a specific mechanism to account for some of these outcomes, is based on "epigenetic" influences on the genome. This is the idea that while the genetic sequence in a person's body does not change, environmental influences can cause "altered gene expression or altered protein regulation associated with altered cell production and cell differentiation that are involved in the interactions between cell types and the establishment of cell lineages."[159] Because of exogenous or other influences, the genes do not express themselves as they normally would or are caused to express themselves at inopportune times, ultimately leading to disease. An epigenetic mechanism may not be the only one to account for predispositions to disease during development, but it appears to be an important one.

How do diseases originate as a result of epigenetic effects during development? Changes in how genes express themselves tend to modify cell production or differentiation. These in turn can induce modified structural or functional changes, as Heindel notes, in the "character of the tissues, organs and systems . . . The end-result is an animal [or human] that is sensitized such that it will be more susceptible to diseases later in life." Traditionally, developmental researchers have focused on exogenous agents that caused premature death of the fetus or birth defects. Recent research has attended to more subtle "changes in tissue function that are not the result of overtly or grossly teratogenic effects, but that result in increased susceptibility to disease/dysfunction later in life."[160]

Several lines of research have further supported the paradigm shift to the developmental origins of disease: nutritional "mismatch" between in utero nutrition and postnatal nutrition, leading to diabetes and heart disease later in life; other epigenetic-induced diseases; and toxicologically induced neurological dysfunction.

Nutritional Mismatch, Metabolic Syndrome, and Epigenetic Effects
Nutritional deprivation of pregnant women followed by adequate nutrition for a child after it is born can result in what has come to be called the "metabolic syndrome." This is a collection of symptoms that includes

higher than normal body weight; increased body-mass index (BMI), that is, the height to weight ratio; glucose intolerance; and a tendency to develop type 2 diabetes.[161] These conditions are now quite common in the United States, usually explained as the result of genetics, family behavior, overeating, lack of exercise, or food producers adding substantial sugars and fats to foods. However, other explanations may partially account for such outcomes, including developmental conditions and exposures to toxicants.

Important data for the developmental origins of metabolic syndrome came from human studies of the "Dutch winter famine" of 1944–1945. From November 1944 to May 1945 the Dutch population experienced a famine imposed by the Nazis' in their final throes of controlling the Netherlands. Women who were pregnant during this period but undernourished in effect deprived their developing children of adequate nutrition and calories in utero. After birth the children then had more or less adequate diets. A subset of this cohort of children who were nutrition deprived in utero showed "an increase in body weight, BMI, and waist circumference at 50 years of age. . . . Those low-birth-weight babies who were most vulnerable to developing obesity were men who had been light and thin at birth and had experienced a period of rapid childhood growth."[162] Thus tendencies to obesity and associated diseases were programmed into people while they were developing in utero.

Researchers came to characterize this phenomenon as "the thrifty phenotype." Children who are undernourished in utero develop a "thrifty" biological system to prepare them for what seems to be a nutrition-limited world following birth. However, if their postnatal nutrition turns out to be adequate or more than adequate, they tend to gain greater weight than do children who had not been exposed to a nutrition-deprived environment. To the extent that there is a "match" between the nutrition the child experiences in utero and the nutrition the newborn finds after birth, the child tends not to manifest substantial weight gains.

The siege of Leningrad by the German army during World War II also created a famine for the populace, but because it lasted from 1941 to 1944, researchers have not found metabolic syndrome as a consequence. Malnourished pregnant women carried developing fetuses, not unlike the Dutch. However, the Russian children were born into a nutrition-deprived world that continued in that state for some time. Thus these children did not show the increase in body mass, tendency to coronary heart disease, and diabetes manifested by some of the Dutch children, probably because in Leningrad there was a "match" between the fetuses' nutrition in the womb and the nutrition in the external world the newborns entered.[163]

Similar effects have been seen in animal models, where conditions can be experimentally controlled and modified. Sheep provide a good model, because their pattern of pregnancy closely resembles humans'. Lambs born to nutrition-deprived ewes but given adequate nutrition after birth showed the sheep equivalent of metabolic syndrome. Even ewes deprived of proper nutrition before conception can give birth to lambs that develop metabolic syndrome with adequate postnatal nutrition. It is as if a biological signal in utero suggests a nutrition-reduced external world awaiting the newborn lambs. However, a world with adequate nutrition poses problems for these offspring. According to Caroline McMillen and others, this "results in poor health outcomes that emerge [earlier] in childhood and adolescence and can potentially impact on the health of the subsequent generation."[164]

Exposures to in utero–induced toxicants can also induce metabolic syndrome.[165] According to Heindel, low doses of the synthetic estrogen DES administered prenatally or immediately postnatally "caused an increase in body weight of CD-1 mice that was not evident at birth but reached significance by 6 weeks of age," the equivalent of early adulthood. Dramatic pictures of genetically identical mice that were given the same diet—and whose only difference was a very low concentration of DES given to one animal but not to the other—show the same consequences. The DES-dosed mouse is obese; the other is quite normal. Genetically identical mice, with the same diet and same living conditions, had different outcomes because of a synthetic estrogen administered in utero. In addition, other natural or synthetic estrogens present at low doses in utero can alter the "set point of body weight," the weight to which an animal will grow with adequate but not excessive nutrition. Bisphenol A and tributyltin appear to contribute to similar weight gain and involve epigenetic phenomena.[166]

Estrogenic compounds are of concern because chemicals with estrogenic activity can cause, in Heindel's words, "oestrogen-sensitive gene expression at the wrong time, if the cells are exposed to the agent when oestrogen levels are usually low and oestrogen-sensitive genes are not active." Either increased gene expression or suppressed expression can affect gene functioning, sometimes adversely. Other hormone mimicking compounds can also interfere with normal functioning. Scientists do not yet fully understand how this occurs, but they have data of epigenetic marking that causes "altered gene expression."[167]

Other Epigenetically Modulated Adverse Effects

Epigenetically induced mechanisms of disease go beyond contributing to metabolic syndrome. In experimental studies, DES present in utero increased uterine carcinogens in adulthood (modeling human effects). Moreover, the epigenetic effect and its consequence are transmitted from one generation to another. Endocrine-mimicking substances bisphenol A and dioxin administered in utero increase the risk of mammary gland cancer in rats (analogous to breast cancer in humans), and the tendency to this cancer is transmitted from one generation to another. However, in utero exposure to the pesticide atrazine (also similar to estrogen) inhibits the development of mammary glands, but transgenerational effects have not been observed.[168] Thus not only do toxicants contribute to disease in one generation, but the gene-exposure-modification effect in some instances can be carried to another generation, depending upon the substance, even though the gene sequence itself has not changed.

Females are not the only toxic target; males can also be adversely affected. Both rats and mice exposed in utero to dioxin at close to human doses exhibit visible changes in the tissue structure of the prostate gland, but not sufficiently to regard it as a teratogen (an agent causing malformed tissues or body parts). However, these exposures and subsequent diseases suggest that low-level dioxin exposures could lead to increased size of the prostate and possibly later to prostate cancer. As researchers state, "This suggests that the gene expression is altered by *in utero* exposure to dioxin and that the alterations are permanent."[169] If dioxin can produce such modifications, related dioxin-like compounds would also be of concern because of their chemical and mechanistic similarity (see Chapter 2).

When administered separately at low concentrations in utero, BPA, the pesticide methoxychlor, and DES—which all possess estrogenic properties—and the natural estrogen estradiol each increase prostate weight in adult mice. These weight increases have not been shown to lead to prostate cancer, but prostate cancer in men often follows a substantial increase in the size of the prostate. It is significant that both estradiol, a natural estrogen, and DES, the well-studied synthetic estrogen, cause such effects. They provide a model of known toxic effects against which to compare adverse outcomes of other estrogenic substances. Of course, men are neither likely to be exposed to estradiol after birth nor to DES, but they are contaminated with BPA, dioxins, and other synthetic estrogens, thus raising the possibility that these exposures can contribute to enlarged prostates or possibly increase prostate cancer risks.[170]

Pregnant rats exposed in utero to some phthalates give birth to pups with defective male reproductive tracts that then lead to functional changes later in life. These changes include, Heindel notes, "decreased sperm production and testicular tumours."[171] Phthalates have also been associated with structural changes in human penises, echoing some of the experimental results.[172]

Finally, in utero exposure to methoxychlor or vinclozolin not only resulted in adverse effects on the male reproductive tract but also caused effects that could be transmitted through the male germ line for at least four generations. Vinclozolin is a fungicide utilized on grapes and other agricultural crops and is an antiandrogenic compound, inhibiting the effects of the male hormone androgen. Similar results have now been demonstrated for several other compounds.[173]

A gestating mother rat was exposed during pregnancy to vinclozolin (and was the only animal exposed). Researchers then bred her offspring and their offspring to wild types through several generations. In order to determine transgenerational effects, researchers must breed animals through three additional generations. A pregnant mother's exposure results in exposure of the first generation offspring (the sons) and of the second-generation germ line (in the grandsons). In order to see transgenerational effects, researchers must examine the third generation. In these experiments the various generations of offspring had lesions of the prostate, spermatogenic cell death, defects in spermatogenesis, and sperm death leading in some cases to infertility. The consequences were not limited to sperm defects and deficiencies. Other adverse effects included quite unexpected breast tumors, prostate disease, kidney disease, and immune abnormalities. Most of these adverse effects were seen out to the fourth generation—the great-great-grandsons—indicating that the effects were clearly preserved genetically. Most likely this occurred by means of an epigenetic mechanism, not a change in the genetic sequence in the animals.[174]

In Utero or Postnatally Caused
Late-Life Neurological Dysfunctions

Diseases that result from in utero or perhaps immediate postnatal exposures are not restricted to endocrine disrupters, dioxins, or nutritional mismatches. Other examples may illustrate somewhat different mechanisms that contribute to adverse neurological effects (the scientific jury is still out), but together all these exposures exemplify, in the words of Brian K. Barlow and colleagues, that "susceptibility to certain diseases can be established by environmental stressors encountered very early in life (in utero or perinatally)." These results are "based on solid rationale and some model systems."[175]

Recall from Chapter 3 some of the studies of neurological dysfunction, including Parkinson's disease. The frozen addicts, who received high doses of MPTP, suffered from degeneration of cells in an area of the brain called the "substantia nigra."[176] The neurons in this area suffered sufficient deterioration that the people manifested almost instant Parkinson's-like conditions. Exposures to toxicants causing lesser damage to such cells can set the stage for later disease or dysfunction long removed from the event that damaged the cells. (MPTP poisoning does not particularly illustrate a developmental problem, but it did provide clues to and a model for more insidious developmental contributors to disease.) Substances causing such effects in utero in experimental studies include the herbicide paraquat and the fungicide maneb. Paraquat's chemical structure resembles MPTP.[177]

The explanation for long-term adverse effects seems to be that areas of the brain can suffer adverse effects and lose early in life some of the reserves needed for lifetime normal functioning. However, deficient functioning typically does not appear until decades later. Researchers proposed the general hypothesis more than twenty years ago. Barlow and others argue that a plausible mechanism appears to be that an environmental insult causes a long period of "silent toxicity" often lasting decades, during which this portion of the brain is operating at a suboptimal level, but there are no overt parkinsonian symptoms. During this time other cells may provide compensatory effects, producing seemingly normal muscle behavior. However, a slow decline in dopamine levels that occurs with natural aging, combined with the original insult that reduced or damaged dopamine cells, crosses a threshold, leading to Parkinson's disease.[178]

Exposures to other neurotoxicants also appear to cause Parkinson's-like conditions, which raise concerns for children should appropriate exposures occur. A recent family-based epidemiological study by Dana B. Hancock and colleagues adds support "to the hypothesis that pesticide exposure is positively associated with risk of PD [Parkinson's Disease]." Exposures to "organochlorines, organophosphorus compounds, chlorophenoxy acids/

esters, and botanicals [are] potential risk factors for PD." A quite recent human study by Sadie Costello and others adds to the evidence: "Exposure to a combination of maneb and paraquat increases PD risk, particularly in younger subjects and/or when exposure occurs at younger ages."[179] Pesticides appear to have such effects because many of them are neurotoxicants.

Finally, quite recently trichloroethylene, the degreaser, anesthetic, neurotoxicant, probable human carcinogen, and contaminant of drinking water, has also been proposed as a contributor to Parkinson's disease, especially as a consequence of workplace exposure. Workers exposed to TCE in a Kentucky plant and who seem to suffer from early Parkinson's show evidence of neuronal damage to the substantia nigra.[180] Other researchers, studying twins one member of which had greater exposure to TCE, have found TCE increases Parkinson's disease up to six-fold.[181] These neurotoxicants thus appear to be possible contributors to the developmental origins of neurological disorders with appropriate exposures.

Scientific Summary

The scientific picture that emerges from the review in this chapter is quite compelling but incomplete in some respects. As a matter of general knowledge, scientists have established a number of important findings in developmental biology.

The womb is not a protected impermeable capsule. Development does not proceed from the carrying out of a fixed genetic program, immune to environmental influences. Instead, the womb is more like an internal environment that provides nutrients and fluids to the developing organism, but that is also subject to any toxicants that reach it through a fairly porous placenta. This internal environment, thus, importantly influences how a person or animal develops in utero. Although a person's genetic code does not change, epigenetic changes in utero can modify its functioning. Chemical contaminations can affect the expression or functioning of a person's genetic sequence and subsequent proper development, and some can damage certain areas of the brain leading to neurological problems later in life.

Developing organ systems tend to be more susceptible to adverse effects from toxicants than the same systems in adults. Developing fetuses and children often have greater exposures on a per body weight basis than adults, as well as higher metabolism, breathing rates, and absorption rates. Behavioral tendencies also increase exposures. At the same time, developing children have fewer and lesser defenses against potential toxicants than do mature adults.

Particular compounds—lead, mercury, methylmercury, DES, PCBs, dioxin-like substances, toluene, arsenic, pesticides, tobacco smoke, water disinfectant byproducts, ethanol, anticonvulsant drugs, and radiation—cause human diseases and dysfunction as a result of *in utero* or prenatal exposures. Researchers have identified two hundred human neurotoxicants alone that could pose problems. Experimental studies have illustrated effects from a wider range of substances, including paraquat, maneb, and vinclozolin. Numerous other compounds appear to be neurotoxic in experimental studies.[182]

Some diseases are exacerbated or triggered sooner as a result of early exposures. Prenatal exposures often resemble time bombs, causing adverse effects much later in life, as we have seen in the case of carcinogens, estrogenic compounds, immunotoxicants, and neurotoxicants. Thus the existence and biological plausibility of developmental effects have been shown for numerous substances. Moreover, researchers are beginning to understand some mechanisms for some of the adverse effects.

Yet there are gaps in the science. The evidentiary picture may be something like a pointillist painting, with parts of the picture filled in with many data points, other parts left blank because they have yet not been explored, and still other sections partially filled in. However, the general background is beginning to take shape.

While scientists have established that some adverse effects occur from in utero, neonatal, and childhood exposures, how many different environmental exposures will contribute to such effects is not known. As developmental toxicology is pursued, it seems likely that other developmental toxicants are likely to be to be revealed. Scientists almost certainly do not yet know all the disease processes that can be potentiated by nontoxic substances or initiated, accelerated, or exacerbated by early exposures to toxicants. They may not know the lowest exposures during development that can pose adverse effects, either immediately or later in life.

As research tools are refined to detect more subtle end points, more adverse outcomes may well be identified, and more subtle effects will likely be detected. When known hazards, such as carcinogens, endocrine disrupters, neurotoxicants, and so on, contaminate a developing fetus, it seems a poor bet merely to *hope* that they will not cause adverse effects.

The resulting picture is troubling. There are substantial scientific concerns that substances similar to those with known toxicity—such as PBDEs, bisphenol A, perfluorinated compounds and phthalates, some classes of pesticides, dioxin-like substances, and various estrogenic compounds— are only now being appropriately appreciated.[183]

Plausibility Arguments

Despite the pointillist nature of the science, it establishes a number of biological plausibility arguments that are pertinent for the law and for the discussion that follows. That is, the science has established certain facts of toxicology in humans. These facts are similar to certain kinds of proofs or showings in mathematics. A mathematician often seeks to show that there is at least one mathematical entity of the kind he or she is considering. After the existence of one example is demonstrated, the question then becomes, how many similar entities are there? This is called an "existence proof in mathematics."[184]

Analogous to such existence proofs are certain results that scientists have discovered for particular substances. Once the existence or plausibility of particular adverse effects has been shown for one substance, how many other substances will show similar or analogous effects? Similarly, once adverse effects have been shown at one concentration, at what other concentration levels might some of the same effects occur? These questions await the progress of science on the issues. Consider three "biological plausibility" claims that the current research has shown.

First, it is biologically plausible that virtually all industrial chemicals enter citizens' bodies and penetrate their tissues and fluids by various pathways. All but the very largest molecules appear to invade in this manner. Second, it is biologically plausible (and highly likely) that industrial chemicals in women's bodies will cross the placenta and enter the tissues of developing fetuses. Recall Schardein's summary: "The vast majority of chemicals given the pregnant animal (or woman) reach the fetus in significant concentrations soon after administration."[185] Similarly, many industrial chemicals will enter the breast milk and be transmitted to nursing newborns. Third, thus there is a live possibility that developing fetuses and newborns can be at risk from toxicants contaminating their bodies. This has been demonstrated in humans for numerous substances at certain exposure levels. Experimental animal studies broaden and deepen these concerns.

From the above plausibility claims and the discussion in Chapter 2, it is clear that what our parents and other individuals expose us to or what we do to ourselves in contaminating our bodies can affect our health and disease status. This contamination may even affect the health of our future generations. Insofar as we are contaminated by others' toxicants and as these substances potentially cause diseases and dysfunctions, this can raise major questions of justice.

These plausibility claims probably have not been appreciated beyond the scientific community. Yet they have profound implications for the law, as we will see in the next two chapters.

RECKLESS NATION:

HOW EXISTING LAWS FAIL TO

PROTECT CHILDREN

Some industrial chemicals, pesticides, pharmaceuticals, and pollutants have already caused harm from in utero exposure. Others pose live risks, especially during development. Some may trigger transgenerational effects and others have been shown to act in concert with other contaminants to pose unexpected risks. And there are more worrisome signs on the horizon. For example, a recent study found polybrominated diphenyl ether fire retardants (PBDEs) in the livers of human fetuses and at higher concentrations than currently exist for polychlorinated biphenyls (PCBs) in some tissues.[1] By themselves, higher concentrations of PBDEs in children appear to reduce mental and physical development.[2] Because the toxicity of these PBDEs resembles that of PCBs, this is of substantial concern. Moreover, the two groups can jointly cause increased adverse effects Another study found newborns begin life already suffused with 200-300 synthetic and industrial chemicals of known and unknown toxicity.[3]

As a matter of biology and chemistry, the invasion of industrial chemicals into our bodies cannot be prevented. Toxic contamination is another story. It is not inevitable that toxicants must contaminate developing children and put them at live risk for developmental diseases and dysfunctions. Currently the law permits toxic contamination and live risks to our children. The law is the culprit; it also holds the solution.

Most people believe that industrial chemicals, like pharmaceuticals and pesticides, have been tested for their safety before people are exposed. After all, haven't our elected representatives understood the possible risks of untested substances and sought to protect us? Sadly, the answer to this is no. Overwhelming numbers of commercial chemical products (80 to 90 percent) are subject to postmarket legislation, entering the market without any required testing for their toxicity. Thus scientists and public health officials have little or no data about hazardous effects of the vast majority of industrial compounds and can remain ignorant for decades after substances have entered commerce and contaminated us. If the products pose risks, children and adults serve as experimental subjects until health protections are increased. Toxic products remain in commerce and in our bodies until legal action forces better public health protections. The motivation to commercialize products and the creativity involved in doing so far outstrip scientific understanding of their risks and our health protection.

Consequently, if substances adversely affect developing children in utero, how can postmarket laws protect them? How can such laws protect children from substances such as phthalates, bisphenol A (BPA), perchlorate or solvents, in utero exposure to which appears to pose developmental or reproductive problems?

It gets worse. Before anyone becomes a fetus, the biological material that surrounds and that eventually becomes part of the fetal tissues has intimate contact with substances of known and unknown toxicity. Men's sperm and women's eggs, the resulting fertilized eggs, developing tissues, and early fetuses are surrounded by these compounds. While nascent organ systems are developing, they too are contaminated. How can postmarket laws prevent harms from occurring, if such substances contribute to harm at early life stages?

Postmarket laws simply cannot protect developing children. They also create incentives to greatly delay protections whenever public health officials attempt protective efforts. The nature of scientific studies and the time needed to conduct them exacerbates this problem. Better public health protections typically become entangled in obfuscation, procrastination, and long disputes about science and legal procedures. Existing laws are misaligned with the science that could help protect us. They are reckless toward our health and our children's, and they permit companies that create such products to be reckless toward us as well. More fundamentally, such laws are based on inappropriate models drawn from older areas of law, and the current legal conception of enforcement is quite inappropriate for toxic molecules.

In order to understand some of these issues, we should ask several questions. How do some of the main environmental health laws aim to protect us from toxicants? How good are these protections? What are some of the consequences of this legislation?

Environmental Health Legislation

In the early 1970s, during the heyday of environmental concerns, the U.S. Congress (and many state legislatures) ushered in the era by passing a wide range of laws aimed at protecting our health and the environment. This spate of legislation sought to remedy shortcomings of existing institutions and to address better and more comprehensively risks to public health and the environment.

The Clean Air Act (1970), the Clean Water Act (1972), and the Safe Drinking Water Act (1976) sought to prevent risks from industrial pollutants in the air, surface waters, and drinking water. The Occupational Safety and Health Act (1970), or OSH Act, aimed to protect employees from workplace risks and harms. The Consumer Product Safety Act (1972) addressed risks from consumer products, for example, toys, hair dryers, and furniture. Congress also amended the pesticide laws (1972) and the laws concerning food additives, food contaminants, and drugs (1968).[4]

By 1976 these initial legislative efforts were seen as inadequate because they tended to focus on pollutants one venue at a time—the air, surface waters, drinking water, or the workplace—or they addressed only a limited group of substances, such as pesticides, drugs, food additives, and consumer products. Collectively, they failed to address toxic substances in a more comprehensive manner. The Toxic Substances Control Act (1976), or TSCA, sought to remedy this gap. There were other inadequacies as well. Toxic wastes dumped into the ground or poorly discarded over a long period needed to be cleaned up. This led to the Superfund legislation, the Comprehensive, Environmental Reclamation and Cleanup Act of 1980 (with some important amendments being passed in 1986). There was other legislation during this period, but these were some of main laws.

Since the decade of the 1970s into the early 1980s, some of these laws have been amended, but the wave of environmental health legislation typical of that decade has not been repeated. Since about 1980 there has been considerable political pressure in support of industries that create products with risks; this pressure has frustrated improved health protections. This political view has been manifested not only in limited legislative action but also in a significant decrease in specific health protections by the delegated agencies and decreased funding for their tasks. At the same time courts have also tended to constrain public health agencies from implementing what limited protections they sought to institute.

Legal Strategies to Protect the Public

How do the various laws seek to provide protection from toxicants? There are two broad approaches. One is to try to prevent outright any harm from occurring by means of administrative, or regulatory laws. Administrative laws are creations of legislatures, either federal or state, that authorize administrative agencies (such as the Environmental Protection Agency, the Food and Drug Administration, or the Consumer Product Safety Commission and some equivalents in the states) to address risks from toxicants before they cause harm. I will focus on federal laws here with one exception.

A second strategy, part of older common law, roughly speaking waits until harm has occurred and then metes out punishment via the criminal law or provides legal procedures by which injured parties may institute action to receive compensation for their injuries through the tort, or personal injury, law. (The criminal receives no discussion in this book; the tort law only modest consideration.)

A preventive strategy is clearly the preferred approach. On this approach, Congress passes laws to address problems that are too particular, require too much scientific expertise, and need too much detailed attention for a legislative body. More particularly, Congress creates legal blueprints for how administrative personnel should address specific risks by regulation in order to provide health or environmental protections. Congress delegates legal authority to administrative agencies to carry out these tasks, in accordance with procedures specified in the blueprint, to achieve the health protections indicated in the general legislation.

There are two broad preventive approaches. A premarket law requires products to be tested for certain kinds of hazards and risks before they enter the market. Companies may not commercialize their products until scientists with the agency are persuaded that they meet the safety standards required by the legislation. A postmarket law permits products to enter commerce without any premarket testing or review of risks that the product might pose. A public health agency reviews the safety of a product in the market only if there are data showing it poses risks in violation of the underlying statute. Premarket laws govern only a relatively small number of substances (about 10 percent, possibly up to 20 percent). These laws apply to pharmaceuticals, pesticides, and new food additives.

Postmarket laws apply to the vast majority of substances to which we are exposed. These laws have also been called "policing" or "cops and

robbers" laws.[5] They are related to a more traditional approach to pollutants and resemble in some respects nuisance laws that owe their origin
to Anglo-American common (judge-made) law.

Postmarket Laws

How can postmarket policing laws "prevent" harms, since they do not
permit legal action to reduce risks until after substances are in commerce
and the public is exposed to them?

Congress spells out general procedures that the agencies must follow
to identify substances that pose risks to the public or workforce, just as a
police officer would identify legal violations, and to issue laws—called
public health "regulations"—to reduce these risks.[6] An agency has a legal burden of proof to find or develop evidence that a substance poses
risks, follow the administrative procedures to ensure that there is a defen-
sible case to be made for regulation, and issue health standards to reduce
the risks before they materialize into harm.

The concept of a risk is critical to this approach. Risks are merely the
chance of harm or some other untoward or undesirable outcome occurring. By definition, a risk has not yet necessarily resulted in injury. Consequently, if there are reliable techniques for identifying risks before they
materialize into harm, agencies would utilize these techniques to identify
risks as early and as quickly as possible and then act to reduce them.

These laws envisioned the use of experimental animal studies and
other nonhuman scientific data to reveal risks so that exposures could be
reduced before harm to people occurred. A new field of risk assessment
burgeoned in support of such legislation. Congress also likely envisioned
that an agency could issue health standards fairly quickly. This approach
seemed to be a clever idea. If the laws and their procedures functioned
well, the identification of risks followed by quick administrative action
could prevent harm to people or the environment. Early public health efforts were much more efficient than those in recent years.

Attractive though it seems, this approach has not prevented the public
from being put at risk. After all, if compounds in commerce actually contribute to harm until they are identified by animal studies, the public remains at risk until research has been completed. At their best, such laws
might prevent harm to most people, but they are not likely to protect some
citizens. Postmarket risk-based health protections were an improvement

over no laws at all concerning toxic threats and were improvements over traditional common tort law actions, which required people to be injured before they could bring legal action to put matters right.

Unfortunately, this bright idea has not been realized and is fundamentally flawed. Postmarket administrative processes are far from quick. Typically public health agencies do not quickly find or develop data from animal or other studies. In recent years they have been poorly staffed and too underfunded even to carry out these tasks, which further hampers their efforts. In addition, there are numerous incentives in postmarket laws and procedures for affected companies to slow the process. Consequently, these efforts have bogged down in scientific and other disputes. Issuing public health standards can easily take several years, sometimes much longer, and occasionally decades. Sometimes standards simply die without being issued. Standards are also subject to review by appellate courts, which not infrequently overturn them.

Such laws are sufficiently cumbersome and slow to provide protections that historically a couple of ad hoc provisions were developed to address these problems. Soon after the Clean Water Act was passed, a federal judge, upon petition from an environmental group, oversaw a consent decree that greatly expedited the issuance of more health protective discharge permits under this legislation in order to better protect public health and the environment. This is known as the "Flannery decree," named for the judge who enforced it.[7] In the same spirit, when Congress passed laws to govern the cleanup of toxic spills, it created stringent "legislative hammers" that set automatic environmental standards necessitating clean up. These remained in effect until the EPA eventually changed them to more practical numbers. Seven years after the legislation passed, the EPA had yet to act.[8] Again, the aim of legislative hammers was to immediately institute public health protections that might be overly stringent instead of permitting toxic exposures to continue until public health agencies addressed them, thus, becoming bogged down for long periods of time. In recent years it appears Congress has forgotten some of the strategies that might assist the implementation of postmarket laws.

Legislation based on postmarket regulation of risks is broken or nearly broken. As Chapter 4 has shown, harms that result from in utero or immediate postnatal contamination cannot be prevented until long after some people have already been exposed and perhaps harmed. Such laws are especially inadequate for protecting our children.

Technology-Based Laws

Some laws have a comparatively simple structure to protect the public. They first require agencies to identify substances as hazardous and then require companies to "do the best they can" with existing or achievable or the best technology (different laws have different standards) to reduce the risks from their products.[9]

For instance, the Clean Air Act requires the regulation of "hazardous air pollutants," such as carcinogens. Congress required the EPA to create and to revise from time to time a list of hazardous air pollutants that it intends to regulate. This is now quite long.[10] If technology-based controls are inadequate to protect the public, the agency must set "residual risk standards that protect *public health with an ample margin of safety.*" For carcinogens the agency must set a residual-risk standard if technology controls leave the most exposed person with a lifetime cancer risk of greater than one in a million.[11]

Amendments to the act in 1990 also required a report on the effect of carcinogens on "hypothetical and actual maximally exposed individuals," including children. It also called special attention to how risk assessments for children should be conducted, given their inherent susceptibilities and differences from adults.[12]

How would the agency carry out this task to reduce a toxicant such as benzene in the air? Once benzene has been identified as hazardous, the agency issues a proposed regulation to reduce benzene in the air by means of technology. The legislation typically provides for a comment period, during which an agency is obliged to hear and take comments from the public. Although the legally mandated comment periods can be relatively short, in practice they are often extended and can take considerable time, sometimes years. The agency must review these comments and take them into account as required by law.

After that, the agency may issue a final standard that creates a law quite specific to the problem at hand. If the technology standard is issued in accordance with lawful procedures and if affected parties do not challenge it in court, it becomes law on that issue. If the regulation is appealed because one or more of the parties believes that it was created by means of improper procedures or that it does not comply with the substantive requirements of the statute, then typically the legal appeals must be exhausted before the regulation becomes binding law. If a health standard or some portion of it were invalidated, then the invalidated sections

138

would not have the force of law. The agency would have to revisit the problem and reissue the regulation or drop it.

Technology-based laws have a number of virtues, compared with other strategies to reduce exposures to toxicants. These laws simplify some public health tasks and ease implementation. According to Wendy Wagner, technology-based laws assume "that there is pollution, that it is undesirable, and that a strong effort to reduce the pollution is needed."[13] They avoid the time and costs of arguing over the appropriate "safe" levels of exposure in the ambient environment. They are comparatively easy to issue and reduce exposures more quickly than some other laws. They are more predictable and are more easily enforced.[14] An agency still has a legal burden to identify toxicants subject to technology controls, to determine which industries would be subject to the controls, and (sometimes) to identify the range of technologies that would satisfy the law. In the end, however, technology laws do not catch toxicants before they enter commerce and expose and contaminate people.

California's Proposition 65 Warning Law

A somewhat analogous law that simplifies governmental tasks to an even greater degree is California's Proposition 65 warning law (not a federal law).[15] It requires the governor (acting through the California Environmental Protection Agency) to identify and list carcinogenic and reproductive toxicants. The toxicants typically are listed by an expert panel, but they can also be listed because certain other nationally or internationally legally recognized scientific committees have identified them.

Once substances are listed, within one year companies that expose the public must issue "clear and reasonable warnings" to those who may come into contact with the toxicants, or the companies must choose to reduce exposures below statutorily required risk levels in some other way.[16] Companies could show that exposures are already below the legally mandated levels, but they would have the legal burden to demonstrate this. They could reduce exposures below the statutory level. They could also phase out the products that cause exposures, or replace toxicants in them with less hazardous materials. Firms thus have considerable leeway in how they comply with the law, but they have the legal burden to do these things.

This law has several strengths. It has some automatic provisions for listing toxicants. The law itself specifies quantitative risk levels that must be met, eliminating arguments about the appropriate degree of health

139

protection and reducing the time before the law applies to a listed substance. It has also some expedited procedures for assisting companies to determine if their products pose risks that would violate the law.[17]

Importantly, once a substance is listed, the law shifts the burden of proof to companies so that they must carry out some tasks ordinarily performed by governmental agencies. Governmental tasks, other than enforcement, cease once the scientific committee has identified toxicants. Compared with laws that place all the legal burdens on governmental agencies, this warning law expedites "protections" for the public. The benefit occurs because companies have private incentives to assist the public process rather than resisting or obstructing it (beyond any debates concerning listing). Moreover, posted warnings identify particular companies as responsible for exposing the public.[18] (The warnings must be located on their facilities or products.) Because many companies choose not to be identified in this undesirable light, they opt to reduce exposures.

The California attorney general enforces the law. Significantly, however, if private citizens are aware of violations of Proposition 65, they may initiate what is called a "private enforcement action" against the offender. The attorney general has a statutory period during which to take the enforcement action, but if that office does not, the private action can continue, with the possibility of the private citizen receiving a portion of the fines upon successful prosecution.

Warnings do not necessarily ensure protections from toxicants; warnings provide people with opportunities for self-protection, should they choose to heed these warnings. Warnings also tend to motivate companies to reduce exposures. How protective warnings can be depends on the risk and how completely it can be avoided. Forewarning about large holes in the ground might be quite effective (though not a Proposition 65 issue). Alerts about toxicants in one's midst likely offer less protection simply because a toxicant cannot be detected by the senses (as a hole can be) and because there may be no clear boundaries beyond which one is reasonably safe.

Proposition 65 appears to motivate firms to remove many toxicants from the market. To date, the state has listed about 887 carcinogens and reproductive or developmental toxicants, adding about 108 toxicants to the Governor's list in the last four years.[19] For many of these substances, companies simply reduce or eliminate exposures. For instance, a private enforcement suit was brought against fast food businesses and companies that sell potato chips and French fries because their processes create a carcinogen, acrylamide. Rather than post warnings about acrylamide in their products, McDonald's, Carl's Jr., and Frito-Lay, among other companies, simply changed cooking processes, to reduce exposure levels to permissible levels.

Other private enforcement actions or suits by the attorney general have reduced exposures to lead, secondhand smoke, toluene, mercury, ethylene oxide, some phthalates, trichloroethylene, and perchlorethylene. This is likely a better record than that of any federal agency. As Ed Weil, a deputy California attorney general put it, "companies have learned that if you keep a chemical off the list, you save a lot of headaches."[20]

California's Proposition 65 has produced substantial benefits and reduces the time for calling attention to toxicants with warning labels. There will still be disputes about whether substances are reproductive or carcinogenic toxicants, but these tend to simpler than arguments about risks from ambient exposures. However, if companies do not care about any bad publicity from exposing the public to toxicants, they simply post a warning and let people respond as they see fit. Does this warning modify people's behavior? There is some debate about this, but a reasonable subset of the public appears to modify its behavior, to try to avoid or reduce risks about which they are warned.

Technology-based and warning laws simplify some government tasks (Proposition 65 being a superior law in terms on this dimension) and require much less governmental information generation than do the ambient exposure laws discussed in the next section of this chapter.[21] However, some disputes remain that can delay protecting the public. Which substances should be subject to technology controls? To which industries should a regulation apply? Do the technologies sufficiently reduce risks so that other procedures do not have to be followed? Answering all these questions can take considerable time.[22]

Under Proposition 65 there can be substantial disputes how much evidence is needed to list substances as carcinogenic or reproductive/developmental toxicants. And the speed with which a science advisory panel acts on such substances can be quicker or slower, depending upon the particular experts serving on the panels and guidance from the governor for implementing the law.

Ambient Exposure Laws

Other postmarket laws have more elaborate blueprints for agencies to follow in order to protect the public. These laws require agencies to identify substances that pose risks after people have been exposed and then to set "ambient-exposure" levels to protect citizens. Ambient exposure levels are levels of toxicants in the environment—the air, water, drinking water, or workplace. If they currently pose health risks, legally they must be reduced.

Such legislation imposes much greater legal burdens on public health agencies than other laws. In order to issue a more protective health standard, agencies first must identify substances that are hazardous. They then must determine what exposure levels can cause adverse effects and what some of these are. Next, they must assess the level of exposures in the relevant environment. Finally, agencies must assess levels of exposures in the relevant environment that would be sufficiently safe as required by the statute. Addressing these issues requires all four stages of a full-fledged risk assessment.

These scientific tasks must be proposed in a regulation, subjected to a comment period, and followed by agency review of the comments before a health standard is issued as a final rule. The final rule may be appealed to a court by one or more parties affected by the regulation. A public health agency has the burden of providing evidence of risk or harm sufficient to justify issuing a health standard, of setting a protective regulation, and of carrying out these tasks by using sufficient evidence and proper administrative procedures to avoid being overturned by a reviewing court.

Occupational Exposures

For instance, under the Occupational Safety and Health Act, if the secretary of labor seeks to reduce exposure to a toxicant in the workplace environment, because of a 1980 Supreme Court decision, the Occupational Safety and Health Administration (OSHA) must show that there is a risk from the toxicant that needs reduction and then show that proposed exposure levels will better protect employees in the workplace.[23] OSHA has regulated few carcinogens since legislation passed in 1970 and even fewer since that 1980 court case.

Before 1986 OSHA's research agency, the National Institute of Occupational Safety and Health, had recommended that 71 carcinogenic chemicals or processes be addressed, but OSHA issued health standards for only 22 (with one group of 14 issued together in 1973).[24] In 1989 OSHA proposed to update health standards for 428 substances by adopting voluntary standards that industry's own health professionals had rec-ommended. While most companies and the unions accepted the rule, a few companies appealed five proposed health standards. The United States Court of Appeals for the Eleventh Circuit struck down the entire agency action. It held that OSHA "failed to establish that existing expo-sure levels in the workplace present a significant risk of material health impairment or that the new standards eliminate or substantially lessen the risk."[25] Only one or two occupational health standards were issued during the eight years of the George W. Bush administration, and one was under court order. OSHA's procedures are quite burdensome, and it is too underfunded and understaffed to carry out its tasks commensurate with myriad workplace risks.

Affected industries strongly resist health standards for employees for various reasons, including requirements to change plant structures or in-dustrial processes, and costs. It is difficult to generate any public political support for the agency's proposed actions, since the general public does not see itself as immediately affected. OSHA's task of protecting the health of the working person is well below or even off any public radar screen.

The level of risks permitted under different laws varies with the statute. One law might require regulations to prevent *unreasonable risks of harm to health*," while another would aim to prevent human health risks "with an *adequate (or ample) margin of safety*."[26] A very protective level of risk regulation was the Delaney Clause of the Food, Drug, and Cos-metic Act. It required that if a "food additive causes cancer in humans or in animal tests, it is declared 'unsafe' and is not allowed as a food addi-tive. The risk to human health is the only factor taken into account. This is a 'no-risk' statute." (That language has now been amended.) Another reasonably health-protective law, at least on paper, is the OSH Act. OSHA must set health standards assuring, "to the extent feasible, that no em-ployee suffer material impairment of health or functional capacity," even if exposed to a toxicant for a working lifetime.[27]

Consumer Products

Consumer products, including plastics in various products as well as lead or cadmium in toys or jewelry, are also subject to postmarket laws. The Consumer Product Safety Commission has the authority to regulate most consumer products except for foods and drugs, pesticides, tobacco, motor vehicles, aircraft, and boats. Many such products contain toxicants about which the public is concerned, including BPA, PBDEs, lead, phthalates, and so on. The agency must ensure that products not pose "unreasonable risks" of injury or illness.[28]

The Consumer Product Safety Commission (CPSC) is quite handicapped in implementing laws for which it is responsible, that is, the Consumer Product Safety Act and the Federal Hazardous Substances Act.[29] In 1985 the Consumer Product Safety Commission was quite small and had only a few toxicologists responsible for carcinogens. It is now even smaller and has fewer scientists to review the thousands of products in the market that might pose risks, although recent legislation and additional funding has bolstered it. Moreover, the agency operates under especially burdensome laws. It must not only show that a substance poses an "unreasonable risk" to the public balancing health risks to the public against various benefits from the product. In addition, the law requires that the agency must take the "least burdensome" regulatory option to protect the public.[30]

The Consumer Product Safety Commission's regulatory activities have been minimal in large part because of legislation that handcuffs its efforts to protect the public. For example, the Commission may not impose mandatory health standards if the industry is willing voluntarily to write its own standards as long as they would "eliminate or adequately reduce the risk of injury addressed and it is likely that there will be substantial compliance with such voluntary standards." In short, the industry can write its own rules for its product! Moreover, although the Commission has had some success in removing hazardous products from commerce by issuing public notices about their risks, it may not do this without first giving the manufacturer the opportunity "to mark such information as confidential and therefore [be] barred from disclosure . . ."[31] That is, the CPSC must seek a company's permission to disclose risks about its products! When it does publicize risks, the mere fact of adverse publicity typically leads companies to remove the product quickly from commerce or to modify it in order to avoid adverse publicity. For instance, in the 1980s the agency found that hair dryers were insulated with asbestos. During use, asbestos fibers would break off and be blown into the room where the dryer was being used; these would then be disturbed the next time the dryer was used. The CPSC entered into a voluntary agreement with hair dryer manufacturers, who then quickly removed the asbestos from new dryers.[32]

Even when the agency sought to set health standards for certain products, some posing carcinogenic risks, courts not infrequently struck down its effort. Courts invalidated warnings for home swimming pools and a regulation seeking to protect homes and schools from the carcinogenic risks in formaldehyde foam insulation. The courts have held the agency to quite particular procedural and substantive rules, although a number of critical decisions have come from a very industry-friendly and antiregulatory Fifth Circuit Court of Appeals.[33]

The Endocrine Disruptor Program

In 1996 Congress authorized an endocrine disrupter program to assess and reduce synthetic endocrines that could adversely affect humans and the environment. However, this authorization would occur under existing, mainly postmarket laws. Before the Obama administration, the endocrine disrupter program had languished for twelve years after its inception, with little progress made since the initial legislation and with no products tested as a consequence. For many of those years, only one person was in charge of developing procedures to address this important group of substances. The new Obama administration has begun to pursue these toxicants.[34] However, since this program appears to envision setting health standards under existing laws, Chapter 4 shows how poorly such an approach will protect developing children from endocrine-disrupting compounds that can adversely affect development and disease status years later.

Court Review

In addition, typically all agency health standards are subject to review by federal appellate courts for procedural and substantive issues. Did the agency issue a regulation that was neither too stringent nor too lax under the law? Did the agency consider all the evidence it should have and otherwise follow appropriate procedures in issuing the rule?

Since 1980 appellate review has been increasingly intrusive, following the Supreme Court's benzene decision, which sought to slow OSHA's activities and succeeded in doing so. Not only did that decision affect the Occupational Safety and Health Administration, but it also produced a "chilling effect" on other regulatory agencies, by signaling a greater willingness to review regulations critically and perhaps overturn them.[35] Since that time, agencies responded by providing greater scientific support and much more substantial documentation for their actions, further burdening their efforts to protect the public. For example, in a 1974 decision, the District of Columbia Court of Appeals was seemingly shocked by the substantial length of a four and one-half page document justifying OSHA's improved health standards for asbestos in the workplace.[36] Subsequent to the benzene decision, agency regulatory justifications can now be hundreds of pages long. Finding appropriate published research, incorporating public comments and writing such long regulatory documents bogs down the process and delays health protections for the workforce.

Legislators envisioned the use of animal studies and other nonhuman evidence for risk assessments. Such evidence is routinely utilized, sometimes over the objections of an affected company. Even risk-based laws utilizing nonhuman evidence, however, have shortcomings. Long-term animal studies commissioned by a government agency can easily take seven years to authorize, conduct, and analyze in order to produce usable results for health standards.[37]

Despite emphasis on use of nonhuman evidence, there is constant pressure from affected industries to challenge the scientific value of animal and other nonhuman evidence, as well as pressure for agencies to support their health standards by epidemiological evidence. For instance, in 2006 a National Academy of Sciences (NAS) committee recommended "the most stringent criteria and . . . [required] epidemiologic evidence for drawing any positive conclusions about potential carcinogenicity." The committee continued, "Animal evidence and other test information are used only to confirm cancer causation once epidemiological associations have been demonstrated."[38] If this guideline were followed, some people would have to suffer disease or death before a public health agency could reduce risks of disease and death to others. This approach poorly protects the public.

Fortunately for public health protections, this view by one NAS committee seems to be an anomaly. More recent NAS committees have disagreed with this view, recognizing, as do other national and international scientific committees, that animal and other kinds of evidence can legitimately provide evidence of disease risks to humans.[39]

The Toxic Substances Control Act

In 1976 Congress understood the need to address chemicals as such in a more comprehensive manner, as it contemplated enactment of the Toxic Substances Control Act. By passing this legislation Congress sought to close gaps left by media-specific laws, to address better total individual exposures, and to better control chemicals at their source. In particular, a U.S. Senate report noted, "The alternative of preventing or regulating the use of [a toxic] chemical in the first instance may be a far more effective way of dealing with the hazards."[40] Companies that make chemical products would also benefit, since substances regulated under several different media-based or other laws are a complicated nightmare compared with regulation under a single statute.

Another goal to be served by TSCA was to avoid "a purely reactive posture with respect to chemical hazards." The Council on Environmental Quality noted, "We need no longer be limited to repairing damage after it has been done; *nor should we allow the general population to be used as a laboratory for discovering adverse health effects.*"[41] Explicitly rejecting the tort system as adequate to control exposures to toxicants as well as rejecting postmarket laws, the council implicitly endorsed "some kind of licensing mechanism of the kind that exists for prescription drugs and pesticides."[42]

However, this was not to be. The TSCA that survived congressional compromises to become law remained largely "reactive." And it continued to permit "the general population to be used as a laboratory for discovering adverse health effects."

Premanufacture Notifications: A tiny but inadequate gesture toward testing chemicals before the public is exposed led to a "pre-market premanufacture *notification*" provision in TSCA for new chemicals.[43] This provision requires companies seeking to manufacture new chemicals or use existing chemicals for substantial new uses to notify the EPA. Manufacturers must then provide any data they have about these chemicals. The EPA is then given ninety days to review each substance for any toxicity properties based on the data it has, with the possibility of time extensions to conduct more careful review or to request more data.[44] If companies are aware of any toxicity properties of their products, they must provide this information.

Companies have no legal duty to conduct toxicity testing under this provision, unless the EPA specifically requires it. They need only report what they know about the product. As John Applegate puts it, the agency "must take what it is given." Typically this includes some chemical and physical properties as well as any biological activity data about it. If the EPA identifies some feature of concern about a substance, it can require other data about or testing of the product to try to determine any risks it poses to humans or the environment.[45] However, the data that the EPA receives is so sparse that it has had to use its own resources to develop extensive scientific tools to try to discern clues to toxicity from chemical structures alone. This process has been helpful, but is a quite distant cousin to more fulsome testing. Moreover the EPA has the burden to justify more tests and legally to require them, instead of a company being automatically required to provide them. The EPA may order a company to submit additional information. However, if the company objects, the agency must take it to court and carry a burden of proof to obtain data or prevent the substance's commercialization.[46] It may also issue a rule to require toxicity data, but this is an even more cumbersome process.

The premanufacture provisions have probably eliminated some hazardous substances before public exposure. Companies that submit substances likely would review them in the same manner the EPA would, and since companies would know that certain chemical structures are likely to pose toxic risks, they would choose not to pursue them. In general, because manufacturers know their own products better than other people, they might become aware of toxicity properties and decide not to commercialize these substances.

Perhaps Congress hoped that firms would voluntarily submit toxicity information about their products. They have not. In 1983 about one-half of the substances submitted for review contained no toxicity information and, according to the Office of Technology Assessment, "only 17 percent of [the chemicals proposed for manufacture had] any test information about the likelihood of the substance's causing cancer, birth defects or mutations—three biological effects that were singled out for special concern in TSCA." Four years later, the same office found that "most PMNs [premanufacture notifications] do not contain *any* toxicity test information."[47] Manufacturers seemed to learn that they did not need to submit toxicity data.

By now the picture seems to be even worse. Approximately 50,000 new substances proposed for manufacture since 1979 (when TSCA was implemented) have been subject to this provision.[48] About eighty-five percent had no data concerning health effects. Moreover, even the chemical identity of a majority of them is concealed by claims of confidential business information. In addition, many compounds already in the market may be more toxic than might appear. For instance, most new chemicals proposed for manufacturing have no data on their ability to cause genetic mutations, a relatively simple and reasonably accurate test to conduct. Yet a sampling of one hundred chemicals in commercial use and a sampling of forty-six chemicals produced in more than a million pounds revealed that about twenty-percent of each group were mutagenic. When a substance is mutagenic it is also highly likely to be a mammalian carcinogen. Consequently not only is little known about products, but they are also likely more toxic than the public has reason believe.[49]

As new technologies enter the market, ignorance about industrial products in our midst will only increase. A new nanotechnology industry is developing products for electronics, medicine, and even pollution control. Their chemical structures are larger than a molecule and smaller than a biological cell, ranging in size from one to one hundred nanometers in size (a nanometer is $1x\,10^{-9}$ meters). A PCB molecule is about 1 nanometer. The number of new nano products has increased forty-fold to more than one thousand in just four years.[50] However, some nanotubes are similar to asbestos, while others can be tiny particles. Both kinds of products have exhibited worrisome toxicity properties. Some nanoparticles cross the placenta and titanium oxide nanoparticles have been found to cause genetic and chromosomal damage in experimental studies.[51]

Other TSCA Provisions: In seeking to protect the public from a substance already in the market under the TSCA, the EPA must show that a compound "may present an unreasonable risk" to human health or the environment.[52] While such a standard may appear quite appropriate, it has a number of problems.

This language makes it difficult to achieve very health-protective goals because the EPA must balance many different considerations against health protection. Congress could have made different choices. For instance, Applegate notes that Congress could have chosen to "classify chemicals as hazardous and ban them or as safe [enough] and allow their sale," much like the Delaney Clause of the Food, Drug, and Cosmetic Act mentioned above.[53] Congress might have endorsed merely classifying substances as toxicants or not and then required companies to "do the best they can" with technology to reduce exposures to the lowest achievable levels.

Instead, the EPA is required to determine what ambient exposures result in a risk that is "unreasonable." The agency must consider the probability and severity of any risks and benefits to the public from the product. It needs to assess the costs of regulation, impacts on small businesses, substitutes for the substance, and so on, and then choose the least burdensome means of reducing risks.

Satisfying these requirements is quite science- and information-intensive. The EPA must develop considerable science to determine levels of exposure that pose health risks. Before agency personnel can issue final standards, they must determine as best they can the likelihood and severity of risks to humans and the environment at different exposure levels, as well as major alternatives to reduce the risks, along with any risks and costs alternatives might pose. In addition, one EPA must generate considerable other information concerning costs of alternative courses of action, effects on the industry, and general economic effects on citizens.

Congress also "intended[ed] that the reviewing court engage in a searching review of the [EPA] Administrator's reasons and explanations for the Administrators conclusions."[54] When Congress required the EPA to greatly restrict the use of asbestos, a notorious killer, the agency's regulation phasing out virtually all uses of asbestos was struck down. As a result, the agency will likely never again use that particular section of the law. The court found that "the more stringent the regulation, the greater the degree of proof to justify regulation is required."[55]

Moreover, if the EPA believes that a substance in commerce poses health risks, it may not simply require a company to test the substance and submit data to assist in determining the safety of the substance. The agency must "issue a regulation" which shows that the chemical needing testing "may present an unreasonable risk," and that rule must have scientific data to support the request. The EPA thus faces what Applegate calls "a regulatory Catch-22." That is, the "EPA must have chemical information in order to prove that it needs it, but it needs the information because it does not have it."[56] Congress's General Accounting Office found that the EPA's authority under the TSCA "for collecting data on existing chemicals [does] not facilitate [the] EPA's review process because [the act] generally place[s] the costly and time-consuming burden of obtaining data on [the] EPA."[57]

As an additional burden, when the agency issues such a rule to require testing of a product, it must be supported by "substantial evidence" in the record, the same burdensome standard to which health protections are held on judicial review. The agency has required the testing of some chemicals under this burdensome requirement. About 140 of 200 substances has been tested as a result of rule making; the remaining 60 were the result of a voluntarily agreement with the affected manufacturer. In short, the legal testing requirement was so burdensome that the EPA had to find a less onerous alternative, namely voluntary testing. This alternative appears to have met with some approval from the EPA and from affected companies, but it reveals the limitations of the law in seeking test data in a timely manner. In order to protect the public's health from an existing chemical, it is as if the EPA must beg a company to do the testing or else engage in a time-consuming, burdensome rulemaking to require the needed data, if a company objects.

The upshot is that testing existing compounds under the TSCA has been extremely sluggish and hopelessly inadequate in the face of enormous ignorance about the toxicity of chemicals in commerce. Each year the National Toxicology Program under the TSCA authorizes only a small number of animal tests on potentially toxic substances, often ten or fewer.[58] At this rate, it would take three hundred to twelve hundred years to test the three thousand to twelve thousand substances thought to pose risks to the public that are already in commerce. This is completely inadequate for determining the toxicity of chemicals to which developing children are exposed.

Moreover, even when there are data from such tests about carcinogens, the assessment and follow-up by the federal agencies lag far behind the rate of studies. Regulatory agencies are so underfunded that they cannot keep pace with existing toxicity data. If federal or state agencies have not assessed the toxicity of substances, it is unlikely that others have.

The failures of the TSCA are traceable to at least three main issues. First, there is no minimum requirement for toxicity testing under the PMN program. Second, the agency has the burden of proof to initiate legal action for ambient environmental health standards and for requiring tests of products that are in the market. Together, these two issues almost ensure that the TSCA will poorly protect the American public from toxic risks. Third, the TSCA is further hampered in its efforts to protect the public by requirements that the agency regulate ambient exposures of toxicants (the most data-intensive health standards) to ensure that there are no "unreasonable risks" and that the agency rule out more burdensome alternatives. Together, these conditions ensure that more protective health standards will be slow, few, far between, and rarely modify the toxicity status quo.

Despite much of the initial motivation for the TSCA, it has utterly failed to measure up to early hopes. New substances are not required to be tested before they enter commerce, and there is no independent certification of some degree of safety. If new substances pose risks, they continue to be used in commerce until the EPA presents a sufficiently strong scientific and legal case under a very burdensome law to establish more protective health standards. In addition, products that are in the market appear to be more toxic than we might think. As already noted in this chapter, it appears that about twenty-percent of chemicals in commerce, including those to which we are most exposed, are mutagenic.

Under postmarket laws developing children and adults alike are certain to remain guinea pigs for the commercial creations of American industry, contrary to the original intentions for a comprehensive chemicals law. Children are contaminated in utero by substances that enter commerce with unknown toxicity. And the contamination will continue until exposures are reduced. If individuals are tainted by persistent substances, however, stopping exposure will not cease contamination. For developmental toxicants that are subtle and take years to manifest, many cohorts of children will be subject to any risks after dangerous substances have been firmly identified and eliminated. This is not a health-protective approach toward chemicals.

Premarket Testing and Licensing Laws

Premarket laws guide the testing and approval of only a small percentage of products.[59] These laws operate in accordance with a different blueprint for protecting the public from that of postmarket laws.

Premarket laws specify that companies must test their products in accordance with FDA or EPA guidance before these products enter the market. Next, the laws authorize scientists to review the test data. Finally, these laws require some level of demonstrated safety (or minimization of risks) and explicit agency approval before the company is licensed to sell its products. The laws aim to identify toxic risks so that they can be intercepted before the public is exposed. Premarket laws address only three generic classes of products—drugs and (to a limited extent) new food additives (under the FDA) and pesticides (under the EPA). I will consider only drugs and pesticides.

Premarket laws resemble other much more mundane legislation that protects us. If I seek to build or remodel my house, I must first submit plans to a building department. It reviews the plans to ensure that the house is properly planned and engineered so that it will be soundly built to withstand earthquakes and other hazards in California. I may not begin construction until the plans are approved. Moreover, at critical stages of the building an inspector is required to review the structure to ensure that it is being erected in accordance with the approved plans, engineering, and standardized building procedures. Finally, he or she must sign off on the final project. If the house is not built according to the plans or is built without approved plans, it is subject to various sanctions. Thus there are substantial "premarket" features to these provisions. I may not erect any old structure I wish to and then wait for some legal authority to find it was improperly built or that it imposes undue risks, to borrow Richard Merrill's metaphor, the way a highway patrol officer might see or discover a violation.[60] Neither may I build any old house I choose, and then provide compensation through the tort law if it collapses and harms someone.

Pharmaceuticals

If Innovative Pharmaceuticals Inc. seeks to introduce a new drug, "Cures X," into commerce, the company must conduct a number of scientific tests and go through a regulatory process in order to license this product for commercialization.

Typically, Innovative Pharmaceuticals would test Cures X for toxicity in experimental animals for shorter or longer periods of time, depending upon the proposed duration of the drug prescription.[61] More extensive animal testing could be required to determine whether long-term use of the drug can cause cancer or birth defects. According to the Food and Drug Administration, "The purpose of preclinical work—animal pharmacology/toxicology testing—is to develop adequate data to undergird a decision *that it is reasonably safe to proceed with human trials of the drug.*"[62]

Next, the company would file an application with the FDA, outlining proposals for human testing in clinical trials. If the results of preclinical testing are sufficiently reassuring that Cures X does not have serious side effects and that the benefits are what the company expects (and what legally must be shown), the company could then proceed with tests on human volunteers in clinical trials: Phase I, II, and III trials. The tests begin with small numbers in Phase 1 (twenty to eighty volunteers) and ultimately use much larger numbers of volunteers in Phase III (several hundred to several thousand).[63]

After clinical trials, Innovative Pharmaceuticals would submit a new drug application for a license to sell Cures X. The FDA would then review the file and assess the research on Cure's X's safety and effectiveness. Next, the FDA would review the information that goes on a label accompanying the drug and inspect facilities where it will be manufactured. If the FDA were satisfied after each of these steps, it would license Cures X for commerce.

Once Cures X is licensed for and enters commerce, the company may be required to conduct further human studies (Phase IV trials) in order to gather additional data about the drug's safety and efficacy in a larger more heterogeneous population.[64] In addition, the company must inform the FDA of any adverse effect reports (case reports) it receives from physicians so that the agency can monitor the collective data on how the drug affects patients.

If a product poses risks, the company has a legal duty to modify its label to put the public on notice about the risks. However, if the risks turn out to be unacceptable the *FDA has the legal burden* to issue a rule to withdraw approval for its marketing. Because this is an onerous task for an underfunded agency, the FDA often tries to persuade the company to withdraw the product voluntarily so that the FDA does not have to go through the elaborate rule-making process.

Just how burdensome this process can be is illustrated by the efforts of FDA to remove from the market Parlodel, a breast milk–suppression drug that posed the risk of hypertension, strokes, and heart attacks to new mothers. Some women choose not to breast feed after giving birth. When they make this decision, the results can be painful as well as unpleasant in other ways. Consequently, Sandoz Pharmaceuticals created a drug, Parlodel, that promised to reduce the swelling, any pain, and other symptoms produced by not breast feeding. Parlodel was approved for this use in 1980.

By 1983 the FDA was receiving case reports that Parlodel was associated with dizziness, hypertension, and strokes. The agency sought to get "Sandoz's agreement to include a warning of these adverse experiences in Parlodel's labeling."[65] The company refused. By 1985 more adverse health effects reports had come in, some suggesting that Parlodel caused heart attacks. The FDA again requested that the company voluntarily label the risks of the drug and inform physicians with a letter. Sandoz refused. In 1987 an epidemiological study failed to allay the FDA's fears about seizures associated with the product, and by then the agency had more reports of seizures, hypertension, myocardial infarctions (heart attacks), and ten deaths associated with Parlodel.[66] Sandoz finally agreed to a change in labeling.

Adverse reaction reports continued to pile up—eighty-five quite serious ones between 1980 and 1989, including ten deaths—leading the agency in 1989 to request Sandoz to voluntarily withdraw Parlodel from the market. The company refused. In 1990 the FDA began the elaborate rule-making procedures needed to withdraw FDA approval for Parlodel. Sandoz filed a contrary petition, protesting the withdrawal.[67]

In 1993 the FDA's proposed rule to withdraw the drug was issued. By then the FDA had received one hundred and ninety-one adverse reaction reports, many quite serious ones, including hypertension, seizures, cerebrovascular accidents, and ten deaths. Sandoz finally voluntarily withdrew the product from the market before legal action could occur.[68] For thirteen years, however, Parlodel produced profits for Sandoz. Hundreds of women experienced adverse effects; some had strokes, others had heart attacks, and at least ten died.

Why did it take so long to remove Parlodel from the market? The FDA has the legal burden to withdraw approval for a drug by means of a rule making, usually a sluggish and labor-intensive process. The agency is likely so shorthanded and underfunded that it could not successfully go through the rule-making process and issue a withdrawal rule any sooner. Thus, even under premarket approval laws, it can be quite difficult to force a harmful product from commerce because of company resistance, burdensome procedures under which the agency operates, and shorthanded agencies.

Pesticides

Somewhat similar but not identical premarket testing must be conducted on pesticides before companies may sell them. Let's suppose Conscientious AgraBusiness Inc. seeks to commercialize a new insecticide, "Stops Pests," designed to reduce pests on food crops.

Conscientious AgraBusiness must test Stops Pests for a number of specific toxic end points before the U.S. Environmental Protection Agency will register the product and license the company to market it. In order for a company to register a new pesticide or to register a new use for a pesticide since the 1996 amendments to the law, the EPA must "ensure that the pesticide, when used according to label directions, can be used *with a reasonable certainty of no harm to human health* and *without posing unreasonable risks to the environment.*"[69] As Tom McGarity has pointed out, it is important to recognize that the "reasonable certainty of no harm" principle seeks to protect against harm to people and may not be balanced against benefits from using the pesticide, as previous formulations of the law permitted.[70] The law also required that, "Where pesticides may be used on food or feed crops, EPA [must set] tolerances (maximum pesticide residue levels) for the amount of the pesticide that can legally remain in or on foods."[71]

The amendments under the Food Quality Protection Act (FQPA) also *"provided heightened health protections for infants and children from pesticide risks;* required expedited review of new, safer pesticides; . . . [and] required reassessment of existing tolerances over a 10-year period."[72] Pesticides are subjected to extensive testing because of human exposures and because of concerns about any effects of pesticides "on infants and children, whose developing bodies may be especially sensitive to pesticide exposure."[73]

Importantly, Congress recognized that a single chemical could expose citizens by multiple pathways, with the body acting as something like a funnel, assimilating substances by several routes. Previously the EPA had treated "exposures to pesticides from different pathways as independent events i.e., [it] only analyzed each individual's exposure to one pesticide via a single pathway. *In reality, however, exposures to pesticides do not occur as single, isolated events, but rather as a series of sequential or concurrent events that may overlap or be linked in time and space.*"[74] Thus, by requiring the agency "to perform *aggregate assessment* (single chemical, multiple pathway/routes), Congress intended that [the agency's] exposure and risk assessments would move closer to describing the pattern of exposure *actually encountered by individuals in the real world.*"[75]

Congress also required the agency to evaluate the cumulative risks to human health that can result from "exposure to pesticides and other substances that are toxic by a common mechanism."[76] Some classes of pesticides act by similar toxic mechanisms. Congress correctly recognized that the human body acts as something like a storage facility, possibly harboring in its tissues (for greater or lesser periods of time) substances that could act by common mechanisms and cumulatively pose greater risks.

For instance, many organophosphate and carbamate insecticides act by a common mechanism. They inhibit a critical nervous tissue enzyme (acetylcholinesterase) that is, according to Donald Ecobichon, "responsible for the destruction and termination of the biological activity of the neurotransmitter acetylcholine (Ach)." If there is too much of this neurotransmitter available at the nerve endings, the result is "continual stimulation of electrical activity," which causes a variety of effects: gastrointestinal cramps, diarrhea, tachycardia, hypertension, tremors, restlessness, emotional liability, mental confusion, and loss of memory.[77] Early versions of organophosphate insecticides were even more toxic, and exposures could easily result in death, among others outcomes.[78] More recent formulations of these insecticides are less potent, but their effects can be additive. In addition, as we saw in Chapter 4 adverse human health effects are beginning to appear.[79]

Toxicity tests for pesticides are somewhat different from those for pharmaceuticals. Since pesticides are considered inherently toxic to some living organism and their selectivity of target-species is not well developed, they can cause harm to nontarget species, including humans. Consequently, pesticides receive scrutiny for risks to humans and the environment. Also, because pesticides are released into the environment, where they will reside for shorter or longer periods of time, they create wider exposures, sometimes by unexpected routes and sometimes by posing surprising risks. They must be reviewed for these possibilities. (Pharmaceuticals typically have not been seen as escaping into the environment. However, they are beginning to appear in drinking water at low levels as a result of poor disposal practices, excretion from patients taking the drugs, and an ability to survive water treatment facilities. While current concentrations are quite low, some officials note that there could be risks to fetuses, to those with compromised health or to wildlife.)[80]

Even though the goal of premarket testing laws is to detect and intercept toxic hazards before they enter commerce, unless the laws or regulation authorize tests specifically for developmental toxicants, they likely will not be detected. Indeed, a number of drugs and pesticides have caused adverse developmental effects, including anticonvulsants and sedatives. For example, more than 45 percent of human neurotoxicants are pesticides.[81] However, until the early 2000s testing did not provide alerts to these effects. Whether or not current tests are adequate is a further issue. There is some concern that the EPA may not have adequate neurotoxicity tests for pesticides and agri-business companies seem opposed to them.[82]

In addition, when Congress amends legislation political disagreements concerning the modified law do not necessarily come to an end. Consequently, interested groups may continue to try to influence the law in their favor as it is implemented in public health protections, either by seeking to modify general principles or nitpicking at particular regulations affecting particular products. The end result can be health regulations that are less protective of the public's health than aimed for in the enacted legislation. This appears to have occurred with the 1996 pesticide amendments. For instance, the safety factors that were designed to protect children were eliminated or reduced about eighty percent of the time for particular pesticides from 1996 to 2001 as the law was implemented.[83]

Finally, premarket testing provisions will not work perfectly, so some toxic effects may be missed despite the best and most conscientious efforts. When this occurs, if manufacturers and agencies have been vigilant, this imperfect discovery of effects may be the best citizens can reasonably hope for. Unfortunately, the regulatory law of the United States has not come close to the goal of adequate testing for the vast majority of substances. In addition, premarket testing will not be free from manipulation. Companies can and have falsified tests to serve their self-interest or have hired others for the task. A major testing scandal that affected pesticides, pharmaceuticals, and other substances involved Industrial Bio-Test Laboratories; this was one of the most significant instances of scientific fraud in the U.S.[84]

Harm-Based Tort (Personal Injury) Law

A major principle of justice is that if one person injures another without legitimate excuse or justification, the harm-doer should "put the matter right" with the injured party.[85] As Tony Honoré notes, this might "require the harm-doer to restore something to the person harmed, or to repair a damaged object, or (when the unharmed position cannot be restored, as it usually cannot be) to compensate the harm-sufferer."[86] This is a matter of corrective, or rectificatory, justice.

The tort law offers some modest protection against risks and harms, because of the effects of general deterrence (the general threat of tort actions) against risky behavior or risky products. That is, the general threat and burden of facing legal action in torts may motivate individuals or companies to exercise greater care to reduce the toxicity of their products than they would without such threats. The success of such deterrence is not clear, since individual companies could choose to respond quite differently to threats of litigation—some fight litigation every step of the way, as the tobacco industry did, hoping in the end never to have to set matters right and pay compensation. Others may settle out of court and modify their behavior or products. Moreover, there are differing views in the applicable literature about how successful deterrence is.[87]

In order to put matters right when a toxicant has harmed someone, the plaintiff in many jurisdictions typically must show that a particular defendant's substance more likely than not *can* cause the kind of injury from which the plaintiff suffers (so-called general causation) and more likely than not *did* cause plaintiff's injuries (so-called specific causation).[88] If these showings are successful to the satisfaction of a jury and if the defendant breached the law, a plaintiff would receive a favorable verdict.

Any "regulation" of toxicants as a result of tort law action would typically not be a direct one (for example, reduction in exposures, product reformulation, or removal from the market). The effect of tort law might occur, first, by means of general deterrence and, second, as a result of postmarket, harm-based and postinjury compensation for those harmed (this would be deterrence by reforming the behavior of the individual company). After unfavorable verdicts against products, a company might then remove or reformulate them to avoid future tort actions. Other companies, seeing results from one tort decision, might similarly modify their products or conduct (this is deterrence by example). Moreover, because the tort law is remedial, any science utilized in this venue and any legal results would come too late to prevent harm to people already exposed.

The tort law provided late compensation for women with vaginal and cervical cancer resulting from prenatal exposure to DES. The regulatory law, researchers on DES, and their doctors failed these women. The threat of tort law actions did not motivate manufacturers to test their products for these adverse effects, nor did it induce manufacturers to take greater precautions. Ultimately women did receive compensation for their injuries. It took an innovative legal theory ensure this, because tracing a plaintiff's particular disease to a particular defendant's product was especially difficult in this case because of the long time lag between exposure and disease. This litigation led to one of the largest products liability line of cases in U.S. history.[89] However, the health of mothers, as well as their daughters and sons (to some extent) was affected because the adverse effects of DES were not identified early enough.

At its best, tort law might have limited effectiveness in reducing exposures to toxicants. Its aim is to correct wrongs for those who have already been harmed and who choose to bring legal action. However, recent legal decisions in the United States have additionally handicapped torts from functioning closer to its corrective justice purpose. Some trial judges and some appellate courts have adopted misleading views of scientific evidence in support of expert testimony to screen expert witnesses. If experts do not pass the screening, they may not testify, and the plaintiff would lose, with the case likely ending without a trial. Such judicial views have asymmetrical effects on injured parties (plaintiffs), since they have the burden of proof to present evidence and persuade judges and juries. Increased judicial scrutiny of expert testimony, correctly keeps out some experts who do not base their testimony on sufficient science, but increased scrutiny based on mistaken views of science also precludes some experts with solid testimony from testifying. Consequently, some meritorious victims are denied the possibility of compensatory justice for their injuries. When judges are mistaken about the science, this also lessens the deterrent message of torts.

There is even greater pressure in the tort law than in regulatory law for legal action to be based on human epidemiological studies. Some courts acquiesce in this.[90] To the extent that legal action is based on such studies, other persons must have already been harmed by exposure to a toxicant before a plaintiff can receive compensation for injuries from a toxicant. In some courts, such as in Texas, the demands for human evidence are particularly strict. In order to bring tort law closer to its potential, judges would need to fully recognize the complexity of scientific evidence, consider expert testimony based on all of an expert's integrated evidence, and then rule on the foundation of testimony.[91]

The tort law can back up failures of environmental health laws and, ideally, provide compensation for those who are injured, as it eventually did for some DES daughters. However, even if the tort law functioned exquisitely well, it is a distant second-best solution for preventing harms to the public, because it merely provides postmarket, harm-based, and post-injury compensation for the victims, with only modest deterence.

Shortcomings of Existing Environmental Health Laws

Premarket laws have the potential of discovering toxic hazards and protecting the public before products enter the market. However, whether this discovery and protection occur depends upon how appropriate the tests are for the end points at issue, what range of tests is required, who does the testing, and how thoroughly the results are reviewed. In principle, such laws do promise greater success on this issue than do postmarket laws.

Postmarket environmental health laws, however, have a number of deficiencies, beyond those already considered even with the 1976 addition of premanufacture notifications under the TSCA.

No Science, No Protections

When there is insufficient scientific data about toxic hazards from products, public health agencies acting under postmarket laws cannot provide better health protections. They need enough data to identify risks before technology-based laws can be issued. And these agencies must have even more evidence to issue protective ambient exposure standards. Under the TSCA, the Environmental Protection Agency must have some data about the potential toxicity of a substance even to begin the complex process of requiring more data about a suspected toxicant. However, the legal fact that there can be no regulatory action without data has much wider ramifications.

Companies don't produce data that can harm their products: Since companies' products create income, any evidence of their toxicity could reduce their value, because of public concerns, the costs of testing, or the availability of safer products sold by competitors. This situation creates temptations for companies not to test their products. If there is no data, a company could say that it had no evidence of adverse effects. If no regulatory activity can occur without data on the risks, for example, of PBDE fire retardants, why test?

The rule of thumb is simple: don't provide evidence that will reduce the value of your product. Thus a company has incentives not to test a substance, even if it believes the product may have toxic properties. But the situation gets worse. Margaret Berger makes the point that frequently a corporation subject to regulation or a tort action took no steps to "test its product adequately initially, failed to impart information when potential problems emerged, and did not undertake further research in response to adverse information." Litigation usually forced companies' hands when "extensive and expensive discovery" revealed documents and witnesses "that showed the corporations' awareness of potential problems.[92] Others echo Berger's concern.[93]

Berger also incidentally illustrates an important virtue of torts. Parties to civil litigation such as toxic torts have the right to discover what each

161

side knows about the issues in a case. As a result, litigants can be legally compelled to reveal any concerns they might have had about the toxicity of a product or any tests they might have conducted. Thus, trial lawyers can obtain information that might well have been precluded from the agencies, but usually well after such data has been denied to them. At any rate, such data would be discovered too late to protect existing plaintiffs and the wider public.

Of course, a company is not the only one that can test chemical substances (although typically it is the only one in a position to investigate its hazards before commercialization). Independent scientists might suspect that a chemical is toxic and begin to explore its properties, as in the case of BPA and PBDEs. This research could enter the public record and assist a scientific case to help establish better health protections. In all likelihood, however, much more evidence would be needed, and someone would have to fund it. However, often such research is not especially interesting and is less likely to be funded through the normal channels of academic science.[94] Scientists would need to find some other form of support. Most importantly, such research is too late; the public and its children will have been treated as experimental subjects between the time the product created exposures and the time risks are verified and exposures are reduced or eliminated.

Companies cast doubt on others' science: A common strategy, first utilized by the tobacco companies but soon followed by other industries, is to raise doubt about the science that shows the toxicity of a product.[95] As a tobacco company internal memo expressed it, "Doubt is our product since it is the best means of competing with the 'body of fact' that exists in the minds of the general public. It is also the means of establishing a controversy."[96] Casting doubt on the science also challenges any attempt to improve health protections, because it gives the appearance that the science is not settled, even if it largely has been. And this tactic can substantially delay better health standards. Moreover, challenging the science can be utilized without disagreeing with standards for protecting the public's health, an important public relations point. However, one upshot is that companies may generate misleading science or denigrate existing science, which, at best, is difficult enough for the broader public to understand. An even greater downside is that this strategy can also contribute to junk science in various scientific fields, degrading them in the interest of commerce.

Companies advocate high standards of proof before regulation: Companies and their public relations firms also try to convince regulatory agencies to demand "proof" of toxicity before substances are subject to regulation. The idea of proof is more naturally at home in the world of deductive math and logic, where certainty is the coin of the realm, not in the biological sciences, where public health officials make decision on the weight of the scientific evidence. This distinction has not stopped advocates from using "proof" to try to shape the terms of debate concerning better health standards. To the extent a party can persuade a public health agency or an appellate court that it must meet high standards of scientific proof before it changes the legal status quo, barriers to better health protections are raised even higher.

It is clear why a company would advocate a quite high standard of legal proof, if its product is threatened. The product could be commercially important to a company's profits. The more central a product is to a company's existence, the harder the firm is likely to fight to prevent regulation or removal. However, legal standards of proof for impartial adjudication of disputes need not be and are not as demanding as a company might want for its particular self-protection. Ordinarily, a public health agency need only show that its health standards are not the result of arbitrary or capricious procedures and the use of science (or, more rarely, that they are supported by substantial evidence in the regulatory record).

Companies demand multiple sources of evidence: A related tactic might be to demand more studies. Companies urge agencies to require different kinds of evidence, for example, mechanistic data about how a substance functions at the molecular level. This is usually evidence that is very difficult to provide and ordinarily simply not known. For instance, researchers understood many of the beneficial and adverse effects of aspirin for about one hundred years without understanding the mechanisms by which either result was achieved.[97] Or companies might urge that there should be more studies of the same kind, such as epidemiological research, to corroborate existing data. Firms might even seek studies of different types of research, such as animal studies, if epidemiological research is already available, in order to have a model for toxicity effects.

If there were world enough and time, some of these requests would be appropriate in the world of academic science, where little of great moment (except perhaps the desire for understanding, advancing the field, and receiving scientific credit) awaited a final judgment about the toxicity of a chemical. In terms of protecting the public's health, unreasonable data demands have bad consequences. Most worrisome is that when a product is hazardous and poses risks, these demands leave the public at risk while more data are generated.

Of course, an agency must have sufficient evidence to justify its health standard. However, legal action is delayed by exaggerated demands for more evidence, by spurious claims that the existing science is not sound, and by unrealistic demands that conclusions must be supported with greater certainty.[98] Sometimes more scientific data is legitimately needed, but too often it is a delaying, obfuscating, or obstructive tactic.

If health agencies acquiesce to unreasonable demands for more studies, or if courts force them to comply with such demands, needed protections can take years or even decades to be instituted.[99] This demand leaves people at risk because substances are already in the market, contaminating them, and these substances will continue to pose these risks while more data are generated. If the products were not dangerous to health, then such delay would not be a problem.

The problems just noted result from a dynamic created by products in commerce whose safety becomes suspect and an agency that has a legal burden to protect the public. This dynamic applies to products subject to both premarket and postmarket laws. Companies have strong self-interested reasons to keep risky products in commerce, and postmarket laws make this easier and make better public health protections more difficult. The products did not enter as a matter of a permission, which in principle could be withdrawn, perhaps on lesser evidence. Instead, it is as if they are there as a matter of *legal right* until there is some demonstration that they violate the law. Moreover, because the product entered commerce without any toxicity testing, greater scientific effort must be devoted to identifying toxic properties because far less is known about it than would be known under a premarket law, under which there are legal duties to report adverse effects. Establishing toxicity under postmarket law is likely to take longer than under premarket laws, which have some data about a product already available, and there is a duty to report adverse effects. And the longer products stay in the market, the better for the firm. Any strategy or tactic that assists in keeping such products in the market longer tempts companies to use it.

One can heighten this point by comparing it with premarket testing and approval laws. If a company seeks to introduce a new drug or new pesticide, it would not insist on the very best science for testing its product. It would not insist on standards of proof analogous to mathematics to ensure their product is safe. It would not argue for the need for multiple kinds of evidence to substantiate risks. It would not seek to slow entry into the market. Quite the contrary, the incentives would be reversed; companies would likely seek to minimize the number and stringency of tests conducted, to lower standards of evidence for safety, and to expedite the tests (as they argued in the late 1990s for expedited approval of pharmaceuticals).

Companies sometimes adopt less honorable tactics and strategies: Companies or their surrogates do not necessarily play by fair rules within science, the law, or ethics. Firms are tempted to bring the "ethics" of the marketplace or even of advertising into the realms of scientific research and public health.

Experts for hire may mislead regulatory agencies and scientific journals about the science involved.[100] Scientists or their employers have sometimes modified studies so that they can claim that there are no or only minimal risks from their products. Recall from Chapter 3 how the chromium industry modified a study to show lesser risks than the data more impartially considered would have shown.

Sometimes companies or their experts, as David Michaels points out, "salt the [scientific] literature with questionable reports and studies . . . which regulatory agencies have to take seriously." But the main effect of such studies "is to clog the [regulatory] machinery and slow down the process."[101] Some research projects have been proposed from the outset to find no adverse effect. Such studies are not the outcome of impartial scientific inquiry, do not follow where the evidence leads, and reach some pre-determined conclusion. Too often they are commercially motivated and the research is designed retain the status quo. The petroleum industry proposed such studies on benzene in order to have data to fight health protections for its workers and to fight possible tort suits concerning blood and bone marrow cancers.[102] The petroleum industry was quite explicit about what the study would show before it had been conducted.

Other studies have been specifically designed to blunt regulation or frustrate toxic tort suits. An obviously angry appellate judge in Pennsylvania described several noteworthy abuses by a pharmaceutical company:

Articles were intentionally inserted in peer review journals for use in court. Studies for publication in peer review journals were tailored to the needs of litigation, and paid for out of defense funds. Most significantly, for the integrity of a judicial system, "scientific" articles for publication in "peer review" journals were edited before publication by lawyers litigating the issues presented in the article. The testimony revealed that "follow-up" studies were solicited by the defendant through intermediaries, funded by the defendant: but the scientific methodology changed, to obscure positive findings.

The practice of ghostwriting scientific papers has not disappeared. In the fall of 2009, editors of medical journals began to place tougher disclosure policies on scientific papers to try to eliminate the practice.[103]

Moreover, John C. Bailar III, a well-known biostatistician, has shown a variety of ways that scientific studies can be designed to frustrate and mislead the impartial assessment of causal associations without actually lying about the science.[104] In Chapter 3, I reviewed various ways studies could be reported that would minimize risks or produce "no effect" results. There are other examples of such responses in the last thirty years, including some high profile examples.[105]

Ignorance of the Toxic Properties of Products

Substantial social and scientific ignorance about chemical products is an important consequence of postmarket laws and companies' responses to them. A 1984 National Research Council committee of the National Academy of Sciences sought to determine "the toxicity testing needs for substances to which there is known or anticipated human exposure, so that federal agencies responsible for the protection of public health [would] have the appropriate information needed to assess the toxicity of such substances."[106] The committee divided the representative toxicity data into several categories: (a) data sufficient to conduct a complete health hazard assessment, (b) data adequate to conduct a partial health hazard assessment, and (c) minimal toxicity data. Data for a partial health hazard assessment would permit a limited characterization of the hazard associated with the safe use of a substance. Minimal toxicity data would be less than this characterization, allowing only the most minimal hazard assessment.

The NRC results were disturbing, even appalling, in revealing how many substances are produced and how little is known about the toxicity of substances with which citizens are contaminated:

- 12,860 substances produced in volumes exceeding 1 million pounds per year—78 percent of which had no toxicity information available.
- 13,911 chemicals produced in volumes of less than 1 million pounds—76 percent of which had no toxicity data.
- 21,752 chemicals produced in unknown volumes—82 percent of which had no toxicity data.
- 8,627 food additives—46 percent of which had no toxicity data.
- 1,815 drugs—25 percent of which had no toxicity data.
- 3,410 cosmetics—56 percent of which had no toxicity data.
- 3,350 pesticides—36 percent of which had no toxicity data.

The vast majority of substances (80 to 90 percent) are subject to post-market regulation. There is no toxicity data for about 70 percent of these substances. At that time about 75 percent of 3,000 substances produced in the highest volume and likely having the greatest exposures had no toxicity data.

Even these results were inadvertently optimistic about needed data. In 1984 data on reproductive and developmental toxicity were not considered necessary for assessing health effects.[107] By now many different substances are known to be capable of posing reproductive and developmental risks. If reproductive and developmental toxicity data had been considered as a necessary part of the toxicity test portfolio, the NRC likely would have noted even greater data gaps. Thus in 1984 the data were even more appallingly incomplete than the report suggested.

In the early 1990s, there were insufficient changes in the data to warrant updating the earlier findings. The three thousand substances produced in the highest volume might have been the most worrisome, but in 1995 the U.S. Congress's Office of Technology Assessment found an additional one thousand to twelve thousand substances for which extensive toxicological information would be important but was not available.[108] In 1998 Environmental Defense found that there remained substantial knowledge gaps for 75 percent of about three thousand substances produced in the highest volume. The EPA has reported a somewhat lower percentage—about 43 percent of the three thousand substances had no toxicity data—but while better, the amount of substances for which there are knowledge gaps remains a problem.[109]

Drugs, pesticides, and new food additives, all subject to premarket testing and approval laws, had somewhat better toxicity data in 1984.

Even this statistic, however, is unimpressive, since premarket testing is now required for these products. If they were tested before entering the market, how could 25 percent to 36 percent of drugs and pesticides, respectively, lack all toxicity data? The answer probably lies in history. Premarket studies were not always required, and this, plus grandfathering in some substances when laws were amended, may account for what appears to be a reasonably poor record under premarket testing laws. Nonetheless, premarket testing will provide much better toxicity data than will postmarket laws. Even if premarket testing of products is required, sensitive tests to detect developmental toxicants will need to be utilized.

Because of toxic ignorance of about 3,000 high production–volume substances, Environmental Defense, the EPA, and industry entered into a voluntary agreement to have producers close the knowledge gaps on these chemicals.[110] Major chemical companies sponsored about two-thirds of 2,800 substances needing data, but about 10 percent of the substances eligible for sponsorship remain "orphans," without sponsors. In 2007, two and one-half years after the deadline for industry to submit final data sets, companies had not provided initial submissions for about one-fifth of the chemicals. Even when some data had been provided, one-third of the companies had not submitted final datasets. Public release of the data from the EPA has been further delayed.[111]

The information reviewed in the previous few pages suggests that any toxic properties of chemicals and any risks they pose are beyond the understanding and control of the current and largely postmarket legal institutions in our free enterprise system. The laws are poorly designed to protect the public.[112] Such laws would not catch substances, such as dioxin or mercury or tributyltin that cause in utero adverse effects lasting a lifetime. Moreover, as we saw earlier in this chapter, existing laws so poorly screen new and even existing substances that there are a reasonable percentage of mutagens in commerce to which we are exposed. In addition, we also saw how poorly they would address new technologies, such as nanotechnology, that already appear to pose health risks. The laws also could not detect and prevent damage to the eggs of a woman or the sperm of a man, damage that might be similar to injuries caused by BPA in experimental studies (see Chapter 4).

Reckless U.S. Industrial Chemicals Policy

Postmarket laws, toxic ignorance, and widespread contamination of the citizenry jointly reveal the substantial recklessness of U.S. laws toward the public's health. Industrial chemicals contaminate people by multiple exposure pathways just as surely as pesticides do. And different substances that affect the same cells, tissues, or organs can produce additive effects. However, the vast majority of chemicals are treated legally quite differently than pesticides are.

The early record of toxicity-induced birth defects and in utero–caused diseases has now been substantially augmented, revealing more widespread but often more subtle adverse effects. Scientists, public health agencies, and companies are now aware of the live possibility of risks to developing children. Research will need to be more sensitive to these risks.

The laws that permit these conditions are reckless toward the public. By this I mean that the policies appear to manifest lack of thought about possible undesirable consequences. Companies that commercialize products without testing for toxicity, fully aware of the contamination of adults and the vulnerability of developing children, also act recklessly.

When researchers, agencies, and companies believed that the womb was a safe, protected capsule, free from external toxic insults, they were perhaps nonculpably ignorant of these possibilities. Before they were aware of in utero and neonatal contamination and the live possibility of risks to developing children, perhaps none acted recklessly in terms of pre- and postnatal contamination. (They might have acted recklessly as a result of exposing the public to particular substances, such as asbestos.) Now however, because there is a far better understanding of the science of perinatal contaminations and their risks, the recklessness of current policies and practices is quite visible.

As a result, citizens are now experimental subjects for the toxicity of products in our chemical society, an outcome that 1970s congressional and presidential committees knew was possible and hoped to avoid. However, in the end Congress failed to enact legislation to prevent it. Moreover, companies have a legal right to contaminate the public until there is sufficient science for a risk assessment and sufficient political will in a regulatory agency to reduce the risks.

More Fundamental Shortcomings

Postmarket laws not only suffer from the deficiencies indicated above, but also have much deeper failings that should be understood.

Postmarket Science Is Too Late to Protect the Public

Despite the theoretical view that risk-based postmarket laws would protect the public, they poorly protect our children and the rest of us. These laws are difficult to improve, given how they can be gamed to frustrate better health protections. The science needed to identify risks and better protect the public is simply produced too late in the life of chemical products under postmarket laws. It is difficult for individuals to take protective self-help steps, and public health agencies need data to provide protections, which are long delayed. Beyond the reasons already considered, why are there such delays?

First, in scientific research there is an endemic (even logical) delay in explanations: understanding and explanation presuppose the existence or truth of something to be explained. This logically must precede explanation. For example, the planet Mercury must exhibit a precession, a shift, in its orbit around the sun before there is something to explain. Then scientists must notice it to be prompted to study it. DES exposure in utero must first cause vaginal cancer before there is something to understand. More importantly, the generic possibility of in utero injury must be known to or conceived by scientists before they can be motivated to consider any other such substances that might have such effects. If they have not yet conceived that such an adverse outcome is possible, it would not occur to them to consider investigating it.

Second, ordinarily there is no great urgency to understand some phenomena scientifically—as in the case of the precession of Mercury or the structure of DNA—beyond the usual scholarly desires to advance the field. In the public health area, the urgency is different. Identifying and understanding risk factors of disease sooner rather later is needed to protect people. It is urgent to shorten the lag between the occurrence of risks and the scientific understanding of them for prevention purposes.

Third, laws can facilitate or frustrate the production of science needed to protect the public. Premarket testing and approval laws facilitate the earlier production of science, because some understanding of a product would have existed before it enters the market. Premarket laws increase the chances, but do not guarantee, that toxic risks will be identified sooner rather than later and, one hopes, before the public is exposed and harms result. These laws will likely also discourage manufacturers from taking chances with risky products. Thus fewer toxic risks are likely to arise in the first place. In addition, some of the science needed to understand the toxicity of substances is unlikely to begin before a wider range of researchers is engaged on the topic. Premarket laws bring data about chemicals into the public domain much earlier than postmarket laws do.

Postmarket legislation results in a misalignment between the laws that aim to protect us and the science that could assist in this. The science will greatly lag toxic contamination and lag the presence of any resulting risks simply because it takes time for the science to reveal sometimes subtle effects. This has occurred even with pharmaceuticals (which are subject to premarket laws) that turned out to be developmental toxicants, for example, anticonvulsive drugs such as diphenylhydantoin and valproic acid, sedatives such as Valium, and antiacne medicine.[113] Postmarket laws are vastly worse in this regard, permitting many annual cohorts to be contaminated and put at risk before scientists can understand the source of harm, delays are overcome, improved health standards are issued, and exposures to harmful substances are reduced. Competitive pressures exacerbate this problem, inviting manufacturers to pursue as little testing as the law requires.

Reducing contamination by and risks from persistent chemicals will be even more difficult. PCBs banned thirty years ago continue to wreak havoc in the environment and with human health.[114] PBDEs will likely pose risks to humans and the environment for decades, by means of many of the same mechanisms as PCBs (see Chapter 4). Insensitive human studies will lag far behind actual but subtle harm to people.[115] Experimental animal research may reveal risks more quickly, but even that takes time from the inception of the study. And skeptics will be ready to issue critiques, usually exaggerated, about animal studies' evidentiary value for humans.

The fourth reason for delays in data and protections is that even if the community sought to expedite the science to identify risks, this is not easy to do. Scientific studies march to their own drummers, based on the kind of study and the effects investigated. Diseases and dysfunctions can have shorter or longer induction and latency periods. Studies undertaken to reveal risk factors of disease simply take time because of the nature of the complex subject.

There is a final point. When there is a substantial lag between exposure and disease, this slows the recognition of the causal relationship between the two. It is quite unlike one billiard ball hitting another, when the effects are immediate. Immediacy facilitates the identification of causal relationships; long latency impedes it.

This issue of long latency periods has important social consequences. It becomes easier for companies to disavow their products' contribution to disease simply because the relationships are more difficult to identify, trace and comprehend. Pinning down causal contributions can be a complicated, subtle, and long enterprise that is difficult enough. Moreover, to the extent that adverse consequences have multiple causes, some of those can be proposed as (and indeed might be) the culprit. Postmarket laws exacerbate the process. Also, the general public may have a false sense of security about exposures to chemical substances simply because the causal relationships are not quickly discernable and certainly not visible. Colliding billiard balls are immediately perceptible and easily understood, but if exposure to BPA can enlarge the prostate and possibly accelerate prostate cancer, as animal data suggest, this causal relationship may be difficult to grasp. It is simpler to ignore or disbelieve such results than to take them seriously.

Cops and Robbers Laws

Postmarket health laws are often characterized as "policing" or "cops and robbers" laws.[116] Firms are permitted to commercialize their products and continue to sell them until a regulatory "policeman" finds that their products pose risks. Richard Merrill notes, "[Agency researchers] look out for products that are [risky] to health, about which it can, like a highway patrol man, do something. That something is to initiate administrative or judicial enforcement action on the premise that the product is not as safe as the law expects it to be."[117]

Many chemicals used as examples in Chapters 2, 3, and 4 and other well-known toxicants are subject to policing laws—PCBs, PBDEs, bisphenol A, phthalates, perfluorinated compounds, chloroform, carbon tetrachloride, toluene, benzene, trichloroethylene, lead, dioxin, mercury, and methylmercury. Reducing or removing exposures to pharmaceuticals or pesticides also resembles a policing action. DES and other pharmaceuticals were missed, sometimes because premarket testing had not been instituted and sometimes because it seems premarket tests were not designed to identify developmental toxicants. When they are missed, they must be addressed via a postmarket police action.

Under postmarket laws an agency has a burden to show that a substance poses risks, as a police department must discover that Joe Smith has robbed a bank. The agency then has the legal burden to take action and to show what should be done to reduce or eliminate the risks posed. It must painstakingly follow all legal procedures and substantive requirements of the law to the satisfaction of a reviewing court, or the action will be overturned and the firm can continue to sell its product and expose the public to the substance.[118]

Policing of chemicals is difficult. Wendy Wagner points out that "the number of chemicals dwarfs the number of regulators by at least two-hundred fold." Thus, she notes, the EPA is at a substantial disadvantage "before [the game] begins."[119] Moreover, vigilant public interest groups, perhaps like vigilant citizens spotting crime, cannot hope to fill data gaps about the toxicity of chemicals.[120] Public interest groups in fact perform a valuable service alerting and assisting public health agencies. However, these groups usually have quite limited resources and personnel. The tort law is also of little effective use, since it suffers from the same (or more difficult) burdens of proof as postmarket laws and probably more burdensome difficulties of proof as tort law is currently administered.[121]

The "cops and robbers" model is insightful, but the metaphor, if pressed, reveals additional shortcomings for public health protections. And there are substantial disanalogies between normal policing activities and the "policing" of toxicants.

Criminal or traffic violations typically occur in a public perceptual space. Highway patrol officers can literally "see" traffic violations: speeding, running red lights or stop signs, or failing to use signal turns. They can receive assistance from citizens who can attest to violations such as failure to stop or driving under the influence. The police might see a criminal rob a bank. Tellers may be able to identify the robber. Witnesses might provide testimony about disguises or masks, testimony that supplies clues for police to follow. Guards or customers who had been held hostage can often provide evidence. Bystanders might identify a getaway car or remember all or part of a license plate number. Television cameras might have recorded the incident. Witnesses might hear robbers boasting about their exploits.

Subtle clues can be left behind: distinctive clothing, fingerprints, or even DNA evidence. In short, most crimes occur in a public world to which other people have epistemic access—they can see and hear robbers and detect other clues. Crimes leave traces that are usually accessible to ordinary citizens.

Molecular invasions and harm-causing, however, are not publicly detectable in the same manner. Molecules are silent, invisible, and usually odorless invaders. If they induce diseases, we are none the wiser. Any molecular-caused disease is typically not immediately known and possibly not discovered for years or decades. Tracing injuries to a chemical invasion may take subtle studies and years of scientific detective work. In addition, molecules may not create an immediate cause-effect relationship as typical crimes do. As a consequence, policing laws seem inadequate for most health protections. To the extent that legislators envisioned laws to prevent harms from toxicants as similar to policing, they seem spectacularly mistaken.

Nuisance Laws Are Poor Models

Nuisance laws, from the older common law, influenced the development of some early postmarket environmental health laws. Alleging a nuisance was a common approach for addressing pollution and its consequences before there were environmental health agencies. However, these earlier laws may have misled legislators because of deep dissimilarities between typical nuisances and molecularly caused harms.

The common law as it is usually understood is the "body of judge-made law that developed during and after the United States' colonial period, especially since independence," some of it inherited from the English legal system.[122] Of course, laws could be made by legislatures as well, but much law in the American legal system was made on a case-by-case basis. An important aspect of the common law is the law of torts, discussed above. Had people been exposed to environmental pollutants of an earlier era, pollutants such as smoke, dust, or arsenic, the tort law might have provided some modest protection by general deterrence and some compensation for nuisances experienced and injuries suffered.

According to *Prosser and Keeton,* in ordinary parlance a nuisance is a "hurt, annoyance or inconvenience." Nuisances in the law referred "to interference with the use or enjoyment of land, and thus was the parent of the law of private nuisance as it stands today."[123] A wide range of pollutants or activities might qualify. They could interfere with the condition of one's land by vibration, blasting, destruction of crops, pollution of a stream or well, and flooding. Smoke, dust, unpleasant odors, excessive light, loud noises, repeated and annoying phone calls might disturb a person's comfort or convenience. "A pond full of malarial mosquitoes" threatening one's health would count.[124] Bawdy houses or funeral homes that perturbed one's peace of mind could be nuisances. "Future threats" could be nuisances: stored explosives, flammable materials, or even vicious dogs.

As noted in *Prosser and Keeton,* as "long as the interference is substantial and unreasonable, and such as would be offensive or inconvenient to the normal person, virtually any disturbance of the enjoyment of the property may amount to a nuisance."[125] The party who has been interfered with has a legal burden to show that an activity interferes with the use and enjoyment of property and that it should cease.[126]Several early antipollution and public health laws—the Clean Air Act, the Clean Water Act, and perhaps the Safe Drinking Water Act—owe some of their inspiration to the law of nuisance. The first two laws also provided some intellectual background for other environmental laws, perhaps including the Toxic Substances Control Act.

William Rodgers traces aspects of both technology-based laws and risk-benefit-balancing laws to nuisance. He argues that the Clean Water Act and the Clean Air Act incorporated both kinds of considerations.[127] Consider technology considerations. A sixteenth-century nuisance case brought against city officials "who allowed swine to run loose distributing their dung in the alleys and lanes" in effect required a low-level technological fix. The judgment directed "that the streets and lanes be cleansed and 'for the future kept clean from dung and dunghills.' "[128] A somewhat similar problem in Michigan in 1970—"offensive odors from a hog finishing barn"—was also dealt with by the best techniques available under the Michigan Environmental Protection Act of 1970.[129]

My concern about similarities between nuisance and environmental health laws rests on two less direct but perhaps deeper points: (1) Common law nuisances are quite different from toxic molecules. (2) The generic procedure of nuisance law within torts and the postmarket procedure of environmental health laws have similarities that may be quite appropriate for nuisances but much less appropriate for molecular hazards.

First, typical nuisances are substantially different from molecules posing risks to people. Vibrations and blasting, unpleasant odors, smoke or dust, excessive light, or high temperatures are all readily cognizable and easily detectable by peoples' normal senses.[130] These nuisances differ substantially from silent, undetectable intrusions by toxicants. Nuisances announce their presence; toxic molecules do not.

The effects of nuisances also tend to be immediate in disturbing one's use or enjoyment of property. The effects of toxicants typically are not.

Substantial latency or induction periods may well delay the appearance of molecularly caused diseases. Immediacy between causes and effects facilitates their identification. Long latency between an initiating event and an adverse outcome hinders recognition of causal connections. In addition, many diseases caused by molecules have undetectable progressions, permitting them to remain hidden for some time before diseases are manifested. Typical nuisances are not hidden—they are apparent at the outset and all along the way.

Second, because nuisances announce their presence, a plaintiff could quickly initiate legal action to remove them. There would be public evidence, shared recognition of nuisances and their effects, and the obvious need to reduce them. Plaintiffs would not have to wait for long induction or latency periods or silent progression before the effects of nuisances would appear. Placing a legal burden on one who experiences a nuisance, thus, seems appropriate. Postmarket approaches to nuisances made and make sense.

It is different for potentially toxic chemicals. They cannot be readily detected. Their causal effects will not be immediately known (except, of course, for quick-acting poisons). Other people can easily doubt whether toxicants are present or what their effects are. Skepticism, thus, can take deeper root toward invisible toxicants than toward public nuisances. It takes time for science to identify risk factors posed by toxicants and perhaps greater time to persuade other scientists and public health officials. For these reasons postmarket administrative action to curtail the risks from toxicants cannot be immediately initiated.

Consequently, if nuisance law provided deep, but possibly unconscious, analogies for environmental health laws, perhaps legislators did not reflect sufficiently on major dissimilarities between typical nuisances and molecular risks and harms. Perhaps legislators were not even fully aware of some of the substantial differences at the time. We now know better.

Conclusion

Enough has been said above to indicate the failure of postmarket laws to protect the public from chemical risks. Nuisances are quite disanalogous to toxic molecules. Policing poorly controls them. Postmarket laws misalign the law with the science needed for health protections. And we have seen the consequences of substantial toxic ignorance as well as of delayed and poor health protections.

Is there a better way to go? Must we rely upon time-consuming, post-market science with imprecise, insensitive studies that are typically too late to prevent live risks to our children? For developing children, many years of exposed cohorts could suffer adverse effects before toxic contaminations can be reduced or eliminated. In the following chapter, I consider some alternatives and their rationales.

6

A MORE PRUDENT APPROACH TO

REDUCE TOXIC INVASIONS

If postmarket laws are reckless toward our health and sufficiently mis-conceived to protect us well from toxicants, what would constitute more farsighted and sensible legal and regulatory safeguards for our health? What kinds of laws could help ensure that our children and we are not contaminated and put at risk by toxicants?

Since most industrial chemicals can contaminate the biological material that goes into creating a child, should public health officials know which substances would disrupt this process? Should they also know whether individual substances could cross the placenta or enter breast milk? Or should they remain ignorant of these issues? Because chemical compounds can put developing children at live risk, what legal/social approach should be taken toward these substances?

Public health officials could continue to remain ignorant of the possible adverse health effects of compounds, unless manufacturers voluntarily choose to conduct tests. But voluntary testing is not likely to work. Com-petition between companies would create a quick race to the bottom of little or no testing. If company A conducts premarket tests, it has higher costs than company B, which does not test its products: why should A test? The consequence of not testing would be continued ignorance about devel-opmental effects from in utero and early childhood exposures.

More prudent, sensible safeguards to protect the public's health from industrial chemicals are needed. I suggest one approach, drawing on cur-rent premarket testing laws, other long-standing legal doctrines, and the ethics of research on human subjects, namely, the testing of new and exist-ing environmental agents for their toxicity before contamination occurs.

Not every substance would need to be tested for every toxic end point; that would be impractical and a waste of resources. Instead a "tiered strategy," suggested by some National Research Council (NRC) committees, "could help to set priorities among environmental agents for screening and could identify end points of mechanisms of action that would trigger more in-depth testing for various end points or in various life stages."[1] Appropriate evaluation should pay particular attention to susceptible subpopulations, as well as to additive and synergistic effects resulting from exposures to multiple toxicants.

This suggestion resembles aspects of U.S. pesticide and drug laws, suggestions made by committees of the National Academy of Sciences, and the European Union's REACH legislation, (REACH is the acronym for the Registration, Evaluation, Authorisation, and Restriction of Chemicals, a regulation that was recently passed by the European Community and that requires specified data on each substance manufactured, used, or distributed in Europe.)[2] In Europe the slogan for existing and new chemical substances emphasizes the need for safety data: "No data, no market."

Premarket Testing and Licensing Laws for Drugs and Pesticides

The U.S. Congress already requires pharmaceuticals, pesticides, and new food additives to be tested for toxicity before the public is exposed; but this requirement has not always been the case. Companies must assess these products in accordance with required protocols and have the results reviewed by independent scientific committees before receiving a license for commercialization.

Pharmaceuticals

Recall that during development pharmaceuticals must be first tested in animals. The period of testing depends upon the likely duration of typical prescriptions.[3] Longer-term animal testing would be required when a drug would be prescribed for a sufficient period of time that cancer or birth defects might become an issue. And historical examples show that exposures need not be extensive to produce adverse developmental effects.

The rationale for such evaluations is to ensure, as much as the best research can, that there are likely to be no risks (or at least only tolerable risks) to any human volunteers during experimental clinical trials and that the drug shows the promised benefits in animal models.[4] Prior testing in animal and other studies must protect volunteers from possible risks from drugs. Although human clinical trials are the gold standard for the safety and effectiveness of drugs not yet approved for use in commerce, the FDA asserts that *"the overriding consideration in these studies is the safety of those in the trials.* [The FDA] monitors the study design and conduct of clinical trials to ensure that *people in the trials are not exposed to unnecessary risks."*[5]

Pesticides

Pesticides must be tested in nonhuman systems before a company is permitted to license a product for commercialization. In order to register a new pesticide or to register a new use for an existing pesticide, the EPA must "ensure that the pesticide, when used according to label directions, can be used with a reasonable certainty of no harm to human health and without posing unreasonable risks to the environment."[6]

We should recall as well that the pesticide laws provide for special protections for "infants and children." The U.S. Congress recognized that children's "developing bodies may be especially sensitive to pesticide exposure." It also appreciated that a single chemical could expose citizens by multiple pathways and required the EPA to conduct risk assessments based on an assumption of aggregate exposures. And since people could be exposed "to pesticides and other substances that are toxic by a common mechanism," it required cumulative risk assessments as well. We have seen, as an example of cumulative risk, organophosphate and carbamate insecticides act by a common mechanism that can inhibit acetylcholinesterase, a critical nervous tissue enzyme.[7]

Although laws for both pesticides and pharmaceuticals require premarket testing of products, it is not clear that the required tests will be adequate to detect substances that are hazardous to developing children. Pharmaceuticals and pesticides have caused disease and dysfunction following in utero or postnatal exposure. Thus, premarket testing by itself is not a panacea; the tests must be appropriate and sufficiently sensitive for identifying adverse effects of concern before they occur.

Moral Requirements on Human Subjects Research

Some of the same concerns to protect people from possible risks are embodied in ethical constraints on conducting medical experiments on adults and children. Research on human subjects need not involve pharmaceuticals, but it is likely the quintessential medical experiment. Research may seek to discover a therapy to benefit people with a particular disease. It might consist of studies to contribute to general knowledge—for example, to understand how long drugs remain in people's bodies—or to test a medical device.[8]

Very strong conventions for such research have been developed as a result of abuses by the Nazis in concentration camps, by the Japanese during World War II, by the U.S. government on black Tuskegee sharecroppers who had syphilis that remained untreated by penicillin, and by the U.S. government in exposing uninformed citizens to radiation. These conventions—the Nuremberg Code (NC), the Declaration of Helsinki (DH), and the Belmont Report—have more than half a century of development and justification behind them.[9]

These conventions have three generic requirements in common: (1) People must consent to participate in the research, and the consent must be based on participants' informed and competent understanding of the risks to which they will be exposed. (2) Researchers must conduct appropriate prior studies so that they and the research subjects have a reasonable assurance of the safety of the experiment and so that participants have the necessary understanding of the risks to give proper consent. (3) A human subjects research project should have appropriate oversight from impartial and independent scientific and ethical committees.

In addition, the Nuremberg Code and Belmont Report require that there must be some overall assessment of the gains in knowledge from the research compared with any risks to experimental participants. Any risks to volunteers, according to the Nuremberg Code, "should never exceed that determined by the humanitarian importance of the problem to be solved by the experiment."[10]

Finally, the Declaration of Helsinki rules out research on children or others who are "physically or mentally incapable of giving consent. . . . These groups should not be included in research unless the research is necessary to promote the health of the population represented and this research cannot instead be performed on legally competent persons."[11] This strong caution recognizes both children's vulnerability and their inability to give informed consent.

Research on human subjects would be morally suspect and probably forbidden if any one of the three general conditions above were not satisfied (provided the conditions were appropriate for the research in question). While informed consent looms as an important consideration, that is not my focus.

Importantly, the conventions specify that there must be prior research on the safety of exposure during research and that there must be exploration of potential risks to participants. One rationale is protection of the participants. Another is the need for participants' informed consent, which cannot occur without knowledge of potential risks and therefore prior research. Moreover, because there have been such abuses concerning human subjects, researchers are not permitted to carry out a project without some significant independent and impartial review of the scientific need for the research and of its design, its scientific soundness, and its safety, as well as its ethical permissibility.

Beyond preparatory research, scientists must protect the experimental subjects against even remote possibilities of injury, disability, or death (NC.7). The codes give substantial priority to and express great concern for protecting those who participate. The well-being of the human subject should take precedence over the interests of science and society (DH.6).

In sum, current laws concerning pharmaceuticals and pesticides as well as the ethics of medical experiments place extremely high priority on conducting research prior to subjecting people to substances or medical experiments, in order to protect them from potential risks. Exposures could result from products (a) that will be injected into or ingested by those using pharmaceuticals or (b) that will be virtually certain to enter humans' bodies by multiple routes and result in cumulative exposures as pesticides do or (c) that will the subject of some other human experiments.

Not every drug, pesticide, new food additive, and food contaminant nor every medical experiment will certainly or likely pose risks to those exposed or participating. Some individual products or exposures probably pose no risks at all. None of this obviates the need under law or ethics for prudent testing for possible risks prior to exposure or contamination. The failure to test contaminating substances for their hazards and risks would clearly be a violation of drug and pesticide laws and contrary to long-established ethical principles concerning medical research.

Battery

An action in battery was brought in 1978 against University of Chicago researchers who, during 1950–1952, gave women diethylstilbestrol (DES) in a double-blind experiment as part of their prenatal care at the university's Lying-In Hospital. The women did not know that they were involved in an experiment and had no knowledge that they were being given DES.[12] At that time DES was not known to be harmful, but its toxicity had been documented by early to mid-1970s. The researchers were seeking to determine whether DES might prevent miscarriages, something manufacturers had asserted. It did not.

A district court considered the case and decided that plaintiffs' action in battery could go forward. According to an eminent legal text, in a battery, "the actor must '*intend* to cause the other, directly or indirectly, *to come in contact* with a *foreign substance* in a manner which the other *will reasonably regard as offensive*."' The aim of the law is to protect one's personal integrity, even if "the plaintiff is *asleep* or *under an anesthetic,* or *other wise unaware of what is going on.*"[13] Thus one need not be aware at the time of exposure to a foreign substance in order to regard it as offensive. In its opinion the court, quoting Prosser's *Handbook of the Law of Torts,* added, "Proof of the technical *invasion of the integrity of the plaintiff's person by even an entirely harmless, but offensive,* contact entitles him to vindication of his legal right by an award of nominal damages, and the establishment of the tort cause of action entitles him also to compensation for the mental disturbance inflicted upon him."[14]

The administration of DES to the plaintiffs was intentional, as required for battery. However, the concept of an intentional act needed for this purpose "is broader than a desire or purpose to bring about physical results. It extends . . . also to those [consequences] which the actor believes are substantially certain to follow from what the actor does."[15] Consequently, even if an agent does not make it his or her purpose or aim to expose others, if contact is substantially certain to follow from the agent's actions, the agent has satisfied the intentionality requirement of a battery. The gravamen of the wrong is violation of one's integrity without permission.

Combining the different elements of an action in battery, if an agent is substantially certain that his or her act will cause foreign substances to come in contact with another person without consent and in a manner that the recipient would reasonably regard as offensive even if at the time the person is unaware of the contact, the agent has committed battery. The person has a right to decide whether and when such "contact" may occur and must consent before it does.

By analogy, this rationale suggests that manufacturers of industrial chemicals, who are substantially certain that their substances will come into contact with citizens in a manner that recipients could reasonably regard as offensive, could be subject to a battery action. Moreover, this view is strengthened when substances that have not been tested for their toxicity contaminate people. Would people reasonably regard as offensive invasion by substances of unknown toxicity? They almost certainly would.

The wrongness of a battery consists in the violation of a person's right to be secure in his or her bodily integrity, subject to his or her decisions, not those who initiate offensive contact. This rationale also resembles principles underlying trespass.

Trespass

The doctrine of trespass provides much of the historical background of torts, but what remains in contemporary tort law as trespass largely concerns real property.[16] According to *Prosser and Keeton on the Law of Torts,* to successfully recover in trespass, a plaintiff must show that there was "an invasion (a) which interfered with the right of exclusive possession of the land, and (b) which was a direct result of some act committed by the defendant." The invasion need not cause harm to the plaintiff or to his or her property, and the defendant need not be at fault in causing it. A defendant would not be held liable "so long as he had done no voluntary act, as where he was carried onto the plaintiff's land by others against his will."[17] In addition, the same text points out, "from every direct entry upon the soil of another, 'the law infers some damage; if nothing more, the treading down of grass or herbage.' " There can be additional liability for any harm "caused to the possessor of the land or to members of the household by any conduct of the trespasser during the continuance of the trespass."[18]

The rationale of a trespass, like that of battery, is "directed at the vindication of the legal right." If someone enters the property without having "authorization," without paying for it, or without being granted permission, that person has trespassed and is liable.[19] Thus one who has the right to exclusive possession of property also has important associated rights: to choose who can be on the property or to authorize the presence of others on it or to license them to be there.

Trespass also encompasses the deposition of molecules and particles, including gases, particulates, and lead on property.[20] Although the trespass doctrine principally protects against invasions of real property, some case law extends that right to persons, for example, when blasting operations on one's property injure parties in public rights of way.[21]

Trespass, like battery, rests on deep considerations related to the integrity of one's person and rights over aspects of one's life. For example, even harmless trespass into your house is a wrong. In a scenario proposed by Arthur Ripstein, suppose I enter your house, doing no damage to it or your door locks, take a nap on your bed, making "sure everything is clean": "I bring hypoallergenic and lint-free pajamas and a hairnet. I put my own sheets and pillowcase down over yours. I do not weigh very much, so the wear and tear on your mattress is nonexistent. By any ordinary understanding of harm, I do you no harm. . . . You[r] objection is to . . . my trespass against your home, not to its effects."[22]

Other violations of one's personal integrity, notes Ripstein, include (1) a dentist's fluoridating a patient's teeth without his or her knowledge when the patient is philosophically opposed to it; (2) a doctor's unauthorized touching of a patient in an intimate place while he or she is sedated; and (3) harmless medical experiments performed on unconscious patients that leave no trace, do no obvious harm, and are done without permission.[23]

Consider a slightly different example. Suppose you dispose of some trichloroethylene (TCE) from home experiments into my hot tub without permission.[24] TCE is a probable human carcinogen and likely neurotoxicant, but diluting small amounts in a hot tub reduces any risk of harm. Have you done anything wrong? You did not actually harm or even pose a risk of harm to anyone exposed to it. But your TCE invaded, or "trespassed" on, my hot tub and my bodily integrity without my consent or license. The *Mink v. University of Chicago* reasoning suggests that the disposal of TCE could also be considered as a battery.

Consider a final example, one similar to invasions of industrial chemicals. Suppose a chemist has an unknown aerosol. He does not know whether it is toxic or not, poisonous or not, or totally benign. It has not harmed him. It might even be beneficial, but he does not know this either. It has a pleasant odor. If he has properly informed the EPA of the chemical properties of his aerosol but has not tested it for health effects, and if the EPA has required neither more data nor any testing under the Toxic Substances Control Act (TSCA), he may introduce the aerosol into commerce. If he sprays it into a conference room as a meeting convenes, knowing that participants will inhale it, but does not inform them or request permission to do this, has he wronged the conferees?

These examples suggest that invasions of people's bodies by chemical substances are trespasses or batteries, just as giving women DES of unknown toxicity without their knowledge is a battery. Common to battery and trespass is a deeper point: a person has considerable rights or authority or "sovereignty" over certain areas of his or her life and property.[25] However, I do not develop this normative view; that must be considered another time.

Battery and trespass are wrongs, but not because the invasions are necessarily risky or harmful. Instead, a wrong in battery results because one could reasonably regard the pertinent contact as offensive or, in trespass, because an unauthorized invasion of a person's bodily integrity occurred.[26] Discoveries in biology have revealed much deeper trespasses by industrial chemicals and pesticides than we might have previously realized. Nearly every substance will contaminate our tissues, cross the placenta, enter the tissues of developing fetuses, and be conveyed via breast milk to nursing infants. They suffuse the biological material that goes into creating a child before it is ever conceived. Children are born tainted with them.

Surely if we understand the biology, invasion by industrial chemicals constitutes battery and trespass. In addition, when substances are untested for their risks and hazards this makes the invasions even more offensive. Trespass would not need this additional consideration.

Property rights advocates often argue that persons may use commercial property in any way they see fit (a more extreme assertion) or with wide degrees of latitude and subject to some constraints (a more nuanced version). Moreover, these advocates claim any interference with such property rights requires justification by those who would interfere. However, if a chemical company, exercising its property rights, manufactures substances that inevitably invade and contaminate persons' bodies in a manner these persons would reasonably regard as offensive and without their permission, this would constitute battery and trespass against their persons. Should not such invasions also require permission and justification by one who would cause the foreign substance to invade, just as trespasses on chemical company property requires permission and justification?[27]

Some Interim Conclusions

Two sets of interim conclusions follow from the discussion above: one concerning exposures regarded as legally worth preventing; and the other, normative rationales.

Exposures from industrial chemicals are quite similar to exposures from pesticides. Both can enter our bodies by multiple routes. Invasion by inhalation or ingestion is obvious, but some chemicals, perhaps many, can also enter through skin contact, with children being at greater risk from this route. For instance, the industrial chemicals carbon disulfide, methylphenyltetrahydropyridine (MPTP) and various pesticides can invade by means of touch in addition to inhalation or ingestion.[28] Benzene, trichloroethylene (TCE) and other organic solvents can also enter by skin contact.

Products that act by toxic mechanisms common to several substances can have additive effects, just as pesticides do. Dioxin-like substances, estrogens, and antiandrogens all have additive effects by attaching to certain (but distinct) cellular receptors. Moreover, as Tracey Woodruff and colleagues have shown, even when toxicants do not affect the same cellular receptors, they may cause the same adverse outcomes (as discussed in Chapter 4).[29] Just as the pesticide laws seek to ensure the protection of children, who are more sensitive and vulnerable than adults, by assuming additive and cumulative effects in children's risk assessments, laws concerning other industrial chemicals should do so as well.

Ethical guidelines and legal requirements for human experiments place substantial constraints on exposing children in experimental settings. In general, it is simply forbidden. If experiments on children are needed because there is no other way to obtain data to protect children, there must be a very strong moral and social case for doing so, and adults authorized to give consent must grant it.[30] Children's contamination by untested industrial chemicals governed by postmarket laws stands in jarring contrast with requirements for medical experiments on children.

Protections from pesticides and protections for people in human subjects research are vastly different from postmarket protections from industrial chemicals or even cosmetics. If a company seeks to introduce nonpesticidal, nonpharmaceutical, non–food additive products, such as PBDE fire retardants, these substances are not subject to premarket testing and licensing (see Chapter 5). No one has to be consulted to give surrogate consent to children's contamination, yet PBDEs are in the highest concentrations in children's bodies.

Should industrial chemicals be treated differently from pharmaceuticals and pesticides? After all, people do not put PBDEs into their bodies by intravenous lines, syringes, pills, or liquids as they do pharmaceuticals. They do not eat foods obviously known to have residues from toxicants on them as fruits and vegetables can have residues from pesticides. We might also think that most people do not inhale or handle industrial chemicals as farmers and farm workers might handle pesticides. And no one is proposing to conduct human experiments with PBDEs.

These appearances are hopelessly misleading. PBDEs are in the dust in our homes, in our food, in the air we breathe, on the floor on which our children play, in our furniture, and on (and in) our pets (see Chapter 2). Moreover, food is a major source of PBDEs. People who eat red meat and chicken are substantially contaminated with PBDEs compared with vegetarians.[31] Lead, PCBs, and older persistent compounds are also ubiquitous, as is BPA.

PBDEs and virtually all industrial chemicals will enter out bodies by ingestion, inhalation, and skin absorption just as surely as pesticides will. PBDEs (along with other persistent substances, such as the perfluorinated compounds) remain in our bodies for much longer periods of time than many contemporary pesticides, often for years. Ingredients in cosmetics will also enter our bodies via several routes. Some are synthetic hormones mimicking toxicants that can contribute to adverse reproductive and developmental effects, while others are carcinogens.[32] (The state of California now requires cosmetic manufacturers to report the use of hazardous substances in cosmetics so that the state can in turn alert consumers.)[33]

The public will not know (and perhaps no one will) what "doses" of the chemicals enter our bodies or what residues are on or in foods, because this information is not legally required and would be nearly impossible to determine in any case. "Dosing" is left to happenstance via the environment, consumer products, and adventitious exposures.

Thus, while there might appear to be superficial differences between pesticide exposures (for which testing is required) and exposures to random industrial chemicals (for which it is not), those vanish upon examination. If pesticides must be tested for their risks and premarket estimates of their hazards provided, why not industrial chemicals? If researchers must carefully test substances before they give them to volunteers in medical research, why should this testing not occur prior to the commercialization of products that will just as surely invade consumers?

Someone might object that pesticides are special, created to be toxic to some living things, and pharmaceuticals must be given in high enough doses to be biologically effective, but such doses can have adverse side effects in some patients. Thus, the argument might run, both classes of substances have the potential to be toxic; this is the reason they are tested.

Industrial chemicals are not necessarily designed to be toxic. There is some truth to these comparisons, but they are misleading. Industrial chemicals may not be designed to be toxic, as pesticides are. However, what they have been designed to do is less important than the live risks they pose or the harms they can cause.

The history of disease and dysfunction caused by industrial chemicals and the live risks they pose to developing children undermine any belief in their toxic innocence. Developing children are exposed in the womb to industrial chemicals just as surely as they are exposed to pesticides. Moreover, the particular susceptibility of developing children reveals that industrial chemicals that might not be toxic to adults can pose risks to children (see Chapter 4). Researchers frequently discover new and unexpected toxic effects from substances as well as new toxicants.

The law in principle seeks to protect developing children from pesticides but not from the wider class of industrial chemicals. There are exquisite and detailed guidelines for protecting fetuses and young children from risks in research on human subjects but not from risks resulting from industrial chemicals. The different protections of children from pesticides, pharmaceuticals, and medical experiments compared with no protections from industrial chemicals will not withstand scrutiny.

The normative rationale underlying the requirements of pesticide and pharmaceutical laws appears to be this: because these substances in fact will enter people's bodies, and because there is a mere possibility of risks, before exposures occur, they should be tested to identify risks, however unlikely, so that health-protective steps can be taken, if needed. If that is correct, it would be as wrong to cause industrial chemicals to enter people's bodies without understanding risks that might result as it would be for pesticides or pharmaceuticals. This consideration places a quite high priority on protecting people from possible risks in products.

Consequently, there are good reasons for testing products, not because every drug, pesticide, new food additive, or medical experiment will certainly or likely pose risks to those exposed. Rather, out of concern for fellow citizens prudent testing is required to identify risks in order to avoid them or to ensure that any residual risks are manageable. Moreover, the tests are typically conducted on surrogates for humans—on whole animals or on biological systems, such as cell cultures, bacteria, and so on—to reveal risks to humans. The wider class of industrial chemicals should be similarly tested.

A second rationale, the one underlying battery, is that the integrity of a person should not be violated by contact, without his or her permission, with an agent or a foreign substance in a manner the recipient would reasonably regard as offensive. This rationale resembles the wrongness of trespass on property or of trespass of a person. Given the public concern highlighted in news reports about exposures to BPA, phthalates, and PBDEs, the public in all likelihood reasonably regards as offensive contamination by substances untested for their hazards and risks. They should so regard them.

I am not arguing that industrial chemicals should not be permitted into the market. Instead, I am asking what steps should reasonably be required of companies who manufacture and distribute chemical substances, to avoid wronging people so invaded. Certain principles underlying the long-standing legal doctrines of injunctions and licenses provide additional rationales for appropriate social policies beyond the analogies considered so far.

Injunctions and Licenses Aim to Forestall Risks and Wrongs

Both licenses and injunctions aspire to forestall risks, harms, or wrongs. Designed to prevent harms or wrongs fairly directly, injunctions are issued to stop conduct that is likely or highly likely to cause harm to or to wrong the party seeking the injunction. Licenses also provide a means to reduce risks, but less directly and less immediately than do injunctions. A person may be required to have a license to enter certain professions, but part of the rationale is to reduce the chances of harms resulting from his or her professional activities. We will consider injunctions and licenses in turn.

Injunctions

An injunction is a court order, issued after a showing by a plaintiff, that commands or prevents an action which legally threatens plaintiff's interests. Typically, to obtain an injunction, "the complainant must show that there is no plain, adequate, and complete remedy at law and that an irreparable injury will result unless the relief is granted." Many common injunctions require a showing of a high probability of an irreparable injury unless an injunction is issued. Prohibiting an activity by an injunction would be seen as serving the overall social interest much better than would providing damages after the fact of harm.[34] An important rationale for an injunction is, according to John Murphy, "to force the defendant *to do the very thing that will actually protect the claimant* . . . . In so doing, an injunction denies the defendant the option of acting in breach of duty and paying (probably inadequate) damages upon the occurrence of harm simply because this will probably yield gains that exceed the claimant's losses."[35]

An injunction traditionally has been seen as a more drastic solution to a legal problem than granting damages, and this likely accounts for typically strict conditions placed on its use—a showing of a high probability of irreparable damage.

What is important for this discussion is an insight central to injunctions: an injunction requires a defendant "to do the very thing will actually protect" a person's interests from risks. Since injunctions have some stringent and fairly onerous features, for the most part I do not argue for their actual use in attempts to prevent risks or harms to developing children. (Of course, in extreme circumstances for which typical injunctions are appropriate, they could and should be utilized.) Instead, I consider the main goal of an injunction more as a metaphor from which we can learn rather than as an actual legal strategy. The injunction suggests an important way to think about protecting people's central interests.

Licenses

Similar to injunctions, licenses aim to reduce the chances of harm but do so less directly and less immediately. They do not forbid an activity the way an injunction might prohibit the nuisance of a howling dog or the invasion of cement dust or molecules. Instead, licenses require a person to have a certain kind of preparation and background to conduct the activity to which the license applies, in order to reduce risks from this activity.

Physicians, dentists, nurses, engineers, and lawyers all need licenses to practice their chosen professions; otherwise, they are prohibited from the practice. Licenses permit behavior that would otherwise be illegal. There are good reasons for not permitting just any person off the street to enter a profession by paying a fee. These unlicensed individuals would not have the appropriate education, training, skill, expertise, and other preparation to carry out the profession's requisite responsibilities and to do so safely for those they serve. By acquiring a license, professionals are permitted to engage in activities that ordinary citizens may not: issue prescriptions, give injections, operate on patients, construct bridges, argue before certain courts of law, and so on.[36]

A license does not ensure that professionals will competently deal with patients or clients without injuring them. However, among other things, it should decrease the chances that professionals would inflict harm in the course of their duties. There can even be failures of licensing. Occasionally professionals might not be capable of doing the job for which they are licensed, but this would be a serious shortcoming on their part and a failure to live up to the standards required by the license. To address such problems, professional organizations that require licenses typically have standards below which professionals must not fall and procedures to adjudicate professional incompetence. If licensees' mistakes or incompetence were serious enough, they could be sanctioned, lose their licenses or be punished in the criminal law.

Of course, there are professions that might also require licenses, but the required competency and training might be missing or inadequate. For purposes of this discussion, I will keep the focus on licenses for physicians and lawyers as paradigm examples.

A Principle Underlying Licenses and Injunctions

Both injunctions and licenses suggest a rationale for guiding regulatory policies toward substances that contaminate humans and pose live risks. Conceiving of a chemical invasion as a battery, an injunction would require a person to do the very thing that would prevent the battery, namely, to stop foreign substances of unknown toxicity from coming into contact with another person in a manner the recipient could reasonably regard as offensive. Regarding invasions as contaminations by untested and possibly risky products, an injunction would require a company "to do the very thing" that would prevent possible risks to others, namely, to test substances used in these products and to have an independent review of the results for safety before there are human exposures. Similar rationales would be reflected in licensing such activities. Companies that introduce industrial chemicals into commerce would be required to do appropriate background preparation that will reduce the live possibility of risks before there are human exposures—that is, to test their products for safety before the public is exposed.

Toward a More Prudent Approach

As we have seen, the U.S. Congress has exercised greater prudence toward potential risks from pesticides and pharmaceuticals than toward risks from industrial chemicals governed by postmarket laws. International conventions and laws similarly protect volunteers in human subjects research, as postmarket laws do not.

The laws concerning the broader range of industrial chemicals and the policies toward them should be changed from postmarket laws, if we are (a) to recognize citizens' rights not to be invaded by untested substances, (b) to treat citizens as other than experimental subjects, and (c) to protect citizens better from possible risks in vulnerable life stages. What would constitute a more prudent approach to protecting developing children?

In the spirit of licenses and injunctions, companies that create, manufacture, import, and distribute industrial chemicals that invariably invade humans should be legally required to conduct their business in a manner that will reduce risks and avoid battery and trespass to the public: They should fund or conduct appropriate tests specified by public health agencies (in consultation with independent scientists) in order to identify risks to developing children and people in other sensitive life stages before exposures occur. The battery of tests should aim to ensure that there is a reasonable certainty of no harm to people as a result of contact with the substances in question. The test results should be reviewed by independent scientists before companies are licensed to introduce their products into commerce or to continue marketing products already in commerce. (The language of "reasonable certainty of no harm" is taken from pesticide laws to guide the requisite degree of protection.) Both new and existing industrial compounds should be evaluated. The same principle should also guide appropriate testing of major pollutants (see Chapter 7).

The goal should be to ensure that there is appropriate testing of new and existing commercial substances in commerce. Whether individual companies should conduct these tests is a separate matter. Some firms have created misleading tests in order to receive approval for pharmaceuticals, for example. The best alternative would be for companies to provide adequate funding to public health agencies for the appropriate group of tests; agencies in turn could find and fund independent laboratories to conduct the needed tests in accordance with appropriate protocols. The next best choice might be for companies to fund independent laboratories licensed by public health agencies that met testing protocols to conduct the tests. This approach, however, has not always been successful, as there have been instances of laboratories committing fraud on the public and public health agencies. Perhaps the least attractive alternative would be for companies to do the tests themselves. There may be other possibilities. Such alternatives would need to be addressed by legislation or administrative regulation to ensure impartial tests to protect the public's health.

To accomplish many of the same goals, three National Academy of Sciences committees have recommended similar conceptual changes in testing chemical substances. These committees support testing "all new and existing environmental agents." However, not all industrial chemicals would need to be evaluated. The decisions about "the intensity and depth of testing should be based on practical needs, including the use of the chemical, the likelihood of human exposure, and the scientific questions that such testing must answer to support a reasonably science-policy decision."[37]

One committee adds the following cautionary note: "It is neither practical nor desirable to attempt to test every chemical (or mixture) against every end point during a wide range of life stages. The committee recommends toxicity screening of every agent to which there is a strong potential for human exposure. A well-designed tiered strategy could help to set priorities among environmental agents for screening and could identify end points of mechanisms of action that would trigger more in-depth testing for various end points or in various life stages."[38] Thus the rationale for testing substances for their risks should not apply so broadly as to be impractical.

In addition, in the legislation the appropriate risk level—a "reasonable certainty of no harm"—should be specified quantitatively, just as risk levels are specified by Proposition 65 (see Chapter 5). Legislation similar to this is now being considered in Congress. The legislation proposes that for nonthreshold toxicants aggregate exposure should not pose "more than a 1 in 1,000,000 risk of adverse effects in the population of concern." For substances that act biologically by means of a threshold effect, "an additional tenfold margin of safety shall be applied to take into account the potential vulnerability associated with in-utero, infant, or childhood exposure to all sources of the chemical substance."[39]

Companies that introduce new chemicals into commerce should be subject to a licensing requirement before the public or workers are exposed. This would require a company that manufactured, distributed, or imported a new chemical in sufficient quantity to provide funding for appropriate tests on it prior to citizens' exposures in order to ensure, as well as a pragmatic battery of tests can achieve, that there is "a reasonable certainty of no harm" to citizens as well as a pragmatic battery of tests can achieve.

One can illustrate some differences between postmarket laws and those that take a more prudent approach toward public health protection by considering the recent "NTP [National Toxicology Program] Brief on Bisphenol A." One conclusion of that report is that the "similarities between exposures of certain human populations and laboratory animals treated with 'low' doses of bisphenol A is supported by multiple approaches. For this reason, the possibility that *human development* may be altered by bisphenol A at current exposure levels cannot be dismissed."[40] Potential adverse effects include predispositions to prostate and breast cancer, as well as neural and behavioral changes.

Under postmarket laws this scientific conclusion would probably be insufficient for a health agency to take regulatory action. More study likely would be needed, and affected companies would surely protest that the science was inadequate. They might even argue that human data must support modification of public health standards. However, the product would remain in commerce until appropriate testing had occurred to support improved public health protections, if they were needed. Under a postmarket law a firm has reasons to strongly contest the science and to keep the product in market as long as it is commercially viable.

If governed by a premarket testing and approval statute, the degree of suspicion expressed in the report about the toxicity of BPA would likely preclude it from the market until these issues were sufficiently resolved scientifically to license or not license the product to enter commerce. In such circumstances a firm would need to decide whether specified testing made commercial sense.

Clearly, not all products proposed for commerce would need to be tested. Some will not be manufactured or imported in sufficient quantity to merit testing. Some may be merely proposed and not developed further. Some may be industrial intermediates with no human exposures. These are issues that would need to be addressed by experts more familiar with products and testing. Proposed tests do not necessarily need to be identical to those used for pesticides. Protocols would need to be designed to account for different levels of production, different amounts of exposure, different degrees of persistence, the extent of toxicity and overall risks, and especially risks to more susceptible subpopulations.

If a company introduces a product with little exposure but then increases production and thus exposures, greater testing should be required.

If the company increases production without informing the EPA or without testing the product appropriately for the production level, upon discovery of this fact an injunction might be automatically authorized to stop the manufacturing and distribution of the product. This injunction should remain in effect until testing is conducted and appropriately verified to the agency's satisfaction. The injunction would require the company to do the very act that would ensure reasonable certainty of no harm from increased production, namely, to test the product that would create wider exposures in order to detect potential risks. Just as pharmaceuticals may not cause harms to small groups of volunteers, so industrial chemicals that have little exposure may not pose problems. However, wider release exposes a more biologically diverse population with likely more susceptible individuals within it. Without such an automatic trigger, companies will have incentives to challenge, resist, delay, and frustrate testing requirements for their products (see Chapter 5). Automatic devices, if they are well utilized, help to overcome such delaying tactics.

Existing substances should also be tested by a certain date in order for them to be licensed to remain in commerce. Conditions similar to those required of new products would apply to existing chemicals. In effect, a company would be prohibited from further marketing the product as of a certain date unless testing had been completed and unless the testing was found by an independent scientific body to appropriately ensure that there was "reasonable certainty of no harm" to children, women of childbearing age, and other susceptible subpopulations.

Some percentage of existing products would not need to be revisited, because they entered the market with good premarket testing, approval, and licensing; this might include many, but not all, pharmaceuticals and pesticides. In the past some pharmaceuticals and pesticides were clearly inadequately tested for developmental toxicity, because so many of them have caused such problems.

Some food additives or food contaminants will need to be reconsidered simply because they have not been adequately tested, as concerns about bisphenol A show. There could also be exceptions for products that had exhibited no toxic properties to date, provided they had been appropriately evaluated. However, epidemiological studies that are too small, too short, or otherwise insensitive or that do not have appropriate confirmation in other studies do not constitute such exoneration.

Still other compounds would need no testing. Some are so toxic that they should not continue in commerce without compelling social reasons for their use. This applies to unnecessary uses of lead, such as in toys, paints, water pipes, electronics, medications and cosmetics, as well as to numerous persistent, bioaccumulating and biomagnifying substances. The use of other known toxicants with considerable exposure should be closely scrutinized and perhaps restricted.

In many respects, the structure of the suggested legal requirements should resemble but not be identical to U.S. pesticide laws, provided companies have good tests to detect a broad range of developmental toxicants. Testing protocols would need to be designed by an agency such as the U.S. Environmental Protection Agency or the Food and Drug Administration; an independent body of scientists would need to review the tests to determine if they provided reasonable assurance that would be no harm from exposures. I suggest a few generic guidelines for testing later in this chapter.

The REACH Strategy

The European Community has adopted a similar approach toward approximately thirty thousand new and existing chemicals, under a compilation and strengthening of existing legislation denoted by the acronym REACH, which stands for the Registration, Evaluation, Authorisation, and Restriction of Chemicals. REACH requires specified data on each substance manufactured, used, or distributed in Europe.[41]

As noted earlier in this chapter, the splendidly simple epigram for this law is "no data, no market." REACH requires premarket testing and approval for new chemicals, but it also requires similar testing and review of existing substances to eliminate asymmetry between new and existing products, a long-standing discrepancy.[42] If no data is produced, a new chemical may not enter the market and an existing one may not continue in commerce.

The law "is based on the principle that it is for manufacturers, importers and downstream users to ensure that they manufacture, place on the market or use such substances that do not adversely affect human health or the environment." The "European Union is aiming to achieve that, by 2020, chemicals are produced and used in ways that lead to the minimization of significant adverse effects on human health and the environment."[43]

REACH seeks to "ensure that substances of high concern are eventually replaced by less dangerous substances or technologies where suitable economically and technically viable alternatives are available." Again, responsibility for testing and ensuring the safety of the products rests with those who "manufacture, import, place on the market or use these substances." The number and generic types of tests needed under REACH for a specific substance are based on the number of tons of production volume as a surrogate for exposure. All substances produced in more than one metric ton per year are subject to REACH. Moreover, the tests are cumulative: any test required for lesser production volume also applies to greater production volumes.[44]

Under REACH companies do not have a right to manufacture, sell, or distribute any goods in the European Union that they think are appropriate. Instead, their products may enter the market only if they have a license from the European Union for them, and that permission is conditional upon testing the products and ensuring their safety to the satisfaction of an EU agency.[45]

A licensing or permission model, such as REACH, treats access to markets as conditional upon satisfying some provisions that aim to reduce risks from a company's products and upon the European Union's granting permission once the conditions are satisfied. This is a much different moral and legal relationship between a country and firms seeking to do business within it. Under U.S. postmarket laws, companies in effect have a legal right to market products unless and until the products cause harms or risks of harm.

The European Union specifies tests that provide reasonable assurance of no significant risks from toxicants.[46] Governmental agencies then have some assurance that the chosen tests were consistent and sufficiently protective for their purposes. REACH does permit firms to present evidence of alternate testing protocols, if it can be shown that they better serve the purposes or are more sensitive and so on.[47] This permits flexibility and perhaps innovation in testing procedures. However, under such a flexible policy it can be difficult to compare test results between chemicals, and it requires greater efforts at validation than if an agency identified specific tests that should be used to try to identify risks.[48] To allow alternate testing protocols, a public health agency must find the proposed alternative acceptable, review the results for safety, and then grant or deny the entry of the chemicals into the market (or permit their continued presence in the market).

The overall goal of REACH is to ensure broad screening of thousands of substances. There are, however, some concerns about REACH. A National Research Council committee expresses worries that the pursuit of breadth under REACH may sacrifice some depth of analysis, which may leave some risks undetected. Of course, it is unrealistic to hope for both broad coverage and in-depth analysis of each substance at a low cost. However, REACH permits greater testing "to be triggered on the basis of initial results," a desirable feature.[49] In addition, REACH protocols may not require tests for some "relevant end points and life stages," such as those of fetuses, children, and the very old, but REACH protocols are an improvement on testing compared with many U.S. laws.[50] REACH should be improved for testing for developmental toxicity for the reasons presented above.

Some Considerations for Testing

What are some critical elements of a reasonable program to identify risks to the public and especially to susceptible subpopulations within it, in order to provide greater protections from industrial substances?

According to the EPA, risks are "equal to toxicity multiplied by exposure." For example, a substance with low toxicity and high exposure could pose a risk similar to that of a substance with high toxicity and low exposure.[51] A strategy for reducing risks from chemical substances must address both the extent of human exposures to them and their toxicity.

Exposures

Exposures have several dimensions.[52] The amount of a substance produced and presumably then utilized within national borders would be one issue. Production volume of a product is a crude measure of exposure. How much of it is used could be another measure of exposure, even if it were not produced domestically. For example, the use of billions of pounds of bisphenol A or millions of pounds of PBDE fire retardants is one measure of exposures to the public and workforce, even though the United States may not be the main producer of either compound. However, production or use volume may overestimate exposures if the substances are industrial intermediates that did not escape and expose the public or workforce.

There can also be high local but not nationwide exposures. Some relatively small populations might be highly exposed. For instance, pollutants from power plants or factories might expose immediate neighborhoods, contaminating citizens, as the Tacoma smelter did. Perfluorinated compounds, used in Teflon and other products, contaminated the air, ground, and groundwater near manufacturing facilities of the 3M Company, substantially exposing immediate neighborhoods. (Unfortunately, there is also a problem with more widespread exposures to these products because they persist and can travel so far in the environment.) Consequently, local populations may be quite contaminated with a product, which in turn could have higher or lower toxicity.

The persistence and bioaccumulation of compounds are other indicators of exposure. These are especially critical because there can be initial exposure to humans or the environment, continued existence in the environment, and then later exposures to humans and the environment that in some cases can be more toxic than the initial exposures.[53] For instance, PCBs biomagnify, that is, they become more toxic as they move up the food chain and bioaccumulate in animal tissues.[54] Thus, the persistence of a chemical can create multiple exposures and sometimes more toxic variations.

Even though persistence and bioaccumulation usually are closely associated, they can be different. Persistence refers to the length of time a chemical remains in the environment or in mammalian bodies. In contrast, a substance bioaccumulates if it "builds up" in the bodies of living entities. It could build up within an individual animal or person, or it could build up in increasingly concentrated amounts as it is transferred from one animal to another, for example, when one animal or fish eats another one.

Cadmium, an extremely toxic metal recently found in low-cost jewelry, persists in the environment and in mammalian bodies, but it does not bioaccumulate. Conversely, polycyclic aromatic hydrocarbons (PAHs) are not persistent in bodies—they break down into metabolites fairly quickly—but humans or mammals could have an increasing body burden of the metabolites with continual exposures to PAHs.[55] PCBs and PBDEs tend to build up in the fatty tissues of animals as they consume foods containing these substances or consume other animals whose tissues contain PCBs and PBDEs. Perfluorinated substances (fluorine-based compounds) attach to proteins instead of fats but then are "recirculated" through the body, reattaching to proteins after going through the kidneys. The half-life of perfluorinated substance is similar to that of PCBs. If compounds have any toxicity properties, which PCBs, PBDEs and perfluorinated compounds all do, they can contributed to adverse effects for decades.

Toxicity

A second element of a risk is the toxicity of the substance, its capacity to cause adverse effects in living organisms.[56] There is no best characterization of toxicity, but the most useful ones, according to David L. Eaton and Curtis D. Klaassen, probably "take into consideration both the chemical and the biological properties of an agent and the exposure characteristics . . . most likely to be useful for legislative or control purposes and for toxicology in general."[57] For our purposes, I have used general terms such as carcinogen, reproductive toxicant, developmental toxicant, and neurotoxicant. These serve sufficiently well to guide discussion.

Risks

As we have discussed, toxicity multiplied by exposure creates a risk. The distinctions above provide some guidance for testing substances for risks. Moreover, there are some issues about risks that are especially critical for protecting the public.

Some of these issues can be illustrated by referring to an older and rejected paradigm. At one time, risk assessors tended to estimate the risks that chemical exposures would cause to a typically healthy seventy-kilogram, possibly white, working, male. That is, assessors would estimate risks to one of the healthiest groups of people and then use this as a basis for projecting risks to other segments of the population, including children. (We should recall the healthy worker effect discussed in Chapter 3.) By now the bankruptcy of this paradigm is clear.

The older model ignores or underestimates risks to women of childbearing age, pregnant women, fetuses, developing children, individuals with preexisting diseases or conditions, people with low incomes, minorities impacted by multiple toxicants, and the elderly, among other groups. Even a recitation of these different groups tends to ignore others who might have greater vulnerability because they have greater body burdens of substances or are both heavily burdened by toxicants and suffering from preexisting diseases. Taking people as they are, with all their susceptibilities, accumulated toxic insults, diseases, and so on, means that they are not necessarily fully healthy, pristinely free of contaminants, and fully captured by risk assessment paradigms of the past.

Consequently, a major testing priority would be testing for, in the words of the NRC, the "more sensitive adverse effects or exposures of greatest

201

concern" rather than testing for all potential adverse effects. A more recent report also calls attention to this issue under the idea of human biological variability.[58] Instead of testing for all end points and all possible life stages, the aim should be to ensure that the broad range of the human population would be protected as a consequence of an assessment. Knowledgeable professionals would need to address these issues for people in sensitive life stages. Obviously, developing children and the very old are starting points, but these groups would likely need amendment.

Another especially important sensitive group would be women of childbearing age (approximately between the ages of fifteen and forty-five) assumed to be contaminated by toxicants. A woman's contaminated body is the developmental environment in which a child comes "into existence," in which it grows and from which it draws nourishment. If that environment is contaminated by lead, halogenated hydrocarbons, endocrine disrupters, and so on, these substances surround the cells and tissues that become a child and also enter the child's body as it develops.

As another important priority in protecting the public, the NAS committee also recommends the use of cheaper and quicker tests to the extent that this is possible. The idea would be to use such tests as indicators of toxicity and then conduct more in-depth analysis if preliminary indications suggested the need for it. At the same time, the NRC recommends "stopping" rules so that there would be a means for preventing all substances from being subject to all tests when there likely would be no need for them. An overriding concern would be to conduct additional testing only if it would be likely to make a difference in assessing the risks from substances and their exposures.

Recent research also has found better ways to assess risks from additive and cumulative exposures for the general population and for especially sensitive subpopulations. Recall that there are some additive effects from substances that attach to similar cellular receptors, substances such as estrogens, dioxin-like compounds, or organophosphate insecticides. Moreover, we should keep in mind more generalized additive effects that Tracey Woodruff and colleagues have identified. Different compounds can interfere with or perturb different "upstream pathways" and cause adverse effects without attaching to identical receptors (see Chapter 4). Researchers have also identified several compounds that jointly cause greater adverse effects than their additive doses would predict.[59] This form of testing would aim to identify substances that disrupt fundamental biological processes and lead to adverse effects, even if they do not arise from an identical mechanism more narrowly construed.

Clearly, chemical compounds should not be tested in isolation. Existing or new chemicals should be tested for toxicity assuming that similarly acting substances already contaminate humans, when there is evidence for it. In experimental studies Andreas Kortenkamp and others have shown the importance of cumulative testing of very low doses of substances, at individual doses of these substances for which no adverse effects have been seen. When doses from four such substances were combined in these studies, there were surprising effects. These researchers found "something [that is, adverse effects] from nothing" (doses below a level that produces adverse effects for each individual substance).[60] A reviewing body should try to take such data into account as best it can. (The reviewers could reference body burdens from the Centers for Disease Control and Prevention biomonitoring program, for example.)

None of us is contaminant free. Developing children are contaminated in the womb and every day after that. Biological material that goes into a child's creation and development is in contact with industrial chemicals before any new life is conceived. Proper assessment of risks should recognize these facts and seek to ensure that risks to citizens are not underestimated. On occasion, some have even suggested that if a substance enters our bodies, accumulates, and is hazardous even in experimental studies, it should be banned.[61] It seems quite clear that if a substance can cross the placenta and has no known safe level, as lead does not, it is a candidate for exclusion from the market.

A related point is recognized in a recent NAS report on a more technical issue. The report urges that "threshold" and "non-threshold" mechanisms of toxicity should be addressed in an integrated manner (see Chapter 3.) The NAS argues that "background exposures and underlying disease processes contribute to population background risk and can lead to linearity at the population doses of concern. Because [threshold models] do not quantify risk for different magnitudes of exposure but rather provide a bright line between possible harm and safety, their use in risk-risk and risk-benefit comparisons and in risk-management decision-making is limited. Cancer risk assessments usually do not account for differences among humans in cancer susceptibility other than possible differences in early-life susceptibility."[62] Thus both threshold models and linear models of toxicity should be modified to take into account cumulative and aggregative exposures plus existing contaminations. When models of toxicity do this, we can reasonably expect that the two ways of assessing risks will tend to be similar. Threshold models will tend to show more linear effects because of preexisting contamination (and may have no identifiable "safe" level), and linear models will need to take into account multiple exposures and additive effects. Testing protocols should recognize these possibilities. When protocols do, they will provide greater public health protections.

Finally, consider a point about endocrine-disrupting chemicals. Theo Colborn, along with an informal group of endocrinologists and the Endocrine Society, the world's largest professional group of endocrinologists, propose legislation that would modify how testing for endocrine disrupters would be done. At present the long-suffering endocrine disrupter program at the Environmental Protection Agency has presupposed that testing would occur within fairly traditional toxicology paradigms. Colborn and colleagues argue that this supposition is fundamentally flawed for hormones, because the timing of exposure and cumulative doses are perhaps even more important for these toxicants than for others.[63]

Consequently, this group urges that a research and testing program be set up not in a regulatory agency but in a scientific governmental research institute, namely, the National Institute of Environmental Health Sciences (NIEHS). The NIEHS, supporters of the bill argue, should convene appropriate independent experts to devise tests that can detect endocrines that can cause adverse effects at extremely low doses, especially when they are ill timed during development. The tests would be well grounded in the best science of the field, substances would be tested in accordance with the new protocols, and when minimal levels of concern are found from the tests, the results would be forwarded to the EPA or other government agencies to utilize for regulatory purposes. A much broader goal would be to more sharply separate the science of testing from the political pressures of regulation. Other aims include closing large data gaps.[64]

None of the above is a detailed prescription for testing; this will need to be devised by others with appropriate expertise. However, I have sought to call attention to some generic and conceptual aspects of testing that must be considered in developing more prudent protocols to protect the citizens' health. Both new and existing chemicals, as well as major pollutants, need to be tested better for toxicity. An independent body should review those results before companies are licensed to commercialize their products.

Costs

Some readers, aware of the costs of licensing a new pharmaceutical or pesticide, will be concerned about potential costs of testing existing and new chemicals. I have not advocated the full battery of tests that are used for pesticides and certainly not the same battery that is used for new drugs (both of which may be quite inappropriate). Instead, I have sketched some considerations that a less comprehensive but more targeted procedure (following suggestions from the NAS) should take into account. I will address the cost issues of testing in the final chapter.

What Social Bet Should We Make?

An alternative title of this chapter might be "Should We Bet Our Children's Health on Postmarket Harm Principles?" This asks a social, not an individual, self-interested question. The arguments of this chapter, along with those of Chapters 4 and 5, however, rest on some quite significant claims, which appear to be correct. If we wish to bet our children's health on one of these claims not being true, those would be especially poor uses of our funds.

Should we bet that there will be no more contamination of our bodies by manmade chemicals? No: virtually every industrial chemical will invade our bodies.

Are we willing to bet that mothers will not share their chemical burden in utero or through nursing? This too would be a bad bet. A few substances may not enter the womb because of size, electric charge, and so on, but most will. Perhaps not all substances present in mothers' bodies will enter breast milk and expose developing children, but lipophilic and other compounds likely will, and probably most chemicals will not follow the "calcium stream" as lead does, but some will.

Should we bet that children will not be more susceptible than adults, be more exposed on a body weight basis, or have fewer defenses against chemical assaults than adults? Should we bet that developing organ systems will not be as susceptible to toxic substances as they have been in the past? These claims would be contrary to scientists' understandings of developmental biology and the stability of biological systems over time.

Should we bet that all developmental toxicants have been identified and that there will be no others? Of course, it is more difficult to know the answer to this question, but it would be particularly unscientific to assume that all developmental toxicants are known and understood. Quite the contrary, the problems just now coming to light are a consequence of better assessments of substances and more subtle science focusing on the developmental origins of disease. As scientists develop the tools to identify more subtle adverse developmental effects, the more they are likely to find. This increase in discovery of effects has occurred with such diverse substances as lead, mercury, DDT, arsenic, various plastics, and endocrine disrupters. Moreover, PBDEs—lipophilic, bioaccumulating substances— will almost certainly cause the same kinds of problems that PCBs do.[65] These chemicals and their problems will persist for decades. In addition, we should be alert to substances that by themselves do not pose adverse effects but can potentiate other substances, that is, we should be alert to nontoxic substances that make a known toxicant more toxic.

Should we bet that research has discovered all the substances that can cause developmental problems at low concentration levels? This, of course, is also difficult to know. As research progresses, it appears that adverse developmental effects are being identified at quite low levels. This is certainly true of lead, radiation, and some carcinogens, for all of which there appears to be no lowest level of toxicity.[66] How many substances are sufficiently similar? The general biological principles that developing fetuses and newborns are much more susceptible than adults argues against such a bet. Moreover, even posing the question ignores the possibility of additive and synergistic effects at very low doses of individual substances.

Some might want to bet that nonhuman mammalian research models will have little biological relevance to humans. This is a point that manufacturers of products and some scientists will argue, even though they utilize the same kinds of studies in their own research. Nonetheless, experimental mammalian studies are the primary means by which scientists come to understand the toxicity of substances. These mammalian studies are genuine scientific experiments and avoid the need to conduct unethical experiments and chemical trespass on humans. Moreover, such animal research will continue to be utilized for the foreseeable future; for the most part, it will be the best science available until superior short-term tests are designed. Finally, if we do not utilize animal studies and other nonhuman tests, as a society we will put at risk many annual cohorts of children.

Does the citizenry wish to bet that postmarket laws, which seem so ineffective in addressing toxic risks to children, will suddenly become strikingly efficacious in some manner preventing risks to children while permitting them into commerce without testing? Just posing the question reveals how bizarre the suggestion is. Business as usual will continue to be risky and will very likely be harmful to our children.

A much better course of action is to develop testing and approval laws based on rationales related to licenses and injunctions, to protections for medical experiments and to existing premarket laws. Companies should do the very acts that will reduce risks to citizens, namely, fund testing of their new and existing products as well as major pollutants for toxicity.

7

WHAT KIND OF WORLD

DO WE WANT TO CREATE?

Imagine a world much like ours but with one major difference. In accordance with the proposal in Chapter 6, companies' new and existing products would be tested to determine hazardous properties so that any risks would be reduced or the products removed from (or prevented from entering) commerce before contamination occurs. Major pollutants would also be tested for their toxicity because these too can pose significant risks. The tests might not be unerringly accurate, but there would be less toxic contamination and many fewer risks because of better and more comprehensive testing than at present.

If such a world had existed in the past, we would have had a world with no thalidomide babies, no women with vaginal and cervical cancer because of in utero exposure to untested diethylstilbestrol (DES). There likely would have been fewer women with breast cancer, because DES, dichlorodiphenyltrichloroethane (DDT), and other synthetic estrogens would have been identified as toxic and prevented from entering commerce. Fewer children's mental functioning and behavioral problems would have been adversely affected by lead, mercury, and other neurotoxicants. Persistent organic pollutants—such as the older polychlorinated biphenyls (PCBs), previously banned fire retardants such as polybrominated biphenyls (PBBs), and existing fire retardants (the PBDEs)—would have affected few or no children. At present men would be at lesser risk from prostate cancer, genital dysfunction and sperm damage, because bisphenol A (BPA) exposures and other antiandrogens would have been identified and caught before commercialization.

We would need to spend less time, effort, and money trying to reduce individual exposures to such substances, because a well-designed, well-funded, and well-implemented effort would have identified chemicals that pose risks before the public was exposed. We would be treated more justly by a legal system less reckless toward our health. Companies would take greater responsibility for the social consequences of their products.

Conceiving such a world helps to visualize one way forward and to provide a foil for the status quo. Invasion by molecules is inevitable. Contamination from toxic substances is not; such contamination is a matter of how we design and implement our legal system. If we continue with the status quo, all of us and our children when we pass through development and other sensitive life stages will be invaded by toxicants.

Can we create and achieve together a more prudent alternative world, to reduce the toxic contaminants in our lives? How might the status quo need to be changed to approach this picture? What would this world be like? What concerns might there be going forward into a different legal world?

Improvements on Existing Legal Strategies

First in an attempt to create a more sensible approach to chemical substances, drugs, pesticides, and new food additives would continue to be subject to premarket testing and licensing laws. However, the Food and Drug Administration (FDA) and the U.S. Environmental Protection Agency (EPA) would need to review existing test protocols, to ensure that they are appropriate to identify likely toxicants for developing children and people in other sensitive life stages. There are several reasons for this review. Some pharmaceuticals have caused birth defects, cancers and various neurological dysfunctions. These data suggest that FDA protocols at the time of product development probably were not sufficiently sensitive to identify toxic effects in developing children.

Numerous pesticides have also been identified as causing neurological or other adverse health effects, either in humans or in animal studies. Moreover, biological mechanisms provide reasons to believe that various classes of pesticides would be developmental toxicants. Those familiar with testing protocols for pharmaceuticals and pesticides would need to review existing procedures to determine whether they are sufficiently sensitive to identify risks to developing children, women of childbearing age, those already substantially contaminated by toxicants, and other sensitive subpopulations. These reviewers would also need to ensure that testing could detect chemicals that might cause diseases with long latency periods or multigenerational effects.

New food additives, in theory subject to premarket testing, would

209

likely need greater scrutiny. BPA, used as a liner in canned foods and beverages, in plastic bottles, and in other products containing foods, appears to have initially been poorly tested and understood. Many experimental studies now identify it as a developmental toxicant, and some human data suggest related problems. Phthalates, which can have intimate contact with our mouths, skin, or blood, have sufficiently worrisome toxicity properties that some states have begun to ban them because of developmental problems. To date, the FDA has not been persuaded that there is an association between phthalates and health risks, although they are listed as reproductive toxicants under California's Proposition 65.[1]

Second, in order to assess the safety of substances with which we are contaminated, we would need to update laws governing testing for hazardous and toxic materials in toys, cosmetics, and other consumer products before they enter commerce; in substances in the workplace before workers are exposed during manufacturing; in packaging for household products; in food, feed, and color additives; and in other products. Substantial amendment of a single law, the Toxic Substances Control Act (TSCA)—along the lines suggested here (see Chapter 6), by the National Academy of Sciences and by pending legislation—might go a long way toward achieving improved testing, since the TSCA was originally designed to address the toxicity of chemicals of different origins before they entered different media.

Of course, testing of industrial chemicals would not need to be as extensive as that for pharmaceuticals. Industrial products will not be directly administered to patients in substantial doses. Thus, a therapeutically potent but safe dose would not need to be determined, as is required for pharmaceuticals. Moreover, human studies would not be needed for industrial chemicals, and these studies would be ethically out of bounds in any case. Nonetheless, there would need to be sufficient research to ensure a reasonable certainty of no harm to humans.

Enhanced testing should be guided by some obvious priority considerations in order to determine toxicity effects of concern as well as guided by characteristic properties of substances to trigger closer scrutiny. Improved testing would be needed to detect developmental toxicants. Toxicity testing would need to be continued for other obvious end points: carcinogenicity, neurotoxicity, immunotoxicity, reproductive toxicity, and so on. Substances to which there is widespread exposure should have priority for testing or perhaps for updated testing if they have been tested in the past but not recently.

Persistent and bioaccumulating substances should be carefully scrutinized because of their long lives, distribution in the environment, and biomagnification, and because of multiple opportunities for exposure. Whether these substances should be permitted into commerce at all is an open question that would need to be addressed. Perhaps there should be a presumption against their use for the above generic reasons unless there are strong social benefits to the contrary. Some developed countries have already banned a number of them.

Safety standards for heavy metals, such as lead, mercury, and arsenic, should be reviewed. There is likely little need for further testing of these substances. They are reasonably well understood. The main issue is likely to be whether health standards for them are sufficient to protect various subgroups in the population and sufficient to protect them based on the assumption that similar acting substances already contaminate them.

Benzene illustrates the point about substances with widespread exposure. This industrial chemical is central to many petroleum-based products and plastics. Most of the public and many in the workforce are exposed to benzene. The current occupational exposure standard of one part per million (ppm) per day for a forty-year working lifetime may not be adequate to protect employees who work closely with it. An older (1976) exposure level of 0.72 ppm, which was voluntarily adopted by many companies, may not even be adequate. Employees have experienced benzene-related adverse effects at these levels.[2] (Typically, voluntary standards have been set at levels that are achievable or that had already been achieved, rather than at levels that protect workers' health.)[3] In addition, while governmental agencies have calculated much lower "minimal risk levels" for the public for noncancerous risks—levels below which adverse effects are unlikely to occur—it is not clear that these new levels have been translated into public health standards and are sufficient to prevent cancers.

Third, to create a more prudent approach to toxic substances, we would need better guidance, and perhaps updated laws, for testing and licensing obvious air, water, and soil *pollutants* before there are substantial exposures. And safety standards for known toxicants should probably be reviewed to determine if they provide sufficient protection. Some pollutants can contaminate adults and children, pose risks, and cause harm. Businesses that create pollutants should take responsibility not only for risks from products but also for risks from major pollutants that arise from product creation. There are several reasons why businesses should take on these responsibilities.

211

For more than one hundred years the American Smelting and Refining Company's Tacoma smelter polluted a thousand-square mile area with the heavy metals lead, cadmium, and arsenic. All are quite toxic, leading to a number of disorders, including mental retardation, loss of IQ, and antisocial behavior (in the case of lead); lung and prostate cancer, kidney damage, emphysema, and bone disease (in the case of cadmium); and lung and other cancers (in the case of arsenic). Lead and arsenic can cause adverse effects from in utero and immediate postnatal exposures; cadmium can cause lung, kidney, bone, and cardiovascular disorders, as well as some cancers.[4] It has also exhibited adverse developmental effects in animals. Air pollution from diesel exhaust can contribute to these adverse health effects. Because pollutants can pose such risks, tests should be conducted on them and reviewed before the pollution is permitted to continue, if it can be safely emitted.

Commercial enterprises are more familiar with their own industrial processes than other organizations or governmental agencies are. Consequently, these businesses should understand what effluents are likely to emerge as waste products in the air, liquid waste streams, or solid waste products. For instance, governmental agencies are just beginning to learn that liquid or solid wastes from coal-fired power plants have a number of toxic pollutants in them.[5] Companies could reasonably have known this, tested them for toxicity, and taken steps to control this waste, if they had been legally required to do so. There are few existing incentives now for companies to take such steps; the law needs to spur this motivation. The EPA in the Obama administration will decide soon whether to treat this waste as a hazardous substance. Rather than wait for a governmental agency to identify the toxicity of "obvious" effluents, a more health-protective strategy would place a legally binding burden on those who are producing the pollutants to identify the composition of their effluents and test them for toxicity, unless they are known to be nontoxic. If studies have revealed adverse effects from such compounds, then public health agencies should ensure that health protections are in accordance with these studies. In addition, as we will consider below, pollutants and accompanying risks are negative externalities imposed on others.

For another example, utilities using coal-fired power plants could and should have been aware that there were small amounts of mercury (as well as larger amounts of other toxicants) coming from their vents. Moreover, these utilities could have determined exposure patterns likely to result from the height of their vents and prevailing wind directions, in order to estimate where pollutants would fall to ground or into water and what could occur as a result. The U.S. Environmental Protection Agency has estimated that about "75 tons of mercury were found in the coal delivered to power plants each year and about two thirds of this mercury was emitted to the air, resulting in about 50 tons [100,000 pounds] being emitted annually."[6] Existing scrubbers screened out twenty-five tons. Discharged mercury can be transported long distances and be deposited on land, can be precipitated out of the air by rain or snow, and can enter lakes, streams, and the ocean via runoff. Bacteria convert the mercury into methylmercury (MeHg), which is absorbed into the tissues of fish and other animals. From there, MeHg tends to concentrate in and migrate up the food chain, bioaccumulating as it does so.[7] Companies should anticipate when industrial facilities would discharge persistent substances, such as metals (lead, cadmium, mercury, arsenic) or some persistent organic pollutants (dioxins, perfluorinated compounds, PCBs, or brominated fire retardants).

Risk assessments are usually conducted after toxicants have affected a geographic area, but they could just as well be carried out in advance of exposures, in order to protect the public health. For instance, the EPA had to conduct a risk assessment around the Tacoma smelter to determine after the population was exposed what concentrations of heavy metals had been left on land and contaminated populations and what risks these concentrations posed.[8] The EPA needed this postdeposition risk assessment for hazard assessment and environmental cleanup. Legislatures, aware of such problems, should require risk assessments before pollution results from new facilities, to determine how populations would be affected, whether the company would need to reduce effluents, and whether improved public health protections would be needed.

Moreover, just because companies have existed for some time does not mean they impose no risks on the public. The Tacoma smelter had been in business for a century, and it was likely contributing to diseases during this period. Sometimes new laws grandfather in existing products or, in this case, an industrial facility. However, if legislators are serious about health protections, they should require companies to identify toxicants in their effluents, to test major ones, and to assess their risks. Ultimately, public health officials acting under improved legislation should update health standards and require modification of individual facilities if need be. Of course, companies would need time to conduct risk assessments and install needed pollution controls. Controls would affect companies' investment and operations, but in principle there appears to be no good

213

public health reason to exempt existing firms from taking actions to protect the public. These actions are of even greater concern when known toxicants are emitted.

Fourth, appropriate and successful on-going testing should reduce toxic contamination; this is an obvious consequence. Testing products before they enter commerce would alert a company and a reviewing agency to a substance's hazards and risks. Faced with such data, either company action or a public health action to preclude the product from commercialization would mean one less toxic substance would contaminate the public. In addition, there is a substantial backlog of industrial chemicals that simply have not been reviewed for toxicity. These will need to be prioritized and a number of them tested and reviewed for toxicity.

More accurate on-going investigation of existing products would reduce live risks to developing children and remove the most worrisome features of unreasonably offensive invasions of our bodily integrity, two underlying rationales for premarket testing. Good-faith testing of a company's products would also eliminate the idea that the business was reckless toward the public.[9] (There may also be ways to reduce risks from an industrial chemical so that it could meet health standards and be commercialized. If so, then the product would still be safer than it would have been without the testing.)

Fifth and a related point, because so few existing products have been tested for toxicity to date and even fewer have been reviewed to prevent risks to developing children, at present there are likely to be numerous "misclassifications" of substances.[10] With improved and on-going testing, the number of these misclassifications would be reduced over time.

Misclassifications result when a substance is regarded as toxic when it is not (a false positive) or as nontoxic when it is toxic (a false negative). A substance can also be regarded as more toxic than it is (overregulated) or regarded as less toxic than it is (underregulated).[11] The National Research Council (NRC), economists, and numerous others regard the economic and social costs of false negatives to be ten to twenty times more serious than those of false positives and overregulation.[12] That is, not knowing about toxic effects or missing them in tests (or underestimating them) is a much more serious social mistake than concluding substances are toxic when they are not. Enhanced testing of products would substantially reduce false negatives.

Improved testing for developmental effects will likely reveal an even greater number of substances that are toxic to developing children, even though these substances are currently regarded as nontoxic or less toxic than they are (and are thus underregulated).

Sixth, several environmental health laws seek to protect susceptible subpopulations; these goals would be better achieved with improved testing. The U.S. Congress has used laudable language in several postmarket environmental health statutes that seek to protect susceptible subpopulations. These quite worthy goals cannot be well achieved under postmarket laws, because susceptible populations only receive protections after they have been exposed for some substantial period of time. Improved testing of new and existing products would better serve to protect vulnerable subpopulations.

The legislative history of the Clean Air Act suggests various protections for sensitive subpopulations. That history focuses attention on children, the elderly, and people with preexisting or special conditions, such as chronic obstructive pulmonary disease, heart disease, or pregnancy.[13] The Safe Drinking Water Act directs the EPA to determine "maximum contaminant level goals" for each contaminant that "may have *any adverse effect* on the health of persons." These goals should be "set at a level at which . . . no known or anticipated adverse effects on the health of persons occur and which allow an adequate margin of safety."[14]

The Occupational Safety and Health Act implicitly mandates protection for susceptible worker subpopulations. It requires health standards which "most adequately [assure] . . . that *no employee* will suffer material impairment of health or functional capacity," even if exposed for a working lifetime.[15] The standards aim to protect any worker during a forty-five-year working lifetime. (However, this protective language can be substantially attenuated by so-called feasibility conditions.)

These laws in principle have a greater commitment than do some other laws to protecting people during sensitive life stages or vulnerable periods. These substantive goals would be much better achieved if products and pollutants were better tested.

Finally, as this book goes to press the U.S. Congress is beginning to consider major amendments to the Toxic Substances Control Act (TSCA), introduced by Frank R. Lautenberg, U.S. Senator from New Jersey. This is known as "The Safe Chemicals Act of 2010."[16] These amendments seek to correct many of the major deficiencies of U.S. chemical policy and TSCA in particular that have been discussed in this work. Some of its major features include the following.

At all times the manufacturers or processors of chemical products bear the legal burden of proving that exposures to chemical substances present a "reasonable certainty of no harm to human health."[17] In determining whether or not this standard has been satisfied, the EPA must find that exposures to substances must not present "more than a negligible risk of any adverse effect" to people's health, taking into account aggregate and cumulative exposures of the general population or any vulnerable populations to the substances. "Adverse effects" include a variety of biochemical or anatomic changes, functional impairments, or pathological lesions, or their precursors that among other things can cause "irreversible change in the homeostasis of an organism," increase its susceptibility to other chemical or biological stressors or reduce its ability to respond to additional health challenges. Susceptible subpopulations are quite similar to those characterized in this book.[18]

New substances must be supported by data sufficient to meet the safety standard. For existing compounds the EPA within 18 months of passage of this law must compile a list of at least 300 chemicals substances for which safer determinations must be made and this list must be updated periodically so that there will always be at least 300 substances on it, until all substances and mixtures in commerce have had a safety standard assessed. The manufacturer has the legal duty to provide data so that the EPA can determine whether new or existing substances meet the safety standard. If a manufacturer or processor violates these provisions, the Administrator of the EPA may by order prohibit the manufacture, processing, or distribution of the chemical in commerce or any mixture or article containing it. The manufacturer has a duty to update toxicity data on any substance every three years at a minimum and whenever it becomes aware of new toxicity information on its product.[19]

There are numerous other provisions, but this brief account shows that the law covers both new and existing substances, uses aggregate and cumulative exposures to compounds as well as their effects on susceptible subpopulations to ensure that the they present a reasonable certainty of no harm to the public's health. All of these factors are important in framing more sensible legal and regulatory safeguards for industrial chemicals and pesticides. Proposed legislation is something of a moving target because it may well change over time. Nonetheless, this appears to be a strong proposed law that will go a long way toward correcting many of the problems discussed throughout this book.

Additional Benefits from On-Going Testing

Beyond some of the above consequences of improved toxicity testing for new and existing commercial chemicals, there are some more subtle consequences that should be anticipated.

Reduction in Postmarket Risk Assessments

Long, slow risk assessments that currently plague and clog regulatory processes under post-market laws would be much less needed. At present public health agencies identify potential adverse health effects from substances then engage in detailed and typically lengthy assessments of risks from exposures. The agencies then employ the scientific community to study health effects in animal or human studies, conduct exposure assessments or biomonitoring, and assimilate the studies to estimate risks to the public or workforce. To the extent that the testing of products before entering the market is successful, such time-consuming studies would be a thing of the past. Of course, it is unlikely that such assessments would disappear because some toxicants would be missed or some of their adverse effects not fully understood initially. Thus, if products were later found to cause diseases or dysfunction, postmarket scientific research needs would remain. And, review of exiting compounds would still require postmarket risk assessment.

Plenty of scientific studies would still be conducted. Most of them would just be carried out at an earlier stage in the life of a product—before it entered the market for new products and beginning in the present for high priority substances currently in the market. Many of the same scientific studies would be needed, e.g., various short-term tests, animal testing to determine carcinogenic, reproductive, immunological and developmental toxicants (until these were replaced by quicker and sufficiently accurate short term tests), and so on. Prospective exposure assessments also would be needed, as well as an assessment of hazards and overall estimation of risks.

It may be difficult to for scientists and the rest of us to understand the paradigm shift implied by the developmental origins of dysfunction and disease along with premarket and other testing aimed at preventing risks and harms before contamination. Postmarket risk assessment is deeply committed to the idea of enlisting scientists to identify risks (or harms) *after* they have occurred, and then finding strategies to reduce or eliminate them. Such assessments are far to late to protect children. And under

217

postmarket laws they clog public health agency processes and slow improvements in health protections. We should no longer think in such terms. Instead, we should ask what tests and procedures should be adopted so that diseases do not result from contamination in utero or perinatally? We should seek to avoid such risks in the first place.

Human epidemiological studies likely would not be used as often. They depend upon humans being harmed and then revealing illnesses or injuries by statistical means. To the extent that humans were no longer harmed, epidemiologists would have fewer studies to conduct. Nonetheless, even the best pre-market and early testing procedures are unlikely to detect all risks from chemical exposures, so epidemiologists will likely continue to have work. Recall that both drug and pesticide laws require pre-market testing, but they sometimes miss adverse health effects that must later be detected by epidemiology.

Improved Comparative Toxicity Data

Better toxicity analysis of new and existing products would provide comparative toxicity data about a company's compounds or their constituents so that safer alternatives could be found. Recently, Lynn Goldman, an academic advisor to the Food and Drug Administration and former administrator in the U.S. EPA, pointed out that parents face a trilemma when choosing dental fillings for children's cavities. If they chose an amalgam, they will expose their children to a compound that contains mercury, a well-known neurotoxicant. If parents choose a composite filling, they will expose their children to a compound that contains bisphenol A, a clear endocrine disrupter in experimental studies. Parents could choose some alternative to these two, but the toxicity of these alternatives is unknown. Goldman correctly identifies this predicament.

However, the problem lies in scientific, public, and public health agency ignorance about the toxicity of alternate substances. Whether or not alternative compounds are toxic should be known—and this can be known with improved testing and review of substances. It currently is not. Better social policy could be made on something as simple as dental fillings if there were good toxicity data on all the plausible alternatives.

In addition, responsible companies have self-interested reasons for an improved understanding of the toxicity of products' constituents. Downstream users would likely place great value on such information, so they could avoid toxicants in their products. When downstream users of products do not have toxicity data on chemical constituents, they cannot make informed choices between substances and markets do not function efficiently.[20]

Moreover, some states legally require businesses to comply with toxic use reduction programs. Early and on-going testing would better support these efforts.

For instance, Massachusetts mandates "companies that use large quantities of specific toxic chemicals to evaluate and plan for pollution prevention opportunities, implement them if practical, and measure and report their results on an annual basis. They must also evaluate their efforts and update their toxics use reduction plans every other year."[21] In order to substitute less toxic products for more toxic ones, a company must understand the toxicity of its favored product or constituents for a particular use, as well as data about the toxicity of reasonable alternatives.

This is especially important for those who purchase constituents for their products from other companies. By conducting an alternatives assessment of chemicals for particular uses, companies can find less toxic alternatives, thus creating a less risky product line and lessening toxic exposures to the public and its workforce. An analysis of one of the phthalates, di(2-ethylhexyl)phthalate (DEHP) illustrates this point.[22] DEHP is a synthetic organic chemical used as an agent in plastics to increase their flexibility. It is utilized as a component of resilient flooring materials, as a constituent of vinyl wall coverings, and as a plasticizing agent in medical devices for neonatal care. The U.S. Environmental Protection Agency has classified DEHP as a probable carcinogen. It is also known to be a reproductive toxicant in animals, and recent results suggest that it is associated with adverse reproductive tract outcomes in humans (see Chapter 4). California has listed it as a reproductive/developmental toxicant under Proposition 65.

The Toxic Use Reduction Institute (TURI) reviewed alternatives to DEHP, assessing them on several dimensions. In assessing health effects for resilient flooring, TURI found that in general some available alternatives— cork, natural linoleum, and polyolefin—were favored over DEHP/PVC flooring. For environmental effects and costs, the other products were quite similar to DEHP. For plasticizing compounds in plastic sheets and tubing applications in medical care, TURI found that four alternatives were superior on health grounds, with three of them less likely than DEHP to migrate out of the plastic tubing into lipid-soluble solutions. One alternative product was similar to DEHP and likely to exhibit such migrations. Several alternatives were superior to DEHP on technical criteria, such as flexibility when cold, clarity, ease of sterilization, and "plasticizer loss during manufacture and use." The costs of two alternative plasticizing agents were similar to those of DEHP, while two others were more expensive.[23]

This particular alternatives analysis could be conducted because the team of researchers had available data on health effects, technical criteria, environmental effects, and costs. Such data is dependent on previous studies having been conducted on the complex agents as well as on the individual materials that constitute them. Better toxicity data for chemical constituents is a prerequisite for such analyses.

Some might protest that alternatives to existing products will be more expensive; the TURI analysis shows that this is not necessarily the case. Several products that could replace DEHP are less expensive for the intended purposes. While no one product may substitute for all possible uses of DEHP, there appears to be one or more products that would substitute for each use, some of which are less toxic and some of which are the same or lower in cost than phthalates while meeting the same technical requirements. While more research may be needed on some of these issues, the report is useful in showing that there are substitutable products that meet technical demands, that are less toxic, and that are available at the same or lower cost.

Greater Justice

Enhanced testing would better secure justice for citizens contaminated by industrial chemicals, compared with the status quo. Diseases that result from poorly tested products can be unjust on several grounds.

We might think that diseases are adverse events that "happen" to us, not matters of justice or injustice. There is something to this, because we do not individually or collectively have control over many diseases. However, even naturally caused illnesses, such as polio or various chronic diseases that induce daily limitations, can raise questions of injustice, depending upon social and institutional responses to the causes or to the treatment of the diseases.

Naturally caused diseases, such as Parkinson's disease, autism or cancer, arbitrarily hinder people's opportunities. By itself this fact does not constitute injustice; that results from institutional responses to the disease. If the diseases could easily have been prevented, or if people cannot access services to ameliorate their illnesses, these are institutional failures. In turn they can pose problems of fair opportunity. Between two people with roughly equal abilities and motivations, opportunities are unfair when one or the other person fails to have approximately equal chances to obtain the same social goals because of how their lives are influenced by alterable institutions. Put another way, opportunities between individuals are fair when they have approximately equal motivation, talents, and abilities for obtaining various life aims (such as becoming a physician, lawyer, symphony conductor, and so on) and when the individuals' chances of success are approximately equal. For example, in earlier decades a man and a woman of similar talents and motivation had quite different opportunities in life because of educational systems, social attitudes and other prejudice barriers. In addition, the community in which one is born and over which one has no choice or control can adversely influence one's prospects because of the community's educational resources or the lack of them. The principle of fair opportunity is central to John Rawls's and other major theories of justice.[24]

Norman Daniels extended this idea to a just health care.[25] Naturally occurring illnesses and dysfunctions constitute arbitrary barriers to opportunity. Such maladies undermine our ability to live normal, fully functioning lives over a life span and to exercise opportunities appropriate to our talents and motivation. If people do not have access to a good health care system that prevents diseases in the first place (via vaccinations, preventive medicine, environmental and workplace health protections, and good prenatal and pediatric care, for example), or if they do not have access to good health care to recover from or ameliorate diseases or dysfunction (to the extent recovery can be achieved), they are treated unjustly when they get sick. For the purpose of our discussion, I will focus on preventive issues. When a community does not design and implement its legal system to prevent illnesses that it reasonably could have, it treats those who contract diseases or dysfunction unjustly.

For instance, the mission statement of Eunice Kennedy Shriver National Institute of Child Health and Human Development expresses a reasonable feature of just legal institutions. Among other goals, the institute seeks "to ensure that every person is born healthy . . . and that all children have the chance to achieve their full potential for healthy and productive lives, free from disease or disability."[26] If children are not born healthy because of diseases or dysfunctions (or predispositions to them) that social or legal institutions could have reasonably prevented, this is a failure of justice. No one need have caused such adverse effects,

221

but failure to provide various preventive services or access to health care creates an injustice.

The discussion in the preceding four paragraphs assumes that the actions of no one initiated disease. However, when institutions permit one group of citizens recklessly to create and distribute toxic substances that contaminate others, causing illnesses and premature deaths, this introduces an additional injustice. The institution permits one group of people with its products and contaminants unjustly to cause other individuals to contract diseases and dysfunctions that would not have occurred without the toxic contaminations. This would be an instance of distributive injustice, a failure of justice between people in the community.

Better detection of hazardous substances and elimination of them before they pose risks or before they have posed too many risks and harms (if possible) helps to address this second aspect of injustice. DES-caused vaginal or breast cancer, TCE- or pesticide-hastened parkinsonism, lead-lowered IQ, lead-induced behavioral problems, or PCB-caused neurological problems can all substantially interfere with fair opportunities that people would have had if they had not been contaminated with toxicants. Postmarket laws, thus, sanction the creators of hazardous products to cause injustices to those who contract disease or dysfunction as a consequence. These injustices are quite preventable with different legal structures.

Do There Need to Be Major Institutional Changes?

The consequences and needed changes that would come with improved on-going testing and review of commercial chemicals seem apparent. However, do we need to go down the path of significant legal change? Might many of the same results be achieved within existing legal structures? Could one avoid risks from chemicals via self-help? Or would laws notifying the public of toxic exposures do the job just as well? By considering these issues, we begin to see why it is so difficult to avoid contamination by potentially toxic substances, why some obvious alternatives will function poorly, and why recommended alternatives to existing institutional arrangements are a substantial improvement.

Self-Help for Reducing Exposures and Contamination

One alternative to wider, more accurate, and timelier testing of chemicals might be to follow individual self-help strategies within existing laws. En-

222

vironmental groups, other organizations, and news outlets could publicize the toxic substances in our midst so that we could choose to avoid them. To some extent, this is already occurring and is laudable. However, a moment's reflection will reveal the limitations of this suggestion, including the need to take a much more systemic approach.

As individuals, we might avoid some of the obvious geographic and occupational circumstances that expose us to toxicants, thus somewhat reducing our exposures and body burdens. We could avoid working as pesticide applicators or as farmers or farm workers. We could avoid working in factories that produce plastics, batteries, petrochemicals, rubber, upholstery, and so on. In order to avoid toxic contamination under present conditions; we should not work in industrial facilities that utilize toxic substances, unless they are used as intermediates within a tightly closed system and there is no exposure of consequence (a scenario that may not be realistic). We could avoid living next to industrial facilities and railroad yards that emit toxic chemicals. These examples are merely suggestive; the reader will think of others.

Philip and Alice Shabecoff's *Poisoned Profits* provides an appendix that extends further into people's lives the idea of making personal choices in order to better protect developing children. Environmental groups recommend similar approaches.[27] For couples seeking to conceive a child, the Shabecoffs recommend eating organic foods. These couples should also eat "low on the food chain—grains, beans, vegetables, and fruits—and [avoid] foods high in saturated fat. [They should] eat fish lowest in mercury." They should also avoid or reduce cosmetics and sunscreens with a variety of toxicants in them, and request phthalate-free intravenous tubes if a newborn needs hospitalization after birth.[28] Prospective parents should also avoid foods and exposures that would contaminate them with lipid-soluble toxicants such as dioxin-like substances, synthetic hormones, and so on.

In addition, women should avoid behaviors that involve ingesting obvious toxicants during pregnancy: smoking, drinking alcohol, and using various illicit drugs. Physicians already advise this. Women might also take some similar steps in anticipation of pregnancy to reduce their own contamination. Similar lifestyle choices may be important for men seeking to conceive children with their partners.

A prospective parent should also try to live in "relatively unpolluted" communities, substantially away from freeways, and consult the EPA's Toxics Release Inventory of chemical substances by locality to identify nearby chemical facilities. Within a home, these individuals can and should finish renovations before newborns are brought home, live elsewhere during renovations if they can afford it, avoid using pesticides, and have clothes "wet-cleaned" with water instead of using solvent-based dry cleaning. It is best if a prospective parent learns the ingredients in manufactured foods (if manufacturers provide such information), avoids "nonstick cookware coated with perfluorochemicals," and replaces it with stainless steel, cast-iron, Pyrex, or ceramic cookware.[29]

People wanting to reduce contamination can find databases that offer guidance for avoiding brominated flame retardants, phthalates, and PFCs in commercial products. Various environmental and public health groups also make generic recommendations about cosmetics and personal care products, cleaning and laundry products, antimicrobial ingredients, air fresheners, and toys.[30]

Limitations to Self-Help

All this is good advice for individuals who are aware of the risks, who have the skills to be employable in many locales, and who have the resources to implement the suggestions. These recommendations are also attractive because Americans may believe that individually we can do much to control our destiny across a variety of dimensions. For some areas of our lives we can, but for others we cannot.

Some of the advice in the previous section would clearly reduce exposures to chemical products that have short half-lives in our bodies and that require continuous exposures to maintain contamination. And exposures to high concentrations of local toxicants are worse than exposures originating farther away. Avoiding certain kinds of occupations will help as well. However, for the most part, substances are sufficiently pervasive and mammalian biology is sufficiently permeable to molecules that much of what we do will have limited success in reducing contamination.

Individual choices will not make occupations or the wider environments safer. My choice to avoid pesticides on farms does not improve the safety of pesticides or farms. I can avoid some toxic exposures, and this likely would reduce my own contamination somewhat, compared with what it would be if I worked in such industries, but my avoidance has little or no broader effect. From a social point of view, the aim should be to create safer occupations and products.

Of course, if enough individuals avoided occupational settings with toxic exposures, this might drive up employment costs in those jobs, and that might motivate firms to improve the safety of the workplace. How-ever, this is probably an unrealistic expectation. Those occupations will still need employees, and adults will need employment. Moreover, even if many people avoided such employment, there likely would be some who would work under risky conditions because they had few choices or were at-tracted by hazard pay, if it exists and is close to adequate. Consequently—to the extent that there is a need for such jobs, a willing workforce, and risky workplaces—people in them will continue to be contaminated and to be put at risk. Even if other individuals can avoid toxic workplaces, if the toxicants are not reduced and if substances continue to travel the globe, as persistent compounds do, exposures and some degree of contamination of the broader public will not cease.

Many of the workplaces described in this chapter are those that have obvious toxic exposures: farms, petrochemical facilities, rubber plants, battery factories, and so on. This might suggest that other workplaces are much different and safer. They are different but not necessarily free of toxicants. In comparison with a farm or factory, "clean" workplaces, such as white-collar institutions, universities, lawyers' and doctors' of-fices, and stores, appear much less contaminated by toxicants. They likely are less contaminated for some substances. Nonetheless, people will be exposed to halogenated fire retardants; contaminants from plas-tics in computers and other equipment; and toxic materials in carpets, chair cushions, drapes, and janitorial products. Homes exhibit similar exposures. Moreover, since offices are typically "closed" environments and poorly ventilated, they can still generate exposures with risks, some of them perhaps concentrated.

As discussed in Chapter 2, a recent study reinforces the limitations of such a self-help strategy and calls attention to some of the contaminants in a home. A sample of fifty-two homes along the Arizona-Mexico bor-der contained measurable amounts of more than four hundred individual chemical substances.[31] Many were persistent organic chemicals, such as PCBs, long-banned pesticides—DDT and its metabolite DDE, hep-tachlor, and dieldrin—as well as many current-use pesticides, such as chlorpyrifos and Diazinon. In general, the researchers found many classes of compounds, most with known toxicants. Another 120 com-pounds "were detected but not identified with any certainty" by the study. In addition, the authors note, "the mixture of airborne chemicals present indoors is far more complex than previously demonstrated," with a "high potential for human exposure to pesticides."[32]

The chemical "soup" in such environments typically has not been tested, is not understood, and may have adverse health consequences, even if some individual compounds are not known to pose risks at low concentrations. We should recall, for instance, the concern about the additive effects of natural and synthetic estrogens, of dioxin-like substances, and of organophosphate pesticides, as well as generalized additive effects (see Chapter 4).

Indeed, the individual concentrations of many of these substances are likely quite small, but the study reveals hundreds of synthetic industrial compounds in an average home. Perhaps not all homes elsewhere will reveal such substances, but the authors provide no reasons to believe that the sample of homes was unrepresentative.

There are other substantial limitations to self-help. For one thing, it does nothing to generate greater information about the toxicity of products. And we already lack considerable data for many chemical products. Self-help is based upon current widespread and available scientific information. If there are no data about particular substances, a person has no reason to avoid them or locations where they might be.

In current circumstances, self-help is also based on studies conducted some time ago. Because scientific studies take time to conduct, exposures will have existed for some period of time before the public is aware of them. The public likely knows little or nothing about some current exposures that have not been or that are just now being studied. If these reveal risks, the findings will take time to penetrate the public domain so that individuals can take preventive actions that might be open to them.

Circumventing toxic products or locations will substantially lag the presence of risks as long as substances and pollutants are tested only well after the public is contaminated. Perforce, when the discovery of risks depends upon postmarket scientific studies being conducted, eluding toxicants will also lag discovery of their risks. The success of avoidance would also depend upon how long it takes for scientific identification of risks to permeate public consciousness so that people can act upon it.

Finally, there is a further difficulty that dashes the hopes of individual self-help. We should recall that we can be exposed by pesticides and other chemicals by ingestion, inhalation, or skin absorption, or all three, with the body acting as something like a funnel. And certain substances will add to the toxic effects of others.[33] Self-help is likely to do little to avoid or eliminate all routes of contamination or additive effects.

Thoughtful choices concerning workplaces, purchases of consumer products, as well as the decorating and constructing of a home—if one has the knowledge and finances to avoid toxicants—can reduce some exposures via some pathways and will likely make some difference. Individualistic strategies will likely succeed with substances, such as BPA or phthalates, that have short half-lives in the body, providing one can reduce contact with these substances. These strategies will do little to reduce exposures to persistent chemicals. (Of course, even if these strategies were better for compounds with transient half-lives, one would still not know which exposures would trigger risks.) The Arizona study concerning homes and possible routes of contamination greatly tempers confidence in the effectiveness of self-help. At the end of the day, much contamination simply cannot be avoided or cannot be greatly influenced by what a person chooses to do in his or her own life.

There are, however, some signs that once the dumping of persistent substances is ended, environmental and wildlife contamination will be reduced over time. After more than forty years, PCB levels in wildlife in some formerly contaminated areas of the Arctic have fallen dramatically. Blood lead levels in children have dropped substantially since lead additives in gasoline were banned.[34] Compared with self-help, however, this outcome takes a more systemic approach of banning the products and stopping concentrated exposures.

A Federal Warning Law Analogous to Proposition 65

An alternative to the status quo, but one short of more extensive testing of new and existing products, would be a federal, nationwide warning law analogous to Proposition 65 but expanded to include other toxicants. This law would require greater governmental involvement than mere publicity from news outlets and environmental groups. Would this new law sufficiently address toxicants, so there would be no need for much more extensive testing?

Recall that the law in California requires the governor to list carcinogenic or reproductive toxicants. Once these substances are listed, any company that exposes the public must post a warning that the public is exposed; take steps to show that the product does not pose risks; or reduce exposures or reformulate the product (a method of reducing exposures). If a business does none of these things, it is subject to daily fines, which can become quite high over time.

Would a more general federalized analogue to Proposition 65 ensure sufficient protections from carcinogenic, reproductive, developmental, and other toxicants? Proposition 65 certainly has its virtues. Listed hazardous products with warnings receive considerable adverse public attention. People are put on notice when they are in the presence of a toxicant, because particular companies or their products in specific locations are identified as responsible for the toxic exposure. Companies, instead of governmental agencies, have the burden of proof to take actions to reduce exposures or to remove the product from the market.

The law has another advantage compared with self-help. Proposition 65 provides a scientific and governmental imprimatur that a substance is a toxic hazard. This endorsement provides better assurance than do newspaper reports about risky products or Web site postings, whose legitimacy may sometimes be in doubt. Moreover, the endorsement is public information and is not left to happenstance.

Unfortunately, Proposition 65 also has some limitations. It too is a postmarket law. A substance subject to the law is already in commerce and posing risks before it is identified by scientific research and listed by a peer review committee. The California Environmental Protection Agency still has the legal burden to make a postmarket legal and scientific showing that substances are toxic hazards before they can be listed (although some are listed if other authoritative agencies have listed them). Of course, listing toxicants under Proposition 65 is easier than setting public health standards under quite onerous postmarket ambient exposure laws or somewhat less onerous technology-based laws. And substances can be listed on the basis of animal studies.

In addition, the hurdles that must be overcome to list a substance tend to vary with the scientific advisory panels—some are more protective of the public's health, others less so. The California EPA is cautious in seeking to list substances via its scientific advisory panel, because adverse legal consequences can occur if a proposed substance is not listed. Listing still takes time, and people are put at risk until it is implemented. On the one hand, bisphenol A and PBDEs have not yet been listed under Proposition 65, even though there is considerable evidence of their toxic effects in animal studies. On the other hand, several phthalates have already been listed as reproductive toxicants in California, while the U.S. Food and Drug Administration continues to study them.

Finally, just because a substance is subject to Proposition 65, it does not *guarantee* that the toxic effects will be removed; the only legal requirement is that a company must post a clear warning about the presence of the toxicant. However, regulation by warnings may or may not be effective in protecting the public from risks. The toxicity of the product remains if a company does not reduce exposures. People may or may not heed the warning. More seriously, even if people respond to a warning, they may or may not avoid contamination. The actual Proposition 65 applies only to people or companies that do "business" in California. Other sources of carcinogenic or reproductive toxicants are not governed by the law. Toxicants in the environment could still contaminate the citizenry. A national law resembling Proposition 65 might well be more effective on this score, but persistent substances will continue to be present in the environment for some time even under such a law, because the source of contamination is not in products sold by or from premises of businesses.

Thus I conclude that within existing laws self-help will do little to dramatically reduce the public's exposure to and contamination by toxic substances. Moreover, even if there were a national analogue to Proposition 65, it would not address well the underlying contamination from toxic products and pollutants, but would reduce them somewhat.

Technology-based Laws

Many of the same concerns apply as well to technology-based laws, under which public health agencies would list toxicants and specify which industries would need to adopt technologies to reduce exposures to the lowest achievable levels. Compared with warning laws these procedures are more time consuming and invite much more contentious discussions. Thus reducing toxicants takes time, perhaps even considerable time. Technology-based laws have one advantage over warning laws: technology-based laws actually reduce exposures; reductions are not left to the happenstance of company choice or consumer efforts to heed warnings.

However, technology-based laws remain postmarket. And given the time it can take under some of these laws to identify toxicants and to reduce exposures, considerable toxic contamination of the public would occur and continue for some considerable period of time.

Muddling Through with Miscellaneous State and Local Laws

Another alternative to more extensive testing would continue a process already begun: permitting states and municipalities to reduce the presence of toxic substances in products or forbid their sales. Some states and municipalities have passed laws concerning bisphenol A, phthalates, brominated fire retardants, and perfluorinated compounds.

Since federal laws are so poorly protective and since federal agencies have temporized for eight years or more, other jurisdictions have responded. States or in some instances cities have passed or are in the process of passing laws to disallow products with such substances within their jurisdictions. This response suggests substantial unease in the body politic about such products. The concern is great enough that voters in some states have convinced majorities of legislatures as well as governors that they should pass laws aimed directly at one or more of the toxicants just mentioned. Legislative action is typically not easily carried out, yet the concerns are so great that some legislatures have been moved to act in this dramatic manner. A related strategy is for Congress to enact laws aimed at particular substances, such as lead in toys or cadmium in inexpensive jewelry, for instance.

Patchwork or substance-by-substance laws are poor strategies to address these issues, but citizens appear to have felt so strongly about the toxicity of products in their midst that these laws have been enacted. Different parts of the country will have different public health standards, however. This will surely complicate the business of firms that seek to sell products nationally. In addition, even if state and local jurisdictions reduce the presence of such products, this restriction will do little to reduce overall contamination from worrisome products. For substances that contaminate citizens as a result of continuous exposures, such actions would help. However, certain substances will remain in commerce and enter from other jurisdictions. Such restrictions will be least effective in reducing exposures to persistent bioaccumulating compounds that remain in and move through the environment, continuing to contaminate people for many months or for years.

Substance-by-substance laws respond to the public's concern. Legislatures, however, may not be well suited to address these issues. It would be much better to put the legislative effort into passing a national law or laws that addressed much more generally toxicants in our midst. Laws passed only after there has been considerable concern mean that people have already been exposed for some period of time and possibly put at risk as a result.

Other Shortcomings of the Status Quo

Hidden social and other costs permitted under existing laws are likely underappreciated. Considering both the alternative world sketched in the introduction to this chapter and some of the consequent changes needed to achieve this world puts in relief some of the "negative externalities" of failing to test new and existing products.

Externalities

Externalities are characterized as "spillover effects," either beneficial or harmful, that are not fully reflected in the market price of a good.[35] A positive externality might be a privately funded lighthouse that guides all ships into a harbor, but some or many of the ship owners may not have paid to construct it—they free ride on the benefits. A negative externality occurs, "when an economic actor produces an economic cost but does not fully pay that cost."[36] Examples might include economic actors that discharge air pollution, harming the air that others breathe, or those that contaminate a river, decreasing downstream water quality and value. Externalities can also be either symbolic or quite concrete. They can range from physical changes to changes that are "essentially psychological and subjective."[37] A negative psychological externality might be raucous rock music in the middle of the night in a quiet neighborhood, while a positive psychological externality might be the quite wafting of Chopin's nocturnes at bedtime through the same neighborhood.

Economists agree that negative externalities can have several undesirable consequences. Ordinarily, a product or activity with an externality is produced in too great a quantity. The reason for this is that the price of a good or activity with a negative externality is too low; thus more of it is sold than otherwise would be. This overproduction occurs because the full social costs of that product are not reflected in market price. For instance, because asbestos-based products caused disease and death among workers that produced and used them, at one time they did not reflect the costs of preventing the illnesses and avoiding premature death. Because disease prevention was not part of the final product costs, firms could sell the product for less and sell more of it.

Asbestos-caused illness, poor health, and deaths of employees subsidized consumers of asbestos-based products. Without putting too fine a point on it, purchasers saved money and asbestos companies made higher profits because many employees were sickened and died, when these illnesses

231

and deaths could have been prevented. The solution is, as Marc Franklin puts it, that "The cost of the product should bear the blood of the workingman."[38] In addition, more-efficient ways of producing products may have been discouraged within a company because there were no incentives to find other ways to lessen production costs: the workforce, not the company, was already bearing the costs of disease and death.[39]

Externalities also distort economic markets for goods and services. Economists are typically concerned with this distortion and whether markets are functioning efficiently and properly. Economists may or may not be as centrally concerned with justice, responsibility, or distributive issues that can arise from externalities (issues that I will discuss in more detail below).

A comparison of the status quo with our imagined, safer state of affairs reveals myriad negative externalities in our current world, because most products and likely no pollutants are routinely tested for their risks and harms before contaminations occur. Obvious externalities can result from in utero exposures. An embryo may die before birth, children may be born with malformations, for example, phocomelia; they may have growth retardation or various functional impairments.[40]

Malformations or functional deficits impose noticeable external costs on the children and on their parents. Physicians and hospitals may have subsidized treatment or long-term care of these offspring. Governmental agencies might have had to provide children with care, rehabilitation, special education, medical devices, treatment, or more general support. The children may need special medical care and education to compensate for functional deficits so that they can more nearly approximate a normal life. Dealing with these dysfunctions can cost considerable money and burden families, divert parents' attention from other children or from spouses, and in general distort what would otherwise be a more normal family life.

DES obviously imposed externalities on the young women who contracted vaginal or cervical cancer or suffered early onset of breast cancer, and it likely imposed externalities on their parents—costs, needed doctors' appointments, lost work time, along with fear and anxiety. Certain externalities may be imposed on the sons and daughters of the DES daughters. Background worry because of the history of DES may be present in these families as well.

If experimental studies reasonably presage some effects in people, there might well be externalities yet to be discovered. For instance, in utero exposures to the pesticides vinclozolin and dioxin in animal studies show that these substances can produce long-term transgenerational effects on the male germ line as well as numerous cancers. If similar effects were to occur in men exposed to these pesticides, these would be quite serious externalities indeed. Recall that PBDEs and perfluorinated compounds (PFCs) have caused additive effects in experimental studies. Recall as well that PBDEs have additive effects with PCBs. Since people are contaminated with all three, they may experience triply additive adverse neurological and other effects.

Lead has imposed uncounted millions of dollars in externalities on children whose IQ was diminished, on families who had increased costs from children's neurological deficits, and on schools that had more children in special education as a consequence. For instance, before lead in gasoline was reduced, the economic costs of diminished productivity caused by decreased IQ alone were estimated to be from $110 billion to $319 billion for each year's cohort of children. Today those costs may have dropped to "only" $43 billion in each birth cohort.[41] Lead also imposes external costs on communities and families when contaminated children tend to engage in antisocial activity, to commit crimes, and to be disruptive in school because they have lesser executive function controls over their behavior.

Other substances have negative externalities that affect children and communities. Valium, valproic acid, and other anticonvulsants to which developing children were exposed produced birth defects, resulting in costs to the children, their parents, hospitals, the educational system, social services, Medicare, and so on.

It is difficult to calculate the costs from chronic diseases, neurological dysfunctions, premature death, and medical treatment associated with various maladies. But these costs are likely substantial. Neurotoxicants, such as lead, mercury, and some pesticides, that contribute to autism, ADHD, and other conditions could have considerable long-term costs.

Philip Landrigan and others suggest some low-end estimates of environmentally induced diseases. They estimate these medical costs to be $2 billion for asthma, "$0.3 billion for childhood cancer, and $9.2 billion for neurobehavioral disorders." Other researchers estimate the costs of children's intelligence loss from methylmercury exposure each year to be $8.7 billion in lost productivity. About $1.3 billion of that may be due to mercury emissions from coal-fired power plants.[42]

The annual health care costs and loss of productivity are approximately

$55 billion for lead poisoning, asthma, cancer, and neurobehavioral disorders combined, with lead poisoning contributing about $43 billion. This total is less than the annual health care costs due to car accidents ($80.6 billion) and somewhat more than the annual health care costs of strokes ($51.5 billion). These costs are greater than military weapons research ($39 billion) and veterans benefits ($39 billion).[43]

More importantly for what follows, removing some of the high dollar costs of disease and lost productivity caused by toxic contamination might well pay for considerable testing. If testing substances before they entered commerce and testing those already in commerce would reduce a substantial portion of such diseases, the testing might well be worth it on economic grounds alone. And these grounds do not even take into account the injustice and the wrongness of diseases inflicted on people as well as misery, anxiety, and other adverse consequences.

Pollutants from industrial processes are long-recognized, classic externalities. Those that contaminate adults and children pose risks, and cause harm are "externalities doubled." They are not merely nuisances needing to be cleaned up or interfering with a decent life; they substantially affect people's lives and opportunities. These long-term effects increase their externality burden. For example, leaded gasoline created a nationwide public health problem from which we have largely emerged, in addition to any other externalities.[44] Moreover, lead likely was not needed in the first place, because ethanol and other additives were attractive alternatives.

Most of these costs we might consider as first-order externalities—they are direct and obvious adverse consequences of the product or pollutant whose costs are not fully incorporated into the product's price. There are others: second-order externalities that are less noticeable than first-order ones because we might not initially think of these less visible consequences as spillover effects. However, when we compare the imagined world with the actual world, it is clear that people respond to both orders of externalities, going to considerable effort to try to reduce or avoid exposures to toxicants and going out of their way to find and purchase less toxic products and perhaps even to make special arrangements in their lives.

Consider a person concerned about toxicants in her life. She would likely spend extra time and perhaps money researching toxic risks: learning the ingredients of foods, cosmetics, furniture, drapes, computers, baby seats and clothes, and building materials and trying to ascertain what toxicants have contaminated certain neighborhoods. She might spend more money to purchase less toxic products, live in less risky neighborhoods, purchase organic foods, buy furniture without chlorinated or brominated fire retardants, buy formaldehyde-free building materials, and so on. The monetary costs of these alternatives would be greater because riskier products, neighborhoods, or activities would have lowered their costs by creating negative externalities.

Thus she would need to be especially alert to toxicants, to be well enough informed to do the research to avoid them, to have sufficient income to pay for less toxicologically risky alternatives, and to have sufficient time to ensure this lifestyle. Others with less time to gather information and lesser incomes may have greater difficulties avoiding exposures. Nonetheless, as already indicated, even these efforts—all constituting externalities imposed by toxic products—are likely to have somewhat limited effects on her contamination.

Beyond these kinds of externalities, some people rightly feel unease caused by their awareness of toxicants in food, air, consumer products, workplaces, water, and living conditions. Anxieties are also externalities. It is difficult to assign a monetary value to psychological concerns, but they still constitute a variety of ways in which people's lives are distorted or disturbed by toxicants in their lives.

Finally, failure to test products for risks is itself an externality. If in the future a substance must be tested for toxicity and if the National Toxicology Program or the National Science Foundation must pay for some of the needed research because the company did not, this too is an externality. Taxpayers would be subsidizing a product's true price, to pay for a company's failure to test.

What is the significance of this discussion of externalities? Just this. Some readers will be concerned that a proposal for greater testing of existing and new products and pollutants will increase the costs of products and activities. This assumption may be correct, but even if it is, its magnitude is unclear. The presence of externalities shows that the status quo has numerous unaccounted costs, likely quite substantial ones. Consequently, any costs of better testing would need to be judged in comparison with costs of current products plus any externalities—monetary, personal, social, and otherwise—of the status quo. Put another way, if the monetary costs of products in the present were merely compared with the monetary costs of products in a world in which there was much greater toxicity testing but fewer externalities, this analysis would be

invalid. Not all current social and monetary costs of products are sub-
sumed in the dollar amounts people currently pay to purchase them. Instead, the total social costs of the status quo should be compared with the
total social costs of a modified world in which products were much better tested.

Responsibility

Externalities also reveal other kinds of social consequences of the status
quo, consequences that could be greatly improved in a world with enhanced, timelier, and on-going testing for toxicants. Companies currently
do not take full responsibility for the social effects of their products—
they off-load some of those consequences on others. Recall how existing
laws are reckless toward citizens. Many companies apparently choose
not to test their chemicals before submitting premanufacturing notices
required by the Toxic Substances Control Act (or the companies have
data and are not providing it to the EPA, as they are legally required to
do). These failures of the status quo would be greatly improved in a
world of enhanced testing and review of products.

Aspects of the Food, Drug, and Cosmetic Act (FDCA) exemplify a
better legal model of responsibility for products. As we have seen, this
law requires a company to test its product in animal studies and on human volunteers once the animal studies reveal that human volunteers
would face no unreasonable risks. If the product poses no unreasonable
risks as a result of increasingly larger clinical trials and is effective for
prescribed uses, it may be licensed for commerce.

Importantly for our purpose, the company has the legal burden to establish its product's safety and effectiveness; the FDA does not have a legal burden to show a drug is not safe or not effective to exclude it from
commerce. This approach was rejected in 1962, when the U.S. Congress
amended the drug laws to require premarket testing and licensing. However, corporate responsibility extends beyond initial testing.

The law also requires a manufacturer to continue to take responsibility for the safety of its product. The FDA may require the manufacturer
to conduct postmarket surveillance to detect any risks that might arise,
especially if the FDA believes that additional data "will improve how a
product is used." If tests reveal risks or the FDA receives a sufficient number of adverse effects reports, the company must revise its label "to include a warning as soon as there is reasonable evidence of an association

of a serious hazard with a drug; a causal relationship need not have been proved." Congressional amendments to the Food, Drug, and Cosmetic Act in 2007 specified that manufacturers "remain responsible for updating their labels," ensuring that warning labels are appropriate to the risks of the product. A company also faces legal penalties if it fails to update its warning label adequately.[45] Thus this law requires companies to take much greater responsibility for the safety of their products than postmarket laws do.

There remains one respect in which pharmaceutical companies are not required to take full responsibility for the safety of their products. The FDA must make a scientific and legal case for withdrawing approval of a pharmaceutical once it has been approved for commerce; this greatly burdens safety and can leave the public at risk for an extended period of time. As we saw in Chapter 5, the agency's efforts to remove the breast milk–suppression drug Parlodel proceeded at a sloth's pace. The agency had to go through nearly all the scientific effort and legal procedures needed to legally force removal of the drug before the company removed it voluntarily. Until then, the drug stayed on the market, despite many cases of strokes, seizures, heart attacks, and some deaths. Some of the burdens under which the FDA worked in the 1980s and 1990s have been eased; for example, companies, not the FDA, must update product warnings when there is new evidence of serious problems. It is not clear that it is now easier for the agency to remove a product.

A health-protective policy would have companies take full responsibilities for their products, as the FDCA largely requires for pharmaceuticals. The assignment of responsibility would need to be supplemented by better, more expeditious mechanisms for removing risky products from commerce. One such legal device might permit a public health agency to establish a rebuttable presumption based on available scientific research (the nature of the data would need to be specified), conducted by the agency or independent scientists, that a product poses risks to the public health. Once that is done, the firm should have the legal and scientific burden to rebut the presumption to show that the product meets statutory requirements or can be made to do so with reformulation or modification. The idea would be to keep companies responsible for ensuring the safety of their products if the firms wish to keep them in commerce. This approach resembles the spirit of the legal burden shifting under Proposition 65, a law that companies in California have adapted

237

to. Legislation should create private incentives for companies to act to protect the public health interest instead of having incentives to resist greater protections.

Except for the difficulty of removing drugs from commerce, under premarket drug laws a company has both initial and ongoing responsibilities to identify risks of its products, to ensure their safety, to continue to monitor them for safety, and to update warnings if needed. One might say that the company must take more nearly full responsibility for the safety of its product both before it enters commerce and after it is marketed, in order to detect any risks the product might pose and to update warning labels if risks appear. This is a much different conception of responsibility for one's products than typically exists under postmarket laws.

Postmarket laws presently permit a company to abdicate responsibility for its products and to abdicate substantial responsibility toward others' welfare. A company neither ensures that its products are safe nor fully assures others that they are free from externalities, especially risks. In effect, the company foists off costs and consequences on others, who must bear them, subsidizing the market price of the product.

Could the tort law assist in internalizing externalities? It likely would not initially cause the externalities to be internalized, but could perhaps correct some of them. If the tort law were fully utilized, were exquisitely accurate in compensating for losses and could be used with minimal transaction costs, there might be justification for using it to compensate for losses from lack of testing. However, this clearly is not the case and was not true even when torts functioned better than they do at present. In addition, the tort law has high transaction costs that greatly reduce citizen access to it, making it an even poorer instrument to address these problems. And since the Supreme Court changed the law concerning the admissibility of scientific experts in toxic tort cases, access has become much more difficult and transaction costs are even higher.[46]

Finally, even if one could correct losses relatively easily and with much lower transaction costs than at present, there is considerable agreement that an ounce of risk prevention via public health laws is better than a later pound of compensation for harms and other losses. I often ask students or audiences whether they would prefer to be injured and receive adequate tort compensation or whether they would prefer never to be injured at all. Invariably, they prefer to avoid injuries in the first place. We should recall that much of the animating motivation for the Toxic Substances Control Act embodied this idea, even though it became lost in the political process that led to the TSCA's enactment.

Concerns about a New World of Enhanced Toxicity Testing

Readers might be concerned about valuable products disappearing from the market or their costing more because of the need to ensure better health protections. How worrisome should such concerns be?

Would We Lose Valuable Products?

In a changed world with improved testing, might valuable or essential products be lost if testing reveals substances toxic to developing children? If some products were no longer in the market, how serious a loss would this be?

Such trade-offs are difficult to assess and people will make different personal assessments, but progress can be made on these issues by distinguishing between two quite different kinds of risk tradeoffs. Life-taking risks are those that threaten a person with loss of life or a shortened life. Thus, rock-climbing, mountaineering, drinking chlorinated water, and being exposed to asbestos all have life-taking risks. The chlorination of drinking water illustrates one point about life-taking risks.

When water is chlorinated, the chlorine disinfects contaminants in the water, killing many bacteria or microbes that carry life-threatening risks. Consequently, chlorine disinfectants provide life-saving benefits for those drinking the water. However, these disinfectants can also pose low probability life-taking risks. When chlorine interacts with biological material, it creates trihalomethanes, known carcinogens. When consumed in drinking water for over forty-five years, trihalomethanes elevate the risk of bladder cancer about 20 percent over background.[47] However, even if chlorination increased the risk of bladder cancer at a somewhat higher rate, we likely should continue to use it (as long as no safer method functioned as well). Chlorination saves people from life-taking diseases that would occur more often without its use and probably would be much worse than a small increase in comparatively rare bladder cancer. In such a case we would be exposed to a life-taking risk in order to avoid a more serious life-taking risk. Such decisions are called risk-risk trade-offs.

Contrast those trade-offs with risks represented by the now-banned pesticide Alar. It was sprayed on the blossoms of red apples and was incorporated directly into the fruit. The resulting apples remained fresh and had a longer shelf life. (Alar was not used on green apples.) Alar was banned because it and some of its metabolites are potent carcinogens, and these posed particular risks to young children who consumed large amounts of apple juice. Those who were exposed to life-taking risks—the adults and children who ate the apples and drank the juice—did not receive life-saving benefits from the product. The producers benefited by having a product that did not spoil as quickly and that could be shipped more easily. Consumers had red apples available for a longer period of time, and the apples might have been marginally cheaper, but these benefits could hardly be represented as life saving. This is a risk-benefit trade-off.

Loss of a product that has toxic properties but that also provides a net life-saving benefit, such as chlorination of drinking water, likely would be a significant loss. However, removal of a toxic constituent from a product that provides only benefits a consumer might be willing to pay for but not life-saving benefits hardly seems in the same category of concern. Not chlorinating drinking water would pose a substantial health threat. Removal of Alar from the market is not a comparable loss to the pubic. Even if red apples were no longer available as a result, we would still have green apples to keep the doctors away. And without Alar, we still have red apples, but they may have shorter shelf lives. They might or might not cost slightly more. Removal of Alar caused some commercial loss to growers.

Brominated fire retardants, PBDEs, appear to represent risk-risk personal and social tradeoffs. The compounds may save lives by temporarily reducing the flames of treated products that catch on fire. These fire retardants do not prevent fires, nor do they suppress them. At the same time, PBDEs should be considered prime candidates for removal from commerce because of their persistence, bioaccumulation, and toxic similarity to PCBs. Brominated fire retardants already pose many of the same kinds of worldwide problems as PCBs.

It appears that most fires for which PBDEs would be particularly effective occur as the result of people smoking in bed or on couches, falling asleep, and causing fires as a consequence. The PBDEs might provide slightly more time for people to escape any flames that erupted from smoldering furniture. However, there may be a lesser need for these compounds if fire-resistant building materials or alternative mattress materials would result in fewer devastating fires. And there might be even better means to prevent mattress fires, for example, quickly dying cigarettes. This would be a much better social choice to make. Do we need to put developing children and the wider environment at risk by using beds, couches, seat cushions, drapes, and car seats with persistent toxic materials when there are likely safer choices available? This discussion should occur with full toxicity data about different alternatives.

In addition, just because a product offers a favorable risk-risk trade-off does not make it the obvious choice. One should consider various alternatives: not using the product, not using it as formulated, using other possible products or processes for achieving the same results, and so on. A wide range of alternatives should be considered before concluding that products must contain toxic ingredients that can easily escape and contaminate us. Although a decision concerning PBDEs in commerce poses more difficult trade-offs than those of some other products such as Alar, lead in lipstick, or phthalates in many products, eliminating PBDEs should thus be very seriously considered and probably should be done, given the compounds' long-term adverse health effects and environmental consequences that will remain for decades.

With regard to products that do not promise life-saving benefits, increasing the costs of these products by testing them and eliminating any toxic ingredients in them does not present such difficult choices. Of course, some people will be concerned about their increased costs (which I consider in next), but one is not incurring life-taking risks by removing toxicants in a product or by removing the product altogether. In addition, we should recall the discussion above concerning toxic use reduction programs. Some products likely would be reformulated to reduce their toxicity without an increase in their costs.

How Burdensome Would Enhanced Testing Be?

Costs for a pragmatically designed and targeted testing regime may not be as great as some would suggest. I do not propose the full battery of tests that are used for pesticides and certainly not the same battery that is used for new drugs, both of which are inappropriate. Details about appropriate testing would need to be developed. I suggest that we follow recommendations from the NAS and consider a procedure for products more targeted than the procedures the FDA or the EPA pesticide office require.

Nonetheless, if one has concerns about costs, it is useful to consider estimates for testing under REACH, for which at least some research has been done. Admittedly, REACH has a different assessment regime from FDA drug testing and EPA pesticide review. Testing under REACH is also different from procedures suggested by the National Academy of Sciences, although since the Academy's suggestions are merely outlined, it is difficult to determine what would be required. Nonetheless, REACH is sufficiently comprehensive that it provides one basis for thinking about the costs of testing new and existing products of concern.

As a preliminary point, we should realize that the costs of environmental health protections are almost always exaggerated before they are instituted as part of the political regulatory dance. Companies overestimate or misestimate compliance costs in order to reduce the extent to which they must modify existing practices. Governmental agencies may rely too much on industry reports and issue high estimates.[48] In what follows I do not seek to discuss possible costs of product testing in any detailed manner. That is not my expertise, and any suggestions the NAS or I have made about testing are too vague for plausible estimates. Those familiar with appropriate batteries of tests would need to provide details and assign costs. Nonetheless, testing under the new REACH legislation offers us some guidance on this issue.

Two studies commissioned by industry estimated that under REACH such testing would devastate the chemical industry—in Germany, according to one study, and in France, according to the other. These two industry reports arrived at direct costs of testing that were reasonably similar to two independent studies. However, the studies commissioned by industry seemed to utilize, in the words of Frank Ackerman, a "creative calculation of indirect costs such as decreases in productivity and delays in innovation," resulting in extremely high and unrealistic estimates.[49]

The European Union commissioned a study that estimated the costs of testing thirty thousand new and existing chemicals to be 2.3 billion euros over eleven years.[50] Independently, the Nordic Council of Ministers, representing the governments of the Scandinavian countries, commissioned a study to estimate the costs of testing the same universe of chemicals. The study found the expense to be 3.5 billion euros over eleven years. Both estimates are large amounts of money. However, upon examination, they are not such obviously high costs.

Rather than just considering monetary costs in the aggregate, we can also appreciate testing costs by assessing the expense per citizen per year over the time period during which testing must be conducted. For example, one penny per day for each person in the United States for an entire year, which does not seem expensive, aggregates to a little more than 1 billion dollars.[51] Both calculations—per person per day for one year and the total for one year—yield the same result, but one does not seem particularly burdensome, while the other appears socially intimidating. Is that its aim?

A similar illustration of testing costs under REACH shows the high-end estimates (3.5 billion euros) to be approximately 1 euro for each of 450,000,000 citizens per year for each year during implementation of the testing. Over an eleven-year period, this would amount to 11 euros for every man, woman, and child in the European Union. Would a European Union family of four pay 44 euros over an eleven year period in order to test for any chemical risks in their lives, to know what these risks were, and to have some assurance that the European Union would not permit such products into the market if need be? This hardly seems to be an unreasonable amount.

Another approach that provides a perspective on aggregate costs is to estimate how much testing might increase the price of chemical products. Ackerman reports that required testing might increase the costs of chemical products one-sixteenth of 1 percent at the low end to one-fifth of 1 percent at the high end.[52] Thus, for example, if a chemical cost 10,000 euros per barrel before testing, an increase of one-fifth of 1 percent (the higher number) would increase the sale price to 10,020 euros per barrel over eleven years. An extra twenty euros does not seem likely to break the social bank, to bankrupt individual companies, or to put downstream users out of business.

One should also compare the total costs of testing the thirty thousand substances over the eleven-year phase-in with the likely total income of the chemical industry during that period. The higher total cost of testing thirty thousand chemicals is 3.5 billion euros. During 2003 the revenues of the chemical industry in Europe were 356 billion euros, or one hundred times greater than estimated total testing costs. When extrapolated over eleven years (not assuming any cost increases or inflation), the total industry income is 3,916 billion, or nearly 4 trillion euros. The high-end testing estimates would be 0.089 of 1 percent of total revenues during that period. Even if the high-end costs were several times higher, they do not seem unreasonable.

However, even these estimates for REACH may be misleading. They appear to compare the increase in euros for testing thirty thousand substances with costs under the status quo. As noted above, this is an invalid comparison when taken by itself, given existing externalities that are un-accounted for in the costs of current products. A better analysis would evaluate the monetary costs of the product plus the added costs of testing along with any remaining externalities and compare those with current costs of the product plus current externalities.

The REACH cost estimates provide one plausible window into these issues. However, this is only a suggestive analysis. Others will need to address these issues more extensively. Moreover, the costs of testing under REACH are likely much less than the costs would be if all substances subject to REACH were tested in accordance with FDA pharmaceutical or EPA pesticide protocols. How REACH tests would compare with recommended testing that the NAS has sketched out is not clear. Its proposals are too vague for detailed guidance. This comparison would require that details be worked out in light of broad testing goals combined with knowledgeable input from scientists, public health officials, test designers, and economists. Nonetheless, insofar as estimates from REACH provide some guidance, enhanced testing costs seem quite manageable.

More importantly, if the costs of disease, medical care, and lost productivity are as substantial as some estimates suggest, reducing those costs alone would appear to more than cover the price of better testing. If only half the $55 billion annual costs of disease and lost productivity from lead poisoning, asthma, cancer, and neurobehavioral disorders could be avoided by better testing and evaluation of substances before there was substantial exposure, this alone would pay several times over for the testing costs under REACH. If the estimates of testing under REACH were twice the high-end amount (up to $5.6 billion–$6 billion), savings from avoided diseases and productivity losses would still comfortably pay for testing. In that case, the amount spent on prevention through testing would be much better than the greater number of dollars and the social costs that we would pay later for care, cure, and rehabilitation.

Conclusion

Existing laws permit the vast majority of industrial chemicals of known and unknown toxicity to contaminate us without public health officials or scientists understanding their toxicity. Failure to test new and existing invisible, tasteless, and odorless molecules in products or pollutants results in a great deal of ignorance about them. This in turn imposes numerous externalities on the citizenry. Under the current system, taxpayers must pay public health agencies to determine that they are contaminated, and frequently their taxes pay to determine the toxicity of products with which they are contaminated.

Individuals must educate themselves about undetectable but possibly risky components of foods, pacifiers, baby bottles, beverage containers, cosmetics, furniture, computers, building materials, offices, and localities. Because some people are legitimately uneasy about their and their children's health, they will spend considerable time and money trying to avoid risky products, homes, and neighborhoods. These efforts too are externalities. The lag time between exposures to hazardous products and any diseases makes tracing their risks much more difficult after products have been commercialized, adding to the externalities.

Of course, the greatest externalities result from unjust, adverse health effects on people's lives from undetected toxicants: birth defects; early onset of vaginal, breast, prostate, lung or bone cancers; possible infertility and decreased sperm health; early parkinsonism or perhaps dementia; increased rates of autism, diminished mental development, reduced IQ or memory, decreased motor skills, and other neurological deficiencies; lifetimes of greater disease because of poorly functioning immune systems; and ongoing costs and anxieties associated with various diseases or dysfunctions. These effects can lead to disrupted, impaired, shortened, and possibly unpleasant lives for those who experience the most serious diseases induced during development. Off-loaded disease and suffering, together with other social and public costs, subsidize the price of current commercial products. These costs are not just economists' externalities; they also represent moral and legal failures.

It is simply morally outrageous to treat citizens as experimental subjects by contaminating them with untested substances. This stands in stark contrast to the way people are treated under the ethical guidelines for medical experiments and the laws concerning pharmaceuticals and pesticides. More seriously, our ethics and laws require even greater protections for children than for adults in experimental circumstances, generally forbidding children to be medical research subjects. However, current postmarket laws are particularly egregious in treating people and their children as the guinea pigs of reckless free enterprise. Legislatures that permit such recklessness are not free from responsibility toward fellow citizens on this issue.

A largely postmarket legal system permits a company to abdicate responsibility for understanding the toxicity of its products and for their consequences in the world. This legal system foists the costs of preventing illnesses on public agencies and citizens. When the products of companies are identified as toxic, they may claim "no human data" show that their products cause diseases, while understanding how insensitive and misleading such studies can be. Or businesses denigrate experimental animal studies instead of acknowledging their scientific value for identifying risks and owning up to the early warnings of toxicity that animal studies provide. Both claims are moral failures approaching deceit.

When substances contribute to actual diseases or dysfunctions that alterable institutions could have prevented, this is a serious injustice toward those who suffer them. Failure to test products and pollutants unfairly undermines opportunities and unjustly distributes monetary costs. It also leads to disrupted, impaired, shortened, and possibly miserable lives for those who experience the most serious diseases induced during development. The same postmarket laws also create substantial barriers to better health standards when they need to be implemented. Public health officials' efforts at improved health protections become delayed in a quagmire of procrastinating disputes about the science; this delay only perpetuates contamination and any resulting injustices.

We can describe and design a much less risky legal and commercial world, but one requiring changes in the law as recommended by the National Academies of Science and by earlier congressional and executive branch committees. In this world, the science would begin earlier in the life of a product or pollutant and be aimed at identifying toxic hazards before contamination occurs. This world would require testing for developmental and other adverse effects during sensitive life stages.

Successful testing would likely reduce some of the risks of the early onset of cancers, neurological dysfunctions of childhood and old age, lifelong illnesses due to impaired immune systems, heart disease attributable to ill-timed toxicants, and attendant misery and costs. Improved testing would better protect susceptible subpopulations and reduce misclassified industrial chemicals. It would also provide downstream purchasers of compounds with better data by which to create their products. Citizens would likely have less anxiety and concern about toxic ingredients in products because the law would require a good-faith effort to identify toxicants before the public could be exposed. People would need to spend less time, effort, money, and mental resources trying to reduce exposures to toxic substances.

Additional testing might increase the costs of some products, but this outcome is not inevitable, as toxic use reduction studies have suggested. However, even if there are some increased costs, these must be appropriately weighed against existing product costs with their associated externalities, which can be substantial.

Analyses of the testing costs of REACH suggest that, in comparison with the costs of the status quo, costs in a world in which there were better testing of products and pollutants would likely be less and perhaps much less than advocates of the status quo assert. The probable incremental costs to treat fellow citizens justly will likely not be as sizeable as some might worry. In addition, in the end these testing costs may well be an excellent bargain to reduce the social, personal, and monetary consequences of diseases, as these can be substantial.

Citizens will continue to be contaminated by molecular products. However, invasions would be much less toxic if there were enhanced and ongoing testing of products, if public health agencies provided well-designed protocols for tests, and if these tests were carried out in good faith. In the end, molecular contamination per se is not the problem we should worry about; that is inevitable. Contamination by toxic substances and reckless contamination by substances of unknown toxicity are the problems.

Existing legislation that permits reckless exposure to industrial compounds of known and unknown toxicity is a serious issue. Laws requiring earlier and better testing of products and pollutants will go a long way toward reducing or eliminating these problems. We can continue to be contaminated together by toxic products as a result of reckless laws, or we can create and achieve a more prudent alternative world together through legal change.

As I indicated in the introduction, while the message in this book can be unnerving, the various aspects of this story should be told and understood. We should be heartened that the scientific community is revealing previously unknown sources of disease and dysfunction so that they can be reduced. We might be angry toward those who are reckless toward our health and toward public officials for failing to protect us. However, we can now address these issues. The advice of Molly Gray, the mother-to-be from Seattle, Washington, who testified before the Washington state legislature, is a good place to begin, ". . . Something is wrong when I, as an educated consumer, am unable to protect my baby from toxic chemicals. I and all other parents should be able to walk into stores and buy what we need without winding up with products that put our families' health at risk. Now that I've learned that companies can put chemicals into products without ever testing for whether they harm our health, I think we need to change our laws."[53] Citizens, educated and uneducated, should be protected. What should we advocate that congress and legislatures to do? What will we choose to do?

I. INTRODUCTION

1. David Biello, "Plastic (Not) Fantastic: Food Containers Leach a Potentially Harmful Chemical: Is Bisphenol A, a Major Ingredient in Many Plastics, Healthy for Children and Other Living Things?" *Scientific American,* February 19, 2008, www.sciam.com (accessed July 25, 2008); Environmental Working Group, "Timeline: BPA from Invention to Phase-out," www.ewg.org (accessed July 22, 2008); U.S. Department of Health and Human Services, Centers for Disease Control and Prevention, National Center for Environmental Health, *Fourth National Report on Human Exposure to Environmental Chemicals,* December 2009, www.cdc.gov/exposurereport (accessed January 13, 2010), p. 4; Andrew Schneider, Report: Hazardous Chemical in Our Canned Food, American Online News, May 18, 2010, located at www.aolnews.com/science/article/bpa-pervasive-in-our-canned-food-national-workgroup-for-safe-markets-says/19482419 (accessed May 19, 2010).

2. On phthalates, James Bothwell, "Toy Story: Timeout for Phthalates," *McGeorge Law Review* 39 (2008): pp. 551–563, at p. 556–560; Rachael Rawlins, "Teething on Toxins: In Search of Regulatory Solutions for Toys and Cosmetics," *Fordham Law Review* 20 (2009): pp. 1–50, at p. 13; on lead, Donald T. Wigle and Bruce P. Lanphear, "Human Health Risks from Low-Level Environmental Exposures: No Apparent Safety Thresholds," *PLoS Medicine* 2 (2005): pp. 1–3, www.plosmedicine.org (accessed August 5, 2009); on lead and cardiovascular disease, Ana Navas-Acien, Eliseo Guallar, Ellen K. Silbergeld, and Stephen J. Rothenberg, "Lead Exposure and Cardiovascular Disease—a Systematic Review," *Environmental Health Perspectives* 115, no. 3 (2007): pp. 472–482; Committee on the Health Risks of Phthalates, National Research Council, *Phthalates and Cumulative Risk Assessment The Task Ahead* (Washington, D.C.: National Academy Press, 2008), pp. 5-6, 58.

3. On 98 percent of the U.S. population, Antonia M. Calafat, Lee-Yang Wong, Zsuzsanna Kuklenyik, John A. Reidy, and Larry L. Needham, "Polyfluoroalkyl Chemicals in the U.S. Population: Data from the National Health and Nutrition Examination Survey (NHANES) 2003–2004 and Comparisons with NHANES 1999–2000," *Environmental Health Perspectives* 115, no. 11 (2007): pp. 1596–1602, at p. 159. On breakdown of perfluorinated substances, John W. Washington, J.

Jackson Ellington, Thomas M. Jenkins, John J. Evans, Hoon Yoo, and Sarah C. Hafner, "Degradability of an Acrylate-Linked, Fluorotelomer Polymer in Soil," *Environmental Science and Technology,* pubs.acs.org/doi/abs/10.1021/es9021238 (accessed July 21, 2009); U.S. Environmental Protection Agency, "Perfluorooctanoic Acid (PFOA) and Fluorinated Telomers: Risk Assessment, www.epa.gov/oppt/pfoa/pubs/pfoarisk.html (accessed June 25, 2010). Per Eriksson, Henrik Viberg, Niclas Johnsson, and Anders Fredriksson, "Effects of Perfluorinated Compounds and Brominated Flame Retardants on Brain Development and Behavior in a Rodent Model" (presentation at PPTOX II: Role of Environmental Stressors in the Developmental Origins of Disease, Miami, Florida, December 7–10, 2009).

4. Julie B. Herbstman, Andreas Sjödin, Matthew Kurzon, Sally A. Lederman, Richard S. Jones, Virginia Rauh, Larry L. Needham, Deliang Tang, Megan Niedzwiecki, Richard Y. Wang, and Frederica Perera, "Prenatal Exposure to PBDEs and Neurodevelopment," *Environmental Health Perspectives,* 118, 5 (2010): pp. 712-719, at pp. 712, 718. Thea Edwards, "Flame Retardants Linked to Longer Time to Get Pregnant, Finds California Study," *Environmental Health News,* June 3, 2010, located at www.environmentalhealthnews.org/ehs/newscience/flame -retardants-associated-with-increased-time-to-pregnancy (summarizing a study by K. G. Harley, A. R. Marks, J. Chevrier, A. Bradman, A. Sjödin and B. Eskenazi, "PBDE Concentrations in Women's Serum and Fecundability," in *Environmental Health Perspectives,* volume 118 (5) (2010): pp. 699-704).

5. U.S. Department of Health and Human Services, *Fourth National Report on Human Exposure to Environmental Chemicals,* p. 7.

6. Myron A. Mehlman, "Dangerous and Cancer-causing Properties of Products and Chemicals in the Oil Refining and Petrochemical Industry: Part XV: Health Hazards and Health Risks from Oxygenated Automobile Fuels (MTBE): Lessons Not Heeded," *International Journal of Occupupational Medical Toxicology* 4(2)(1995): pp. 219-236; F. Belpoggi, M. Soffritti, C. Maltoni, "Methyl Tertiary Butyl Ether (MTBE)-A Gasoline Additive-Causes Testicular and Lymphohaematopoietic Cancers in Rats, *Toxicology and Industrial Health* 11(2)(1995):119-149 (1995).

7. Nancy Langston, *Toxic Bodies: Hormone Disruptors and the Legacy of DES* (New Haven: Yale University Press, 2010), p. 115, and Michael Balter, "TRADE POLICY: Scientific Cross-Claims Fly in Continuing Beef War," *Science* 284 (1999): pp. 1453-1454.

8. On 212 substances CDC reliably identifies, U.S. Department of Health and Human Services, *Fourth National Report on Human Exposure to Environmental Chemicals,* "Executive Summary, p. 1; on "built-in abilities," Birger G. J. Heinzow, "Endocrine Disruptors in Human Breast Milk and the Health-Related Issues of Breastfeeding," *Endocrine-Disrupting Chemicals in Food,* ed. I. Shaw (Cambridge: Woodhead Publishing, 2009), pp. 322–355, at pp. 324–325. See also Elaine M. Faustman and Gilbert S. Omenn, "Risk Assessment," in *Casarett and Doull's Toxicology,* 6th ed., ed. Curtis Klaassen (New York: Pergamon Press, 2001), pp. 83–104, at p. 84.

9. Sally Deneen, "Infuriated Mom: Why Can't I Protect My Body? Study Pinpoints Chemicals in Moms-to-be," *Seattle Globe,* November 18, 2009, www .seattlepostglobe.org (accessed November 18, 2009).

10. Ibid.

11. Calafat et al., "Polyfluoroalkyl Chemicals in the U.S. Population," p. 1599.

12. Richard W. Stahlhut, Wade V. Welshons, Shanna H. Swan, "Bisphenol A Data in NHANES Suggest Longer Than Expected Half-Life, Substantial Non-Food Exposure, or Both," *Environmental Health Perspectives,* 117, no. 5 (May 2009): pp. 784–789,; on 93 percent of people older than six, Antonia M. Calafat, Xiaoyun Ye, Lee-Yang Wong, John A. Reidy, and Larry L. Needham, "Exposure of the U.S. Population to Bisphenol A and 4-tertiary-Octylphenol: 2003–2004," *Environmental Health Perspectives* 116, no. 1 (January 2008): pp. 39–44, at p. 39.

13. Peter Fimrite, "Study: Chemicals, Pollutants Found in Newborns," *San Francisco Chronicle,* December 3, 2009, www.sfgate.com (accessed December 3, 2009); Marla Cone," Innovations in Designing Green Chemicals Are Emerging in Nearly Every U.S. Industry, from Plastics and Pesticides to Toys and Nail Polish," Los Angeles Times, September 14, 2008, located at http://www.latimes.com/news/local/la-me-greenchem12008sep14,0,1331640,print.story, last visited 4/21/10.last visited 4/21/10.

14. On known human neurotoxicants being likely toxic to developing children, Philippe Grandjean and Philip Landrigan, "Developmental Neurotoxicity of Industrial Chemicals," *Lancet* 368 (December 2006): pp. 2167–2178; on a possible one thousand additional neurotoxicants, ibid., p. 2175; on there being no safe level of carcinogens, lead, etc., Wigle and Lanphear, "Human Health Risks from Low-Level Environmental Exposures," pp. 1–3.

15. On lead and Parkinson's Disease, Marc G. Weisskopf, Jennifer Weuve, Huiling Nie, Marie-Helene Saint-Hilaire, Lewis Sudarsky, *et. al.,* "Association of Cumulative Lead Exposure with Parkinson's Disease," *Environmental Health Perspectives,* online, 31 August 2010, located at doi: 10.1289/ehp.1002339, (accessed September 29, 2010); on BPA, Biju Balakrishna, Kimiora Henare, Eric B. Thorstensen, Anna P. Ponnampalam, and Murray D. Mitchell, "Transfer of Bisphenol A Across the Human Placenta," *American Journal of Obstetrics and Gynecology,* 202, 4, (April 2010): pp. 393.e1-393.e7, at pp.393.e6; Jason G. Bromer, Yuping Zhou, Melissa B. Taylor, Leo Doherty, and Hugh S. Taylor, " Bisphenol-A Exposure in Utero Leads to Epigenetic Alterations in the Developmental Programming of Uterine Estrogen Response," *The FASEB Journal* 24 (2010): pp. 1-8, located at www.fasebj.org (Accessed online February 20, 2010); Jerrold J. Heindel, "Animal Models for Probing the Developmental Basis of Disease and Dysfunction Paradigm," *Basic and Clinical Pharmacology and Toxicology* 102 (2008): pp. 76–81, at p. 78.

16. Philip Landrigan, "What's Getting into Our Children?" *New York Times,* August 4, 2009, www.nytimes.com (accessed August 5, 2009).

17. On illnesses of increasing concern, Bruce P. Lanphear, "Origins and Evolution of Children's Environmental Health," in "Essays on the Future of Environmental Health Research: A Tribute to Kenneth Olden," special issue, *Environmental Health Perspectives* (August 2005): pp. 24–31, quotation at p. 24; on various morbidities, ibid (Lanphear is citing dozens of scientific papers).

18. Randy L. Jirtle and Michael K. Skinner, "Environmental Epigenomics and Disease Susceptibility" *Nature Reviews* 8 (2007): pp. 253–262; Heindel, "Animal

Models for Probing the Developmental Basis of Disease and Dysfunction Paradigm," at p. 77.

19. National Research Council, Steering Committee on Identification of Toxic and Potentially Toxic Chemicals for Consideration by the National Toxicology Program, *Toxicity Testing: Strategies to Determine Needs and Priorities* (Washington, D.C.: National Academy Press, 1984), p. 48.

20. Larry D. Claxton, Gisela de A. Umbuzeiro, David M. DeMarini, "The *Salmonella* Mutagenicity Assay: The Stethoscope of Genetic Toxicology for the 21st Century," forthcoming in *Environmental Health Perspectives,* online 2 August 2010 located at doi: 10.1289/ehp.1002336 (accessed August 3, 2010). On current rates of testing, Ronald Melnick, Senior Toxicologist and Director of Special Programs, Environmental Toxicology Program, National Institute of Environmental Health Services, personal communication about how few substances are tested for toxicity each year on animals, October 25, 2002; Joseph H. Guth, Richard A. Denison, Jennifer Sass, Require, Comprehensive Safety Data for All Chemicals, New Solutions 17(3) (2007): pp. 233-258, at pp. 237-238, (on veil of confidential business informration, p. 241).

21. Claxton, Umbuzeiro, DeMarini, "The *Salmonella* Mutagenicity Assay."

22. Lucio G. Costa and Gennaro Giordano, "Developmental Neurotoxicity of Polybrominated Diphenyl Ether (PBDE) Flame Retardants," *NeuroToxicology* 28 (2007): pp. 1047–1067, at p. 1048.

23. Amber R. Wise, Jackie Schwartz, Tracey J. Woodruff, *Policy Recommendations for Addressing Potential Health Risks from Nanomaterials in California* (Draft report for the California Science Advisory Panel on nanomaterials, June 14, 2010), p. 27; Peter Wick, Antoine Malek, Pius Manser, Danielle Meili, Xenia Maeder-Althaus, Liliane Diener, Pierre-Andre Diener, Andreas Zisch, Harald F. Krug, and Ursula von Mandach, "Barrier Capacity of Human Placenta for Nanosized Materials, *Environmental Health Perspectives* 118, 3 (2010): pp. 432–436; Benedicte Trouiller, Ramune Reliene, Aya Westbrook, Parrisa Solaimani, and Robert H. Schiestl, "Titanium Dioxide Nanoparticles Induce DNA Damage and Genetic Instability in vivo in Mice," *Cancer Research,* 69(22) (2009): pp. 8784-8789.

24. T. J. Woodruff, L. Zeise, D. A. Axelrad, K. Z. Guyton, Guyton KZ, et al., "Meeting Report: Moving Upstream-Evaluating Adverse Upstream End Points for Improved Risk Assessment and Decision-making," *Environmental Health Perspectives,* 16(11) (2008): pp. 1568-1575, and Sofie Christiansen, Martin Scholze, Majken Dalgaard, Anne Marie Vinggaard, Marta Axelstad, Andreas Kortenkamp, and Ulla Hass, "Synergistic Disruption of External Male Sex Organ Development by a Mixture of Four Antiandrogens," *Environmental Health Perspectives* 117 (2009): pp. 1839-1846.

25. Eunice Kennedy Shriver National Institute of Child Health and Human Development, "Institute Mission and Accomplishment Highlights," www.nichd.nih.gov/about/overview/mission/index.cfm (accessed December 21, 2009).

26. Philip J. Landrigan, Clyde B. Schechter, Jeffrey M. Lipton, Marianne C. Fahs, and Joel Schwartz, "Environmental Pollutants and Disease in American Children: Estimates of Morbidity, Mortality, and Costs for Lead Poisoning, Asthma, Cancer, and Developmental Disabilities," *Environmental Health Perspectives* 110,

no. 7 (2002): pp. 721–728, at p. 727; Guth, Denison and Sass, "Require Comprehensive Safety Data for All Chemicals," p. 242 and Carl F. Cranor, *Toxic Torts: Science, Law and the Possibility of Justice* (New York: Cambridge University Press, 2006), pp. 160-164.

27. Frank Ackerman, "The Unbearable Lightness of Regulatory Costs," *Fordham Urban Law Journal* 33 (2006): pp. 1071–1096,1076–1077.

28. Sally Deneen, "Frustrated local mom testifies to Senate: Why can't I protect my baby from chemicals?" Seattle Globe Reporter, Friday, February 05, 2010, located at seattlepostglobe.org/2010/02/04/infuriated-local-mom-testifies-to-senate-why-cant-i-protect-my-body-from-chemicals, (accessed 4/29/10).

29. Arlene Blum and Bruce N. Ames, "Flame-Retardant Additives as Possible Cancer Hazards," *Science* 195 (2007): pp. 17–23; Philip J. Hilts, *Protecting America's Health: The FDA, Business, and One Hundred Years of Regulation* (New York: Alfred A. Knopf, 2003), pp. 144–165.

30. Hilts, *Protecting America's Health*, pp. 148–149.

31. Ibid., p. 145.

32. Ibid., on the drug being used as a sedative and treatment for nausea, quotation at p. 150; on the 1938 law, quotation at p. 152.

33. On numbers of thalidomide birth defects worldwide, James L. Schardein and Orest T. Macina, *Human Developmental Toxicants: Aspects of Toxicology and Chemistry* (Boca Raton, Fla.: Taylor and Francis, 2007), p. 131; on avoidance of thalidomide birth defects in the United States, Hilts, *Protecting America's Health*, p. 158, and see also Schardein and Macina, *Human Developmental Toxicants*, p. 131; on Grünenthal's withdrawing the drug from the German market, Hilts, *Protecting America's Health*, quotation at p. 155.

34. Ibid., pp. 155–156.

35. Ibid., pp. 132–140.

36. Ibid., on Republicans attack on legislation, quotation at p. 140; on rights of businessmen and doctors, quotation at p. 161.

37. Ibid., on senators fearful of blame, quotation at pp. 159–160; on comment by official of the Pharmaceutical Manufacturers Association, quotation at p. 161.

38. Ibid., p. 160.

39. Nancy Langston, *Toxic Bodies*, p. 115.

40. U.S. Environmental Protection Agency, "Appendix: The Toxic Substances Control Act: History and Implementation," p. 111, located at www.epa.gov/oppt/newchems/pubs/chem-pmn/appendix.pdf (accessed 4/22/10).

2. NOWHERE TO HIDE

1. Alicia J. Fraser, Thomas F. Webster, Michael D. McClean, "Diet Contributes Significantly to the Body Burden of PBDEs in the General U.S. Population," *Environmental Health Perspectives* 117, no. 10, (2009): pp. 1520–1525.

2. Charles Duhigg, "That Tap Water Is Legal but May Be Unhealthy," *New York Times*, December 17, 2009, www.nytimes.com (accessed December 17, 2009).

3. Barbara A. Cohn, Mary S. Wolff, Piera M. Cirillo, and Robert I. Sholtz, "DDT and Breast Cancer in Young Women: New Data on the Significance of Age

at Exposure," *Environmental Health Perspectives* 115, no. 10 (October 2007): pp. 1406–1414.

4. Sarah Schmidt, "Fish Oil Supplements May Bring PCB Compounds: Study," *Canwest News Service*, March 2, 2010, located at www.canada.com/health/Fish%20supplements%20bring%20compounds%20Study/2634004/story.html (accessed June 3, 2010).

5. Marla Cone, *Silent Snow: The Slow Poisoning of the Arctic* (New York: Grove Press, 2005), pp. 21–23, 28–29, 30–33, and 142–143.

6. Quoted in Josh Mankiewicz, "It's Not Easy Being Green: Ever Worry That Your Family Uses Stuff Every Day That Contains Potentially Harmful Chemicals? Two Very Different Families Face a *Dateline* Test," *NBC Dateline,* www.msnbc.msn.com/id/24228167 (accessed July 23, 2009).

7. National Research Council, Steering Committee on Identification of Toxic and Potentially Toxic Chemicals for Consideration by the National Toxicology Program, *Toxicity Testing: Strategies to Determine Needs and Priorities* (Washington, D.C.: National Academy Press, 1984), p. 3.

8. Joseph H. Guth, Richard A. Denison, Jennifer Sass, Require, Comprehensive Safety Data for All Chemicals, *New Solutions* 17(3)· (2007): pp. 233-258, at p. 234.

9. U.S. Congress, Office of Technology Assessment, "Screening and Testing Chemicals in Commerce" (Background Paper, Office of Technology Assessment, OTA-BP-ENV-166, September 1995), p. 1. According to this report, "some 15,000 chemicals . . . are produced in significant volumes, with approximately 3–4,000 produced in excess of 1,000,000 lbs/year." Subtracting that 3,000 produced in excess of 1 million pounds leaves up to 12,000 substances about which there appears to be some concern produced in "significant volumes." On 6,000 chemicals of concern for exposure, Harvey Black, "New Frontiers—and Limitations—in Testing People's Bodies for Chemicals," *Environmental Health News*, December 2, 2009, www.environmentalhealthnews.org/ehs/news/new-horizons-in-biomonitoring (accessed December 2, 2009).

10. Elaine M. Faustman and Gilbert S. Omenn, "Risk Assessment," in *Casarett and Doull's Toxicology,* 6th ed., ed. Curtis D. Klaassen (New York: McGraw-Hill, 2001), p. 97.

11. On PBDEs, U.S. Department of Health and Human Services, Agency for Toxic Substances and Disease Registry, "Public Health Statement of Polybrominated Diphenyl Ethers," http://www.atsdr.cdc.gov/toxprofiles/phs68-pbde.html (accessed August 25, 2008); on MTBE, U.S. Department of Health and Human Services, Centers for Disease Control and Prevention, National Center for Environmental Health, *Fourth National Report on Human Exposure to Environmental Chemicals,* December 2009, p. 7, www.cdc.gov/exposurereport (accessed December 25, 2009); on bisphenol A, Antonia M. Calafat, Xiaoyun Ye, Lee-Yang Wong, John A. Reidy, and Larry L. Needham, "Exposure of the U.S. Population to Bisphenol A and 4-tertiary-Octylphenol: 2003–2004," *Environmental Health Perspectives* 116, no. 1 (January 2008): pp. 39–44, quotation at p. 39.

12. Ibid., p. 39.

13. On biomonitoring, Ken Sexton, Larry L. Needham and James L. Perkle, "Human Biomonitoring of Environmental Chemicals: Measuring Chemicals in Human Tissues Is the 'Gold Standard' for Assessing People's Exposure to Pollution," *American Scientist* 92 (2004): pp. 38–45, at pp. 38–39; quotation on biomonitoring, U.S. Department of Health and Human Services, Centers for Disease Control and Prevention, National Biomonitoring Program, www.cdc.gov/biomonitoring/about.html (accessed August 20, 2008).

14. U.S. Department of Health and Human Services, Centers for Disease Control and Prevention, National Center for Environmental Health, *Third National Report on Human Exposure to Environmental Chemicals,* September 2005, www.cdc.gov (accessed August 20, 2008).

15. Sexton, Needham and Perkle, "Human Biomonitoring of Environmental Chemicals," both quotations at p. 41.

16. Prasada Rao S. Kodavanti, "Neurotoxicity of Persistent Organic Pollutants: Possible Mode(s) of Action and Further Considerations," *Dose-Response* 3 (2005): pp. 273–305, at p. 274; Arnold Schecter, Olaf Papke, Kuang Chi Tung, Jean Joseph, T. Robert Harris, James Dahlgren, "Polybrominated Diphenyl Ether Flame Retardants in the U.S. Population: Current Levels, Temporal Trends, and Comparison With Dioxins, Dibenzofurans, and Polychlorinated Biphenyls," *Journal of Occupational and Environmental Medicine* 47, no. 3 (March 2005): pp. 199–211, at p. 209. On lead see Robert A. Goyer and Thomas W. Clarkson, "Toxic Effects of Metals," in *Casarett and Doull's Toxicology,* 6th ed., ed. Curtis Klaassen (New York: Pergamon Press, 2001), pp. 811–867, at p. 829.

17. Calafat et al., "Exposure of the U.S. Population to Bisphenol A," pp. 39–44.

18. Sexton, Needham, and Perkle, "Human Biomonitoring of Environmental Chemicals," 41.

19. U.S. Department of Health and Human Services, *Third National Report on Human Exposure to Environmental Chemicals;* U.S. Department of Health and Human Services, *Fourth National Report on Human Exposure to Environmental Chemicals.*

20. Environmental Defense, "Toxic Nation: A Report on Pollution in Canadians," 2005, pp. 1–42, at p. 1, www.toxicnation.ca (accessed September 29, 2008).

21. Ibid., p. 1.

22. Ibid., pp. 20–30.

23. On conclusions of report, ibid., quotation at p. 1; on expected results of evapotranspiration and ascent up the food chain, ibid., p. 29.

24. Ibid., p. 1.

25. Ibid., p. 15.

26. Body Burden Work Group and Commonweal Biomonitoring Research Center, "Is It In Us? Chemical Contamination in Our Bodies: Toxic Trespass, Regulatory Failure and Opportunities for Action," 2007, pp. 1–52, 18.

27. Mankiewicz, "It's Not Easy Being Green."

28. Ibid.

29. Ibid.

30. Sally Deneen, "Infuriated Mom: Why Can't I Protect My Body? Study Pinpoints Chemicals in Moms-to-be," *Seattle Globe,* November 18, 2009, www .seattlepostglobe.org (accessed November 18, 2009).

31. Peter Fimrite, "Study: Chemicals, Pollutants Found in Newborns," *San Francisco Chronicle,* December 3, 2009, located at www.sfgate.com (accessed December 3, 2009); Environmental Working Group, "Pollution in People: Cord Blood Contaminants in Minority Newborns," December 2, 2009, www.ewg.org/minoritycordblood/fullreport (accessed 12/20/09).

32. U.S. Department of Health and Human Services, *Fourth National Report on Human Exposure to Environmental Chemicals,* p. 4.

33. Jia Kyunghee, Lim Khob Young, Parka Yoonsuk and Choi Kyungho, "Influence of A Five-Day Vegetarian Diet on Urinary Levels of Antibiotics and Phthalate Metabolites: A Pilot Study with "Temple Stay" Participants." *Environmental Research,* 110, Issue 4 (2010): PP. 375-382.

34. U.S. Department of Health and Human Services, *Third National Report on Human Exposure to Environmental Chemicals,* p. 309.

35. Andreas Kortenkamp, Michael Faust, Martin Scholze, and Thomas Backhaus, "Low-Level Exposure to Multiple Chemicals: Reason for Human Health Concerns?" *Environmental Health Perspectives* 115, no S-1 (2007): pp. 106–114.

36. Gardiner Harris, "Ban Sought on Cold Medicine for Very Young," *New York Times,* September 29, 2007, www.nytimes.com (accessed October 15, 2007); Gardiner Harris, "Makers Pull Infant Cold Medicines," *New York Times,* October 11, 2007, www.nytimes.com (accessed October 15, 2007); Editorial, "Children and Cold Medicines," *New York Times,* November 3, 2007, www.nytimes.com (accessed November 4, 2007).

37. U.S. Department of Health and Human Services, Agency for Toxic Substances and Disease Registry, "Public Health Statement for Lead," August 2007, www.atsdr.cdc.gov/toxprofiles/phs13.html (accessed October 21, 2007).

38. Jack Lewis, "Lead Poisoning: A Historical Perspective," *EPA Journal,* May 1985, www.epa.gov/history/topics/perspect/lead.htm (accessed August 20, 2007).

39. Quotation on use of lead 1950–2000, U.S. Department of Health and Human Services, "Public Health Statement for Lead"; on use of lead in gasoline and its effect on automobile market and innovations in automobiles and airplanes, Lewis, "Lead Poisoning: A Historical Perspective"; on use of tetraethyl lead despite less toxic alternatives, Gerald Markowitz and David Rosner, *Deceit and Denial: The Deadly Politics of Industrial Pollution* (Berkeley: University of California Press, 2002), pp. 17–18; on United States' consuming of 40 percent of the world's supply of lead, Lewis, "Lead Poisoning: A Historical Perspective."

40. U.S. Department of Health and Human Services, "Public Health Statement for Lead," p. 2.

41. Ibid.

42. U.S. Environmental Protection Agency, Region 10, "Health Effects of Lead," yosemite.epa.gov/R10 (accessed July 17, 2009).

43. On lead remaining in the environment, ibid; on worldwide use increasing, Collegium Ramazzini, "Collegium Ramazzini Lead Statement: Call for Worldwide

Reduction in Exposure to Lead," p. 2, www.collegiumramazzini.org/publications
.asp (accessed December 20, 2009).

44. Mireya Navarro, "Lead Poisoning Is in Decline, but Proves a Stubborn
Nemesis," *New York Times,* April 21, 2010, located at www.nytimes.com/2010/
04/22/nyregion/22lead.html?hp (accessed 5/2/2010).

45. U.S. Department of Health Services, Centers for Disease Control, Agency
for Toxic Substances and Disease Registry, "Toxicological Profile for Mercury," http://www.atsdr.cdc.gov/toxprofiles/tp46.pdf, at p. 3 (accessed July 20,
2007).

46. U.S. Environmental Protection Agency, "Mercury: Consumer and Commercial Product: Energystar Fact Sheet," www.energystar.gov/ia/partners/promotions/
change_light/downloads/Fact_Sheet_Mercury.pdf (accessed November 10, 2009).

47. Robert A. Goyer and Thomas W. Clarkson, "Toxic Effects of Metals," in
Casarett and Doull's Toxicology, 6th ed., ed. Curtis Klaassen (New York: Pergamon Press, 2001), pp. 811–867, at p. 834.

48. U.S. Department of Health Services, Centers for Disease Control, Agency
for Toxic Substances and Disease Registry, "Toxicological Profile for Mercury,"
Ibid., p. 3.

49. Ibid., p. 4.

50. Ibid., quotation at p. 5; on persistence of methylmercury, Committee on the
Toxicological Effects of Methylmercury, Board on Environmental Studies and Toxicology, National Research Council, "Executive Summary," in *Toxicological Effects of Methylmercury,* books.nap.edu/catalog/9899.html (accessed July 29, 2009).

51. Goyer and Clarkson, "Toxic Effects of Metals," p. 834.

52. On cautionary advisories to limit intake of seafood, U.S. Department of
Health Services, "Toxicological Profile for Mercury"; on content of MeHg in pilot
whales and certain fish, Pal Weihe ("The Faroe Islands in Biomedical Research,"International Conference on Fetal Programming and Developmental Toxicity,
Torhavn, Faroe Islands, May 20-24, 2007).

53. U.S. Department of Health Services, "Toxicological Profile for Mercury,"
pp. 11–12.

54. ABC News, "Report: Toxins Found in Whales Bode Ill for Humans," located at abcnews.go.com/print?id=11003954 (accessed June 25, 2010).

55. Poul Harremoës et al., eds., *Late Lessons from Early Warnings: The Precautionary Principle, 1896–2000* (Copenhagen: European Environment Agency,
2001), p. 64; quotation on no natural sources of PCBs, U.S. Department of Health
and Human Services, Public Health Service Agency for Toxic Substances and Disease Registry, "Public Health Statement: Polychlorinated Biphenyls (PCBS)," November 2000, p. 1.1 (hereafter, ATSDR, "PCBs"), www.atsdr.cdc.gov (accessed
June 13, 2008).

56. Arlene Blum and Bruce N. Ames, "Flame-Retardant Additives and Possible
Cancer Hazards," *Science* 195 (1977): pp. 17–23, at p. 22.

57. On hexagonal structure of PCB molecules, U.S. Environmental Protection
Agency, Office of Research and Development, *PCBs: Cancer Dose-Response Assessment and Application to Environmental Mixtures,* APA/600/P-96/001F, Washington, D.C. 1996, p. 1; on 209 variants of PCB, U.S. Department of Health and

Human Services, Public Health Service Agency for Toxic Substances and Disease Registry, "Toxicological Profile: Polychlorinated Biphenyls," September 2004, p. 443, www.atsdr.cdc.gov/ToxProfiles/tp17-c4.pdf (accessed September 23, 2008).

58. On PCBs being "dioxin-like," Ted Simon, Janice K. Britt, Robert C. James, "Development of A Neurotoxic Equivalence Scheme of Relative Potency for Assessing the Risk of PCB Mixtures," *Regulatory Toxicology and Pharmacology* 48 (2007): pp. 148–170, at p. 149. See also U.S. Department of Health and Human Services, "Introduction to PCBs and other Dioxin-Like Substances," in *Third National Report on Human Exposure,* p. 139. and p. 135; on mechanism being conserved through evolution, Simon, Britt, and James, "Development of A Neurotoxic Equivalence Scheme," p. 151.

59. Andreas Kortenkamp, "Ten Years of Mixing Cocktails: A Review of Combination Effects of Endocrine-Disrupting Chemicals," *Environmental Health Perspectives* 115, no. S-1 (2007): pp. 98–105, at p. 99.

60. On half-life of dioxin-like compounds in people, U.S. Department of Health and Human Services, *Third National Report on Human Exposure to Environmental Chemicals,* p. 135; on half-life for PCBs in both adults and children, Philippe Grandjean, Esben Budtz-Jorgensen, Dana B. Barr, Larry L. Needham, Pal Weihe, and Birger Heinzow, "Elimination Half-Lives of Polychlorinated Biphenyl Congeners in Children," *Environmental Science and Technology* 42, no. 18 (2008): pp. 6991–6996, at p. 6995; on estimates of half-lives likely being underestimated, U.S. Environmental Protection Agency, *PCBs: Cancer Dose-Response Assessment,* pp. 23–24.

61. On different toxicity properties of noncoplanar PCBs, Harremoës et al., *Late Lessons from Early Warnings,* p. 68; on persistence of noncoplanar PCBs in environment and their neurotoxicity, Kodavanti, "Neurotoxicity of Persistent Organic Pollutants," pp. 274, 283, and 297–298; Simon, Britt, James, "Development of A Neurotoxic Equivalence Scheme," pp. 149–150.

62. Harremoës et al., *Late Lessons from Early Warnings,* p. 64.

63. ATSDR, "PCBs," p. 1.2.

64. Fox River Watch, "The History of PCBs: When Were Health Problems Detected?" www.foxriverwatch.com (accessed August 15, 2008).

65. Ibid., p. 1.2.

66. ATSDR, "PCBs," quotation at p. 1.3; on bioaccumulation in water and persistence in animals, U.S. Department of Health Services, "Toxicological Profile for Mercury," p. 3.

67. M. J.Brunner, T. M. Sullivan, A. W. Singer, M. J. Ryan, J. D. Toft II, R. S. Menton, S. W. Graves, and A. C. Peters, "An Assessment of the Chronic Toxicity and Oncogenicity of Arochlor-1242, Aroclor-1254 and Aroclor-1260 Administered in Diet to Rats" (Batelle Study no. SC920192, Chronic Toxicity and Oncogenicity Report, Columbus, Ohio, 1996); T. C. Hornshaw, R. J. Aulerich, and H. E. Johnson, "Feeding Great Lakes Fish to Mink: Effects on Mink and Accumulation and Elimination of PCBs by Mink," *Journal of Toxicology and Environmental Health* 11 (1983): pp. 933–946; R. J. Aulerich, R. K. Ringer, and J. Safronoff, "Assessment of Primary and Secondary Toxicity of Aroclor 1254 to Mink," *Archives of Environmental Contamination and Toxicology* 15 (1986): pp. 393–399.

68. Hornshaw, Aulerich, and Johnson, "Feeding Great Lakes Fish to Mink," pp. 933–946, and Aulerich, Ringer, and Safronoff, "Assessment of Primary and Secondary Toxicity of Aroclor 1254 to Mink," pp. 393–399, both as cited in U.S. Environmental Protection Agency, *PCBs: Cancer Dose-Response Assessment*, p. 3.

69. On double whammy of more-concentrated and more-toxic congeners, ATSDR, "PCBs," p. 1.4; on breast milk of Arctic mothers, Cone, *Silent Snow*, pp. 30–31; on nursing polar bear cubs, Cone, *Silent Snow*, p. 55.

70. Department of Health and Human Services, *Third National Report on Human Exposure to Environmental Chemicals*, p. 135.

71. Ibid., pp. 201, 204–249, 137.

72. U.S. Department of Health and Human Services, Public Health Service Agency for Toxic Substances and Disease Registry, "Public Health Statement: Polybrominated Diphenyl Ethers," September 2004, p. 1.1 (hereafter, ATSDR, "PBDEs"), www.atsdr.cdc.gov (accessed September 23, 2008).

73. U.S. Environmental Protection Agency, *Polybrominated Diphenyl Ethers (PBDEs) Project Plan*, March 2006, p. 1, www.epa.gov (accessed July 15, 2009).

74. Kim Hooper and Thomas A. McDonald, "The PBDEs: An Emerging Environmental Challenge and Another Reason for Breast-Milk Monitoring Programs," *Environmental Health Perspectives* 19, no. 85 (2000): p. 388; U. S. Environmental Protection Agency, "Furniture Flame-REtardant Partnership: Environmental Profiles of Chemical Flame-Retardant Alternatives for Low-Density Polyurethane Foam, p. 2-2, located at www.epa.gov/dfe, (accessed July 15, 2009).

75. U.S. Environmental Protection Agency, *Polybrominated Diphenyl Ethers (PBDEs) Project Plan*, p. 3.

76. On PBDEs being "added" or "blended," Mehran Alaee, Pedro Arias, Andreas Sjodin, and Ake Berman, "An Overview of Commercially Used Brominated Flame Retardants, Their Applications, Their Use Patterns in Different Countries/ Regions and Possible Modes of Release, *Environment International* 29, no. 6 (2003): pp. 683–689, at p. 685; on "leaking" of PBDEs into the environment, L. G. Costa and G. Giordano, "Developmental Neurotoxicity of Polybrominated Diphenyl Ether (PBDE) Flame Retardants," *NeuroToxicology* 28 (2007): pp. 1047–1067, quotation at p. 1048; on leakage being more likely than if PBDEs were chemically bound, Alaee et al., "An Overview of Commercially Used Brominated Flame Retardants," p. 685.

77. Joseph G. Allen, Michael D. McClean, Heather M. Stapleton, and Thomas F. Webster, "Linking PBDEs in House Dust to Consumer Products Using X-ray Fluorescence, *Environmental Science and Technology* 42, no. 11 (2008): pp. 4222–4228, at p. 4226.

78. Fraser, Webster, McClean, "Diet Contributes Significantly to the Body Burden of PBDEs in the General U.S. Population," pp. 1520, 1574.

79. U.S. Environmental Protection Agency, *Furniture Flame Retardancy Partnership: Environmental Profiles of Chemical Flame-Retardant Alternatives for Low-Density Polyurethane Foam*, EPA 742-R-05–002A, Washington, D.C., 2005, pp. 3–15–16, www.epa.gov (accessed September 29, 2008).

80. U.S. Environmental Protection Agency, "Polybrominated Diphenyl Ethers (PBDEs) Project Plan," p. 3; Costa and Giordano, "Developmental Neurotoxicity of Polybrominated Diphenyl Ether (PBDE) Flame Retardants," p. 1048.

81. Schecter et al., "Polybrominated Diphenyl Ether Flame Retardants in the U.S. Population," p. 199.

82. Matthew Denholm, "Cancer Agents Found in Tasmanian Devils," *News .Com.AU,* January 22, 2008, www.theaustralian.com.au/news/cancer-agents-in -tassie-devils/story-e6frg6ox-1111115 (accessed May 13, 2008); Dave Hansford, "Flame Retardants Found in Rare Tasmanian Devils," *National Geographic News,* news.nationalgeographic.com/news/2008/01/080128-devils-cancer.html (accessed December 29, 2009).

83. Parks and Wildlife Service Tasmania, "Wildlife of Tasmania: Mammals of Tasmania: Tasmanian Devil," www.parks.tas.gov.au (accessed May 13, 2008).

84. June-Soo Park, Arthur Holden, Vivian Chu, Michele Kim, Alexandra Rhee, Juja Patel, Yating Shi, et al., "Time-Trends and Congener Profiles of PBDEs and PCBs in California Peregrine Falcons *(Falco peregrinus),*" *Environmental Science and Technology* 43, no. 23 (2009): pp. 8744–8751.

85. Melissa Rose, Debora H. Bennett, Ake Bergman, Britta Fangstrom, Issac N. Pessah, and Irva Hertz-Picciotto, "PBDEs in 2-5 Year-Old Children from California and Associations with Diet and Indoor Environment," *Environmental Science and Technology,* 44 (2010): pp. 2648–2653.

86. On 50 to 97 percent of U.S. citizens, Andreas Sjodin, Lee-Yang Wong, Richard S. Jones, Annie Park, Yalin Zhang, Carolyn Hodge, Emily Dipietro, et al., "Serum Concentrations of Polybrominated Diphenyl Ethers (PBDEs) and Polybrominated Biphenyl (PBB) in the United States Population: 2003–2004," *Environmental Science and Technology* 42, no. 4 (2008): pp. 1377–1384, p. 1378; on body burdens in the United States compared with those reported in European studies, Sjodin et al., "Serum Concentrations of Polybrominated Diphenyl Ethers (PBDEs) and Polybrominated Biphenyl (PBB) in the United States Population," p. 1382; Costa and Giordano, "Developmental Neurotoxicity of Polybrominated Diphenyl Ether (PBDE) Flame Retardants," p. 1049; on serum and breast milk concentrations in 2003, Schecter et al., "Polybrominated Diphenyl Ether Flame Retardants in the U.S. Population," p. 205; on PBDEs with half-lives estimated at two years, Sjodin et al., "Serum Concentrations of Polybrominated Diphenyl Ethers (PBDEs) and Polybrominated Biphenyl (PBB) in the United States Population," pp. 1377–1384.

87. Schecter et al., "Polybrominated Diphenyl Ether Flame Retardants in the U.S. Population," quotation at, p. 203; p 207.

88. Sjodin et al., "Serum Concentrations of Polybrominated Diphenyl Ethers (PBDEs) and Polybrominated Biphenyl (PBB) in the United States Population," pp. 1380–1381, and U.S. Department of Health and Human Services, *Third National Report on Human Exposure to Environmental Chemicals,* pp. 174–197 and 204–249.

89. Linda S. Birnbaum and Daniele F. Staskal, "Brominated Flame Retardants: Cause for Concern? *Environmental Health Perspectives* 112, no. 1 (January 2004): pp. 13–15, and Costa and Giordano, "Developmental Neurotoxicity of Polybrominated Diphenyl Ether (PBDE) Flame Retardants," p. 1048.

90. Sjodin et al., "Serum Concentrations of Polybrominated Diphenyl Ethers

(PBDEs) and Polybrominated Biphenyl (PBB) in the United States Population," pp. 1379–1383; Per Eriksson, Eva Jakobsson, and Anders Fredriksson, "Brominated Flame Retardants: A Novel Class of Developmental Neurotoxicants in Our Environment?" *Environmental Health Perspectives* 109, no. 9 (September 2001): pp. 906–907; Schecter et al., "Polybrominated Diphenyl Ether Flame Retardants in the U.S. Population," pp. 205–207; quotation from Sjodin et al., "Serum Concentrations of Polybrominated Diphenyl Ethers (PBDEs) and Polybrominated Biphenyl (PBB) in the United States Population," pp. 1382–1383.

91. Costa and Giordano, "Developmental Neurotoxicity of Polybrominated Diphenyl Ether (PBDE) Flame Retardants," p. 1050.

92. Rose, et al., "PBDEs in 2-5 Year-Old Children from California," pp. 2648–2653.

93. Ibid., p. 2652.

94. On disease mechanisms of PBDEs likely being similar to that of PCBs, Kodavanti, "Neurotoxicity of Persistent Organic Pollutants," p. 297; on risks of PBDEs for neurological diseases or damage, Birnbaum and Staskal, "Brominated Flame Retardants: Cause for Concern?" pp. 13–15; Kodavanti, "Neurotoxicity of Persistent Organic Pollutants," pp. 296–298;

95. Christopher Lau, Katherine Anitole, Colette Hodes, David Lai, Andrea Pfahles-Hutchens, and Jennifer Seed, "Perfluoroalkyl Acids: A Review of Monitoring and Toxicological Findings," *Toxicological Sciences* 99, no. 2 (2007): pp. 366–394, at p. 366; Environmental Working Group, "PFCs: Global Contaminants," p. 1, www.ewg.org (accessed July 31, 2009).

96. Lau et al., "Perfluoroalkyl Acids: A Review of Monitoring and Toxicological Findings," p. 366.

97. Ibid.

98. Environmental Working Group, "PFCs: Global Contaminants," p. 2.

99. Lau et al., "Perfluoroalkyl Acids: A Review of Monitoring and Toxicological Findings," p. 367.

100. Ibid.

101. Antonia M. Calafat, Lee-Yang Wong, Zsuzsanna Kuklenyik, John A. Reidy, and Larry L. Needham, "Polyfluoroalkyl Chemicals in the U.S. Population: Data from the National Health and Nutrition Examination Survey (NHANES) 2003–2004 and Comparisons with NHANES 1999–2000," *Environmental Health Perspectives* 115, no. 11 (2007): pp. 1596–1602, at pp. 1599, 1596.

102. Ibid., p. 1599.

103. Ibid. pp. 1599, 1596.

104. U.S. Environmental Protection Agency, "Basic Information on PFOA," www.epa.gov/oppt/pfoa (accessed July 31, 2009).

105. Arthur C. Guyton, *Textbook of Medical Physiology* (Philadelphia: W. B. Saunders, 1986), quotation at pp. 876 (emphasis in original), 972–973.

106. Dolores Ibarreta and Shanna H. Swan, "The DES Story: Long-Term Consequences of Prenatal Exposure," in Harremoës et al., *Late Lessons from Early Warnings,* p. 84.

107. Guyton, *Textbook of Medical Physiology,* pp. 974–975.

108. On creation of DES, Ibarreta and Swan, "The DES Story," p. 84; on FDA approval in 1943 and 1947, Edward W. Lawless, *Technology and Social Shock* (New Brunswick, N.J.: Rutgers University Press, 1977), pp. 71–72; Philip J. Hilts, *Protecting America's Health: The FDA, Business and One Hundred Years of Regulation* (New York: Alfred A. Knopf, 2008), p. 162; Nancy Langston, *Toxic Bodies: Hormone Disruptors and the Legacy of DES* (New Haven: Yale University Press, 2010), quotation at p. 60; more generic point pp. 55-60.

109. Lawless, *Technology and Social Shock*, p. 72.

110. Ibid., p. 72.

111. Ibid., pp. 72–73.

112. Ibid., pp. 73–75.

113. Quotation and FDA and USDA not testing until 1965, both ibid., at p. 75.

114. Ibid., pp. 76–77, 80.

115. Langston, *Toxic Bodies*, p. 118.

116. David Biello, "Plastic (Not) Fantastic: Food Containers Leach a Potentially Harmful Chemical: Is Bisphenol A, a Major Ingredient in Many Plastics, Healthy for Children and Other Living Things?" *Scientific American,* February 19, 2008, www.sciam.com (accessed July 25, 2008), and Environmental Working Group, "Timeline: BPA from Invention to Phase-out," www.ewg.org (accessed, July 25, 2008).

117. Biello, "Plastic (Not) Fantastic"; E. C. Dodds and W. Lawson, "Molecular Structure in Relation to Oestrogenic Activity: Compounds without A Phenanthrene Nucleus," *Proceedings of the Royal Society:* London B. 125 (1938): pp. 222–232, as cited in Environmental Working Group "Timeline: BPA from Invention to Phase-out."

118. Biello, "Plastic (Not) Fantastic."

119. Ibid.

120. U.S. Environmental Protection Agency, High Production Volume (HPV) Challenge Program," 2004, www.epa.gov/chemrtk/index.htm (accessed May 11, 2008].

121. Calafat et al., "Exposure of the U.S. Population to Bisphenol A," p. 39.

122. Ibid.

123. Ibid.

124. Richard W. Stahlhut, Wade V. Welshons, and Shanna H. Swan, "Bisphenol A Data in NHANES Suggest Longer Than Expected Half-Life, Substantial Non-Food Exposure, or Both," *Environmental Health Perspectives* 117, no. 5, p. 3; Susanne Rust, "BPA Lingers in Body, Study Finds," *Milwaukee Journal Sentinel,* January 28, 2009, www.jsonline.com/watchdog/watchdogreports/38515489 .html (accessed January 28, 2009).

125. Donald J. Ecobichon, "Toxic Effects of Pesticides," in *Casarett and Doull's Toxicology,* 6th ed., ed. Curtis D. Klaassen (New York: McGraw-Hill, 2001): pp. 763–810, p. 763.

126. Ibid., p. 769.

127. Ibid. pp. 769, 771.

128. Ibid., p. 770.

129. Department of Health and Human Services, *Third National Report on Human Exposure to Environmental Chemicals,* p. 309.

130. Heather Hamlin and Wendy Hessler, "Sleeping with the Enemy: Indoor Airborne Contaminants," www.environmentalhealthnews.org/ehs/newscience/indoor-measure-of-airborne-contaminants (accessed April 28, 2009); Robert W. Gale, Walter L. Cranor, David A. Alvarez, James N. Huckins, Jimmie D. Petty, and Gary L. Robertson, "Semivolatile Organic Compounds in Residential Air along the Arizona-Mexico Border," *Environmental Science and Technology* 43, no. 9 (2009): pp. 3054–3060, pubs.acs.org/doi/pdfplus/10.1021/es803482u (accessed on May 28, 2009).

131. Gale et al., "Semivolatile Organic Compounds in Residential Air," quotation at p. 3054; on many compounds being part of our body burdens, some below limits of detection, Department of Health and Human Services, *Third National Report on Human Exposure to Environmental Chemicals,* p. 349.

132. Ecobichon, "Toxic Effects of Pesticides," p. 775.

133. Ibid., pp. 772–773.

134. Department of Health and Human Services, *Third National Report on Human Exposure to Environmental Chemicals,* p. 349.

135. Ibid.

136. On concentrations of workers being fifty times higher, ibid; on Arizona home study, Gale et al., "Semivolatile Organic Compounds in Residential Air," p. 3056.

3. DISCOVERING DISEASE, DYSFUNCTION, AND DEATH BY MOLECULES

1. Michael Waalkes, "Fetal Arsenic Exposure and Adult Cancer" (presentation at PPTOX II: Role of Environmental Stressors in the Developmental Origins of Disease, Miami, Florida, December 7–10, 2009).

2. Jack Lewis, "Lead Poisoning: A Historical Perspective," *EPA Journal,* May 1985, www.epa.gov. (accessed July 20, 2008).

3. Andreas Kortenkamp, "Breast Cancer and Exposure to Hormonally Active Chemicals: An Appraisal of the Scientific Evidence," background paper published by the Health and Environment Alliance and CHEM Trust, www.chemtrust.org.uk (accessed August 19, 2008); Barbara A. Cohn et al., "DDT and Breast Cancer in Young Women: New Data on the Significance of Age at Exposure," *Environmental Health Perspectives* 115, no. 10 (2007): pp. 1406–1414, at p. 1406.

4. On the IBM study being too small, John C. Bailar III, "How to Distort the Scientific Record without Actually Lying: Truth, and the Arts of Science," *European Journal of Oncology* 11 (2006): pp. 217–224, quotation at p. 219. See also researchers who first called attention to the issue: John C. Bailar III, M. Bobak, B. Fowler, et al., "Open Letter to the Health and Safety Executive," *International Journal of Occupational and Environmental Health* 6 (2000): pp. 71–72; Joe LaDou, "Occupational Health in the Semiconductor Industry," in *Challenging the Chip: Labor Rights and Environmental Justice in the Global Electronics Industry,*

ed. T. Smith, D. A. Sonnenfel, and D. N. Pellow (Philadelphia: Temple University Press, 2006). On elevated risks of cancer among employees, Richard W. Clapp, "Mortality among US Employees of a Large Computer Manufacturing Company: 1969–2001," *Environmental Health: A Global Access Science Source* 5 (2006): pp. 1–10.

5. International Agency for Research on Cancer (hereafter, IARC), "Preamble," *IARC Monographs on the Evaluation of Carcinogenic Risks to Humans*, section B(2)(f), monographs.iarc.fr/ENG/Preamble/CurrentPreamble.pdf (accessed December 26, 2009).

6. On studies designed to minimize or not discover adverse outcomes, David Michaels, *Doubt Is Their Product: How Industry's Assault on Science Threatens Your Health* (New York: Oxford University Press, 2008), pp. 60–78. See also Carl F. Cranor, *Toxic Torts: Science, Law and the Possibility of Justice* (New York: Cambridge University Press, 2006), pp. 220–264. On changes to chromium industry study, Michaels, *Doubt Is Their Product*, pp. 105–108. On the history of misleading reports, Margaret A. Berger, "Eliminating General Causation: Notes towards a New Theory of Justice and Toxic Torts," *Columbia Law Review* 97 (1997): pp. 2117–2152, at pp. 2135–2136; Nelson Wyatt, "Cigarette Maker Destroyed Studies on Tobacco, Researchers Say," *Star*, www.thestar.com/news/canada/article/710458—cigarette-maker-destroyed-studies-on-tobacco-researchers-say (accessed December 31, 2009).

7. Jason Dearon, "Calif. Regulators Will Not List Bisphenol A under Prop. 65, Call for More Study," *Los Angeles Times*, July 7, 2009, www.latimes.com (accessed July 7, 2009).

8. J. William Langston, Phillip Ballard, James W. Tetrud, and Ian Irwin, "Chronic Parkinsonism in Humans Due to a Product of Meperidine-Analog Synthesis," *Science* 219 (1983): pp. 979–980, quotations at p. 979. On symptoms of parkinsonism appearing within a week or less, "The Case of the Frozen Addict," *NOVA*, WGBH and BBC, February 18, 1986, www.pbs.org/wgbh/nova/listseason/13.html (accessed January 10, 2010); Langston, Ballard, Tetrud, and Irwin, "Chronic Parkinsonism in Humans due to a Product of Meperidine-Analog Synthesis," pp. 979.

9. IARC, "Preamble," *IARC Monographs on the Evaluation of Carcinogenic Risks to Humans*, section B(2); Abraham G. Hartzema, Miquel Porta, and Hugh H Tilson, eds., *Pharmacoepidemiology: An Introduction*, 3rd ed. (Cincinnati, Ohio: Harvey Whitney Books, 1998), p. 77.

10. Institute of Medicine and National Research Council, Committee on the Framework for Evaluating the Safety of Dietary Supplements, *Dietary Supplements: A Framework for Evaluating Safety* (Washington, D.C.: National Academy Press, 2005), pp. 131–132; J. P. Collet et al. "Monitoring Signals for Vaccine Safety: The Assessment of Individual Adverse Event Reports by an Expert Advisory Committee," *Bulletin of the World Health Organization* 7 (2000): pp. 178–185.

11. U.S. Food and Drug Administration, Center for Drug Evaluation and Research, "Adverse Event Reporting System," www.fda.gov/cder/aers/default.htm (accessed June 20, 2008).

12. IARC, "Preamble," *IARC Monographs on the Evaluation of Carcinogenic Risks to Humans,* section B(2)(a).

13. Clark W. Heath, Jr. and Henry Falk, "Characteristics of Cases of Angiosarcoma of the Liver Among Vinyl Chorlide Workers in the United States," *Annals of the New York Academy of Sciences,* (January 31,1975): pp. 231-236, at p. 233.

14. U.S. Department of Health and Human Services, Food and Drug Administration, "Sandoz Pharmaceuticals Corp.; Bromocriptine Mesylate (Parlodel) for the Prevention of Physiological Lactation; Opportunity for a Hearing on a Proposal to Withdraw Approval of the Indication," *Federal Register* 59 (August 1994): p. 43347.

15. Ibid.

16. Fabrice Larrazet, Christian Spaulding, Henri J. Lobreau, Simon Weber, and Francois Guerin, "Bromocriptine-Induced Myocardial Infarction," *Annals of Internal Medicine* 118 (1993): pp. 199–200, at p. 199.

17. Ibid., p. 200.

18. Ibid.

19. Kenneth Rothman, *Modern Epidemiology* (Boston: Little, Brown, 1986): quotations at p. 51.

20. Ibid., p. 52.

21. "The Nuremberg Code" (1947), and World Medical Association, "Declaration of Helsinki: Ethical Principles for Medical Research Involving Human Subjects," both in *Contemporary Issues in Bioethics,* 6th ed., ed. Tom L. Beauchamp and LeRoy Walters (Belmont, Calif.: Wadsworth Publishing, 2003), pp. 354–363.

22. Rothman, *Modern Epidemiology,* p. 55.

23. Ibid., quotation at p. 56; on case-control studies, p. 74.

24. Carl F. Cranor, *Regulating Toxic Substances: A Philosophy of Science and the Law* (1993; repr., New York: Oxford University Press, 1997), p. 29.

25. U.S. National Institutes of Health, National Cancer Institute, "Soft Tissue Sarcoma," www.cancer.gov (accessed September 23, 2008); U.S. National Institutes of Health, National Cancer Institute, "Non-Hodgkin Lymphoma," www.cancer.gov (accessed September 23, 2008).

26. Manolis Kogevinas, Timo Kauppinen, Regina Winkelmann, Heiko Becher, Pier Alberto Bertazzi, H. Bas Bueno-de-Mesquita, David Coggon, et al., "Soft Tissue Sarcoma and Non-Hodgkin's Lymphoma in Workers Exposed to Phenoxy Herbicides, Chlorophenols, and Dioxins: Two Nested Case-Control Studies," *Epidemiology* 6 (1995): pp. 396–402, at pp. 396, 401.

27. On follow-up studies, Rothman, *Modern Epidemiology,* p. 57; on suggestion made by physicians who first identified MPTP-caused parkinsonism, William J. Langston, Elizabeth Langston, and Ian Irwin, "MPTP-Induced Parkinsonism in Human and Non-Human Primates: Clinical and Experimental Aspects," *Acta Neurologica Scandinavia* 70 (1984): pp. 49–54, at p. 53.

28. Gwen Collman, "Overview of Developmental Basis of Disease: Environmental Impacts" (presentation at PPTOX II: Role of Environmental Stressors in the Developmental Origins of Disease, Miami, Florida, December 7–10, 2009).

29. David Bellinger and Herbert L. Needleman, "The Neurotoxicity of Prenatal Exposure to Lead: Kinetics, Mechanism and Expressions," in *Prenatal Exposure to*

Toxicants, ed. Herbert L. Needleman and David Bellinger (Baltimore: Johns Hopkins University Press 1994): pp. 89–111, at p. 92.

30. Rothman, *Modern Epidemiology,* pp. 58, 72.

31. "The Case of the Frozen Addict."

32. Arthur L. Herbst, Howard Ulfelder, and David C. Poskanzer, "Adenocarcinoma of the Vagina: Association of Maternal Stilbestrol Therapy with Tumor Appearance in Young Women," *New England Journal of Medicine* 284 (1971): pp. 878–881, at p. 880.

33. On long induction period for cancers after DES exposures, ibid; on relatively shorter latency period, Rothman, *Modern Epidemiology,* p. 14.

34. D. B. Calne, A. Eisen, E. McGeer, and P. Spencer, "Hypothesis: Alzheimer's Disease, Parkinson's Disease, and Motoneurone Disease: Abiotropic Interaction between Ageing and Environment?" *Lancet,* 328, no. 8515 (1986): pp. 1067–1070, at p. 1068.

35. Jerrold J. Heindel, "Animal Models for Probing the Developmental Basis of Disease and Dysfunction Paradigm," *Basic and Clinical Pharmacology and Toxicology* 102 (2008): pp. 76–81, at pp. 79–80.

36. Manilas Kogevinas and Paolo Boffetta, letter to the editor, criticizing a study by O. Wong ("A Cohort Mortality Study and a Case Control Study of Workers Potentially Exposed to Styrene in Reinforced Plastics and Composite Industry," *British Journal of Industrial Medicine* 47 [1990]: pp. 753–762), *British Journal of Industrial Medicine* 48 (1990): pp. 575–576.

37. On the need to be alert to errors or deliberate strategies, ibid. On a study being conducted for a sufficient period of time, David Michaels, *Doubt Is Their Product.* See also Carl F. Cranor, "The Tobacco Strategy Entrenched," review of Michaels's *Doubt Is Their Product, Science* 321 (2008): pp. 1296–1297. On conditions under which "no effect" studies correctly show no evidence of adverse effects, IARC, "Preamble," *IARC Monographs on the Evaluation of Carcinogenic Risks to Humans,* section B.2 (f).

38. Philippe Grandjean and Philip Landrigan, "Developmental Neurotoxicity of Industrial Chemicals," *Lancet* 368 (December 2006): pp. 2167–2178.

39. National Institutes of Health, National Institute of Child Health and Human Development, National Children's Study, www.nationalchildrensstudy.gov (accessed September 23, 2008).

40. J. J. Schlesselman, "Sample Size Requirements in Cohort and Case-Control Studies of Disease," *American Journal of Epidemiology* 99 (1974): pp. 381–384, at pp. 382–383, as cited in Cranor, *Regulating,* p. 36.

41. On case-control studies being better at finding causes of rare adverse effects, Rothman, *Modern Epidemiology,* p. 68; on early research on DES, Herbst, Ulfelder, and Poskanzer, "Adenocarcinoma of the Vagina," p. 878.

42. Rothman, *Modern Epidemiology,* pp. 58–60.

43. James Huff and David P. Rall, "Relevance to Humans of Carcinogenesis Results from Laboratory Animal Toxicology Studies," in *Maxcy-Rosenau-Last Public Health and Preventive Medicine,* 13th ed., ed. J. M. Last and R. B. Wallace (Norwalk, Conn.: Appleton and Lange, 1992): pp. 433–452, at p. 433.

44. Elaine M. Faustman and Gilbert S. Omenn, "Risk Assessment," in *Casarett and Doull's Toxicology,* 6th edition, ed. Curtis Klaassen (New York: Pergamon Press, 2001), pp. 83–104, at p. 86.

45. Vincent James Cogliano, Robert A. Baan, Kurt Straif, Yann Grosse, Marie Beatrice Secretan, Fatiha El Ghissassi, and Paul Kleihues, "The Science and Practice of Carcinogen Identification and Evaluation," *Environmental Health Perspectives* 112, no. 13 (2004): pp. 1269–1274, esp. p. 1270.

46. Richard L. Canfield et al., "Low-Level Lead Exposure and Executive Functioning in Young Children," *Child Neuropsychology* 9 (2003): pp. 35–53, and David Bellinger and Herbert L. Needleman, "Intellectual Impairment and Blood Lead Levels," *New England Journal of Medicine* 349, no. 5 (2003): pp. 500–502; R. I. Canfield, C. R. Henderson, D. A. Cory-Slechta, C. Cox, T. A. Jusko, et. al, "Intellectual Impairment in Children with Blood Lead Concentrations below 10 Micrograms per Deciliter," *New England Journal of Medicine* 348 (2003): pp. 1517–1526, as cited in Bruce P. Lanphear, Charles V. Vorhees, and David C. Bellinger, "Protecting Children from Environmental Toxins," *PLoS Medicine* 2 (2005): pp. 199–204, at p. 201, www.plosmedicine.org (accessed July 18, 2009).

47. James L. Schardein and Orest T. Macina, *Human Developmental Toxicants: Aspects of Toxicology and Chemistry* (Boca Raton, Fla.: Taylor and Francis, 2007), pp. 133–134.

48. M. Carbone, G. Klein, J. Gruber, and M. Wong, "Modern Criteria to Establish Human Cancer Etiology," *Cancer Research* 64 (2004): pp. 5518–5524, at p. 5519.

49. Brenda Eskenazi, Lisa G. Rosas, Amy R. Marks, Asa Bradman, Kim Harley, Nina Holland, Caroline Johnson, et al., "Pesticide Toxicity and the Developing Brain," *Basic and Clinical Pharmacology and Toxicology* 102 (2008): pp. 228–236, at p. 235.

50. Cogliano et al., "Science and Practice," p. 1270.

51. IARC, "Preamble," *IARC Monographs on the Evaluation of Carcinogenic Risks to Humans,* section B.2(b), "Quality of Studies Considered"; on volunteers in a colon cancer study, Kenneth J. Rothman, *Epidemiology: An Introduction* (New York: Oxford University Press, 2002), pp. 96–97; on Down's syndrome, Rothman, *Epidemiology: An Introduction,* pp. 101–102.

52. Ibid., pp. 96–99.

53. Michaels, *Doubt Is Their Product,* pp. 105–108.

54. On "healthy worker effect," Rothman, *Epidemiology: An Introduction,* pp. 97–98. See also Rothman, *Modern Epidemiology,* pp. 83–84. On workers being healthier and having better health care, Russellyn S. Carruth and Bernard D. Goldstein, "Relative Risk Greater than Two in Proof of Causation in Toxic Tort Litigation," *Jurimetrics Journal* 41 (2007), pp. 195–209, at pp. 207–208.

55. Rothman, *Epidemiology: An Introduction,* p. 113.

56. Rothman, *Modern Epidemiology,* pp. 119–120.

57. Cranor, *Regulating,* pp. 31–38.

58. On assessing whether the observed relationship is causal, Austin Bradford Hill,

"The Environment and Disease: Association or Causation?" *Proceedings of the Royal Society of Medicine* 58 (1965): pp. 295–300, repr. in *Evolution of Epidemiologic Ideas: Annotated Readings on Concepts and Methods,* ed. Sander Greenland (Newton Lower Falls, Mass.: Epidemiology Resources, 1987), pp. 15–20. For extensive discussions of these issues, see Cranor, *Toxic Torts,* pp. 102–105, 170–192, 227–259.

59. John Rogers, "Animal Models for Testing the DOHaD [Developmental Origins of Health and Disease] Hypothesis" (presentation at PPTOX II: Role of Environmental Stressors in the Developmental Origins of Disease, Miami, Florida, December 7–10, 2009).

60. Langston, Langston, and Irwin, "MPTP-Induced Parkinsonism in Human and Non-human Primates," pp. 49–54.

61. Bernard D. Goldstein, "Toxic Torts: The Devil Is in the Dose," *Journal of Law and Policy* 16 (2008): pp. 551–587, at pp. 556–557.

62. Jane Kay, "NIEHS Director Birnbaum: 'We Kind of Jump from the Proverbial Fry Pan into the Fire' When Replacing Chemicals," *Environmental Health News,* November 9, 2009, http://www.environmentalhealthnews.org/ehs/news/q-a-with-linda-birnbaum (accessed November 9, 2009).

63. Heindel, "Animal Models for Probing the Developmental Basis of Disease and Dysfunction Paradigm," p. 79.

64. On typical exposure levels of maneb in food and on paraquat in contrast with maneb in the environment, Extension Toxicology Network, "Pesticide Information Profile for Maneb," pmep.cce.cornell.edu/profiles (accessed September 23, 2008).

65. B. K. Barlow, D. A. Cory-Slechta, E. K. Richfield, and M. Thuruchelvam, "The Gestational Environment and Parkinson's Disease: Evidence for Neurodevelopmental Origins of a Neurodegenerative Disorder," *Reproductive Toxicology* 23 (2007): pp. 457–470, at p. 463.

66. Goldstein, "Toxic Torts" p. 556.

67. Huff and Rall, "Relevance to Humans," p. 433.

68. Eric Plutzer and Michael Berkman, "Trends: Evolution, Creationism, and the Teaching of Human Origins in Schools," *Public Opinion Quarterly* 72 (2008): pp. 540–553.

69. Dearon, "Calif. Regulators Will Not List Bisphenol A under Prop. 65" (emphasis added).

70. Hill, "The Environment and Disease: Association or Causation?"

71. Goldstein, "Toxic Torts," quotations at p. 557.

72. David P. Rall, M. D. Hogan, James E. Huff, B. A. Schwetz, and R. W. Tennant, "Alternatives to Using Human Experience in Assessing Health Risks," *Annual Review of Public Health* 8 (1987): pp. 355–385, quotation at p. 356; Goldstein, "Toxic Torts," p. 556.

73. Huff and Rall, "Relevance to Humans," quotation at p. 434. Significant scientific understanding of neural transmission, renal function, and cell replication and development of cancer have come from nonhuman species, often species far removed phylogenetically from humans. On scientists having confirmed the similarity at the gene and protein levels, Mark S. Boguski, "Comparative Genomics: The

Mouse that Roared," *Nature* 420 (2002): pp. 515–516; The Mouse Genome Sequencing Consortium and Alec MacAndrew, "Comparison of Mouse and Human Coding Genes," *Nature* 420 (2002): pp. 520–562.

74. James Huff, "Chemicals and Cancer in Humans: First Evidence in Experimental Animals," *Environmental Health Perspectives* 100 (1993): pp. 201–210, at p. 204; U.S. Environmental Protection Agency, "Proposed Guidelines for Carcinogen Risk Assessment," *Federal Register* 61 (1996): p. 17977. (According to this report, "there is evidence that growth control mechanisms at the level of the cell are homologous among mammals, but there is no evidence that these mechanisms are site concordant." That is, that they must be in the same tissue in rodents and humans.) Bernard D. Goldstein and Mary Sue Henifin, "Reference Guide on Toxicology," in *Reference Manual on Scientific Evidence,* 2nd ed., ed. Federal Judicial Center (New York: LEXIS, 2000), p. 419.

75. Huff, "Chemicals and Cancer," pp. 201, 204. Huff states here that the array and multiplicity of carcinogenic processes are virtually common among mammals, as, for instance, between laboratory rodents and humans.

76. See, generally, U.S. Department of Health and Human Services, Taskforce on Health Risk Assessment, *Determining Risks to Health: Federal Policy and Practice* (Dover, Mass.: Auburn House Publishing, 1986), esp. pp. 10–13, for information on testing in the Food and Drug Administration. For more general discussions, see U.S. Congress, Office of Technology Assessment, *Identifying and Regulating Carcinogens* (Washington, D.C.: Government Printing Office, 1986), and National Research Council, *Risk Assessment in the Federal Government* (Washington, D.C.: Government Printing Office, 1983).

77. Cogliano et al., "Science and Practice," p. 1270.

78. IARC, "Preamble," *IARC Monographs on the Evaluation of Carcinogenic Risks to Humans,* section B(3) (emphasis in original).

79. Huff and Rall, "Relevance to Humans," quotation at p. 437. On other distinguished scientists and scientific communities in agreement, see National Academy of Sciences, "Pest Control: An Assessment of Present and Alternative Technologies," in *Contemporary Pest Control Practices and Prospects: The Report of the Executive Committee,* Washington, D.C., 1975, pp. 66–83, at p. 66, as cited in Huff, "Chemicals and Cancer," p. 205; see also Victor A. Fung et al., "The Carcinogenesis Bioassay in Perspective: Application in Identifying Human Cancer Hazards," *Environmental Health Perspectives* 103, nos. 7–8 (1995), pp. 680–683, at pp. 682. See also Lorenzo Tomatis, James Huff, Irva Hertz-Picciotto, Dale P. Sandler, John Bucher, Paolo Boffetta, Olav Axelson, et al., "Avoided and Avoidable Risks of Cancer," *Carcinogenesis* 18 (1997): pp. 95–105, esp. p. 97.

80. Institute of Medicine and National Research Council, *Dietary Supplements,* p. 157. (The quotation is set in boldface in the original.)

81. Kenneth S. Korach, preface to *Reproductive and Developmental Toxicology,* ed. Kenneth S. Korach (New York: Marcel Decker, 1998): pp. v-vi, at p. v.

82. John M. Rogers and Robert J. Kavlock, "Developmental Toxicology," in *Casarett and Doull's Toxicology,* 6th ed., ed. Curtis Klaassen, (New York: Pergamon Press, 2001): pp. 351–386, quotation at p. 374 (emphasis added).

83. Schardein and Keller as quoted in Rogers and Kavlock, "Developmental Toxicology," at p. 374; Jelovsek as cited in Rogers and Kavlock, "Developmental Toxicology," p. 374.

84. Cited in Rogers and Kavlock, "Developmental Toxicology," p. 374

85. Michael Skinner, "Epigenetic Transgenerational Actions of Endocrine Disruptors on Reproduction and Disease: The Ghosts in Your Genes" (presentation at PPTOX II: Role of Environmental Stressors in the Developmental Origins of Disease, Miami, Florida, December 7–10, 2009).

86. Institute of Medicine and National Research Council, *Dietary Supplements,* p. 157; National Academy of Sciences, "Pest Control, " quoted in Huff and Rall, "Relevance to Humans," p. 437. The IARC concurs with the National Academy of Sciences: "It is biologically plausible . . . to regard agents and mixtures for which there is sufficient evidence of carcinogenicity in experimental animals as if they presented a carcinogenic risk to humans" (IARC, "Preamble," *IARC Monographs on the Evaluation of Carcinogenic Risks to Humans,* section B[2][f]).

87. See National Research Council, *Risk Assessment,* pp. 24–27; see also U.S. Congress, Office of Technology Assessment, *Identifying and Regulating Carcinogens,* pp. 39, 46.

88. On mechanisms that may not operate in humans, see IARC, "Preamble," *IARC Monographs on the Evaluation of Carcinogenic Risks to Humans;* Jerry M. Rice, "Editorial: On the Application of Data on Mode of Action to Carcinogenesis," *Toxicological Sciences* 49 (1999): pp. 175–177. On basing assertions on evidence that different mechanisms are acting in animals than in humans, see James Huff, "Response: Alpha-2u-Globulin Nephropahty, Posed Mechanism, and White Raves" [letter]," pp. 1264–1267, and R. L. Melnick, M. C. Kohn, J. Huff, "Weight of Evidence versus Weight of Speculation to Evaluate the Alpha2u-Globulin Hypothesis," *Environmental Health Perspectives* 105(9) (1997): pp. 904–906.

89. National Academy of Sciences, Committee on Improving Risk Analysis Approaches Used by the U.S. EPA, National Research Council, "Science and Decisions: Advancing Risk Assessment" (free executive summary), pp. 7–8, www.nap.edu/catalog/12209.html (Accessed July 27, 2009).

90. National Academy of Sciences, Committee on Improving Risk Analysis Approaches Used by the U.S. EPA, National Research Council, *Science and Decisions: Advancing Risk Assessment* (Washington, D.C.: National Academy Press, 2008), pp. 126–129.

91. Rogers and Kavlock, "Developmental Toxicology," p. 374.

92. On regarding experiments as strongly presumptive evidence of toxicity in humans, Institute of Medicine and National Research Council, *Dietary Supplements,* p. 157. On not contravening the well-established use of animal studies, Huff, "Alpha-2u-Globulin," pp. 1264–1267, and Melnick, Kohn, Huff, "Weight of Evidence versus Weight of Speculation," pp. 904–906.

93. Huff and Rall, "Relevance to Humans," p. 440. See also T. Woodruff, L. Zeise, D. A. Axelrad, K. Z. Guyton, S. Janssen, M. Miller, G. G. Miller, et al., "Meeting Report: Moving Upstream—Evaluating Adverse Upstream End Points for Improved Risk Assessment and Decision-Making," *Environmental Health Perspectives,* 116, no. 11 (2008): pp. 1568–1575.

94. Faustman and Omenn, "Risk Assessment," quotations at p. 86.

95. Food and Drug Administration, "Final Rule Declaring Dietary Supplements Containing Ephedrine Alkaloids Adulterated Because They Present an Unreasonable Risk," *Federal Register* 69 (2005): pp. 6787–6854; I. Furuya and S. Watanabe, "Discriminative Stimulus Properties of Ephedra Herb *(Ephedra sinica)* in Rats, *Yakubutsu Seishin Kodo* 13 (1993): pp. 33–38; C. R. Lake and R. S. Quirk, "CNS Stimulants and the Look-Alike Drugs," *Psychiatric Clinics of North American* 7 (1984): pp. 689–701, all as cited in Institute of Medicine and National Research Council, *Dietary Supplements,* pp. 205–206.

96. On molecular structures similar to valproic acid, retinoic acid, and glycol ethers providing clues, Faustman and Omenn, "Risk Assessment," p. 86. On relationships providing strong but not always infallible reasons, ibid. pp. 83–104; J. Ashby and R. W. Tennant, "Chemical Structure, Salmonella Mutagenicity and Extent of Carcinogenicity as Indicators of Genotoxic Carcinogenesis among 222 Chemicals Tested in Rodents by the U.S. NCI/NTP," *Mutation Research* 204 (1988): pp. 17–115. On industry utilizing such information during product development, Richard Dennison, "U.S. HPV Challenge and Beyond" (presentation to Ward, Kershaw, and Center for Progressive Regulation Environmental Law Symposium: The Data Gaps Dilemma: Why Toxic Ignorance Threatens Public Health, Washington, D.C., May 6, 2005).

97. On the Ah receptor tending to increase the toxicity of certain substances and sometimes decreasing others, Faustman and Omenn, "Risk Assessment," p. 86. IARC, "Polychlorinated Dibenzo-*para*-Dioxins," *IARC Monographs on the Evaluation of Carcinogenic Risks to Humans,* p. 343, monographs.iarc.fr/ENG/ Monographs/vol69/mono69–5E.pdf (accessed December 16, 2008).

98. Andreas Kortenkamp, "Low-Level Exposure to Multiple Chemicals: Reason for Human Health Concerns?" *Environmental Health Perspectives* 115, no S-1 (2007): pp. 106–114, at p. 113.

99. On certain substances binding to molecules by electronic transfer and interfering with DNA, Zoltan Gregus and Curtis D. Klaassen, "Mechanisms of Toxicity," in *Casarett and Doull's Toxicology,* 6th ed., ed. Curtis Klaassen (New York: Pergamon Press, 2001): pp. 28–76, at pp. 45–46; on other compounds possibly activating or inactivating molecular pathways, Carbone et al., "Modern Criteria," p. 5519.

100. D. B. Calne, J. William Langston, W. R. Wayne Martin, A. Jon Stoessl, Thomas J. Ruth, Micahel J. Adam, Brian d. Pate, & Michael Schulzer, "Positron Emission Tomography after MPTP: Observations Relating to the Cause of Parkinson's Disease," *Nature* 317 (1985): pp. 246–248.

101. Langston, Langston, and Irwin, "MPTP-Induced Parkinsonism in Human and Non-Human Primates," p. 51.

102. Heindel, "Animal Models for Probing the Developmental Basis of Disease and Dysfunction Paradigm," pp. 79–80.

103. Beate Ritz, Angelika Manthripragada, Sadie Costello, Sarah Lincoln, Matthew Farrer, Myles Cockburn, and Jeff Bronstein, "Dopamine Transporter Genetic Variants and Pesticides in Parkinson's Disease," *Environmental Health Perspectives* 117, no. 6 (2009): pp. 964–969, at p. 964; Barlow, Cory-Slechta,

Richfield, and Thuruchelvam, "The Gestational Environment and Parkinson's Disease: Evidence for Neurodevelopmental Origins of a Neurodegenerative Disorder," p. 257.

104. On molecular-level data assisting the discovery of other toxic effects, Cogliano et al., "Science and Practice," p. 1270; Carbone et al., "Modern Criteria," pp. 5519–5520; IARC, "Preamble," *IARC Monographs on the Evaluation of Carcinogenic Risks to Humans,* at section B(4)(b). On research based "primarily [on] mechanistic information," Carbone et al., "Modern Criteria," quotation at p. 5519 (emphasis added).

4. CAVEAT PARENS:
A NATION AT RISK FROM CONTAMINANTS

1. Prue Talbot, Department of Cell Biology & Neuroscience at the University of California, Riverside, and developmental biologist, in a personal communication (December 10, 2009) reports that this was routinely taught to graduate students in that era; : Herbert L. Needleman and David Bellinger, eds., preface to *Prenatal Exposure to Toxicants* (Baltimore: Johns Hopkins University Press 1994), p. ix.

2. Talbot, ibid., recalls that this was a recommendation to pregnant women.

3. American Pregnancy Association, "The Fetal Life-Support System: Placenta, Umbilical Cord, and Amniotic Sac," www.americanpregnancy.org/duringpregnancy/fetallifesupportsystem.html, (accessed November 7, 2008).

4. Ronald D. Hood, "Principles of Developmental Toxicology Revisited," in *Developmental and Reproductive Toxicology: A Practical Approach,* 2nd ed., ed. Ronald D. Hood (Boca Raton, Fla.: CRC Press, 2005): pp. 3–17, at p. 7 (emphasis added).

5. Herbert L. Needleman and David Bellinger, eds., preface to *Prenatal Exposure to Toxicants* (Baltimore: Johns Hopkins University Press 1994), p. ix.

6. U.S. Environmental Protection Agency, Water Quality Criteria, "Fish Tissue Criteria," www.epa.gov (accessed September 10, 2009).

7. Shun'ichi Honda, Lars Hylander, and Mineshi Sakamoto, "Recent Advances in Evaluation of Health Effects on Mercury with Special Reference to Methylmercury—A Minireview," *Environmental Health and Preventive Medicine* 11 (July 2006): pp. 171–176, at p. 176.

8. Masazumi Harada, Minamata Disease: Methylmercury Poisoning in Japan Caused by Environmental Pollution, *Critical Reviews in Toxicology,* 25(1) (1995): pp. 1-24, p.3. On symptoms of MeHg poisoning, J. McCurry, "Japan Remembers Minamata," *Lancet* 367, no. 9505 (2006): pp. 99–100; on extreme cases, Honda, Hylander and Sakamoto, "Recent Advances in Evaluation of Health Effects on Mercury," pp. 171–176; on animal behavior, McCurry, "Japan Remembers Minamata," p. 100.

9. On MeHg poisoning in children, Bernard Weiss, "The Developmental Neurotoxicity of Methyl Mercury," in *Prenatal Exposure to Toxicants: Developmental Consequences,* ed. Herbert L. Needleman and David Bellinger (Baltimore: Johns Hopkins University Press, 1994), pp. 112–129, at p. 116; National Institute of

Neurological Disorders and Stroke, Cerebral Palsy Information Page, "What is Cerebral Palsy?" www.ninds.nih.gov (accessed July 20, 2009); Harada, "Minamata Disease," p. 8, 16.

10. On the increased level of MeHg concentration in fetal blood, Honda, Hylander, and Sakamoto, "Recent Advances in Evaluation of Health Effects on Mercury," p. 173; on the fivefold level of concentration, Weiss, "The Developmental Neurotoxicity of Methyl Mercury," p. 113.

11. Weiss, "The Developmental Neurotoxicity of Methyl Mercury," p. 116; Philip J. Landrigan, Carole A. Kimmel, Adolfo Correa, and Brenda Eskenazi, "Children's Health and the Environment: Public Health Issues and Challenges for Risk Assessment," *Environmental Health Perspectives,* 112, no. 2 (2004): pp. 257–265, p. 260.

12. Weiss, "The Developmental Neurotoxicity of Methyl Mercury, p. 116.

13. National Research Council, Committee on the Toxicological Effects of Methylmercury, *Toxicological Effects of Methylmercury* (Washington, D.C.: National Academy Press, 2000), p. 18.

14. On thalidomide entering the market, James L. Schardein and Orest T. Macina, *Human Developmental Toxicants: Aspects of Toxicology and Chemistry* (Boca Raton, Fla.: Taylor and Francis, 2007), p. 129; Philip J. Hilts, *Protecting America's Health: The FDA, Business, and One Hundred Years of Regulation* (New York: Alfred A. Knopf, 2003), quotations at p. 147.

15. On thalidomide being more like a mild sedative, Hilts, *Protecting America's Health,* p. 147; ibid., quotation at p. 149; on estimates of peripheral neuropathy, ibid., p. 154.

16. Ibid., p. 149, pp. 149, 151, and p. 150.

17. Ibid., p. 154; Schardein and Macina, *Human Developmental Toxicants,* p. 131.

18. Hilts, *Protecting America's Health,* p. 153; Schardein and Macina, *Human Developmental Toxicants,* on functional deficits, pp. 131–132, on spontaneous abortions and affected children dying within the first year, pp. 130–134, on period after conception when birth defects were induced, p. 132.

19. On effects of a single dose, Luz Claudio, Winston C. Kwa, Allison L. Russell, and David Wallinga, "Testing Methods for Developmental Neurotoxicity of Environmental Chemicals," *Toxicology and Applied Pharmacology* 164, no. 1 (2000): pp. 1–14, at p. 6; on disease and dysfunction rate, Hilts, *Protecting America's Health,* pp. 154–155.

20. Hilts, *Protecting America's Health,* p. 158. See also Schardein and Macina, *Human Developmental Toxicants,* p. 131.

21. Schardein and Macina, *Human Developmental Toxicants,* p. 284.

22. Ibid., pp. 284–285. Schardein and Macina estimate that the number of women who took the drug could be as low as 1 million and as high as 10 million, with a best estimate of 3 million. On age of cancer onset, John M. Rogers and Robert J. Kavlock, "Developmental Toxicology," in *Casarett and Doull's Toxicology,* 6th ed., ed. Curtis Klaassen (New York: Pergamon Press, 2001), pp. 351–386, at p. 354; Arthur L. Herbst, Howard Ulfelder, and David C. Poskanzer, "Adenocarcinoma

of the Vagin: Association of maternal Stilbestrol therapy with Tumor Appearance in Young Women," *The New England Journal of Medicine,* 284, 18 (1971): pp. 878–881.

23. U.S. Department of Health and Human Services, Centers for Disease Control and Prevention, DES Update: Consumers, "Recent DES Research," located at www.cdc.gov/DES/consumers/research/recent.html, citing E.E. Hatch, J.R. Palmer, L. Titus-Ernstoff, K.L. Noller, R.H. Kaufman RH, et al., "Cancer Risk in Women Exposed to Diethylstilbestrol in Utero," *Journal of the American Medical Association,* 280 (1998): pp. 630–634.

24. On DES cancer risk, U.S. Department of Health and Human Services, Centers for Disease Control, "DES Update for You, Your Family and Your Health Care Provider," www.cdc.gov/DES/consumers/download/cdc_des_update.pdf, pp. 1–119, quotation at p. 19 (accessed July 20, 2008); on product liability cases, Schardein and Macina, *Human Developmental Toxicants,* p. 28.

25. On autopsies of 281 neonates, Lorna D. Johnson, Shirley G. Driscoll, Arthur T. Hertig, Philip T. Cole, and Rita J. Nickerson, "Vaginal Adenosis in Stillborns and Neonates Exposed to Diethylstilbestrol and Steroidal Estrogens and Progestins," *Obstetrics and Gynecology* 53 (1979): pp. 671–679; Medicine for the Family, "Vaginal Adenosis," www.medfamily.org (accessed August 3, 2008).

26. Retha Newbold, "DES Model: Cancer in Animals and Humans" (presentation at PPTOX II: Role of Environmental Stressors in the Developmental Origins of Disease, Miami, Florida, December 7–10, 2009).

27. On latency periods for cancers, Malcolm A. Smith, L. Rubinstein, R. S. Ungerleider, "Therapy-Related Acute Myeloid Leukemia Following Treatment with Epipodophyllotoxins: Estimating the Risks," *Medical and Pediatric Oncology* 23, no. 2 (1994): pp. 86–98, at p. 87; U.S. Department of Health Services, Centers for Disease Control, Agency for Toxic Substances and Disease Registry, "Toxicological Profile for Asbestos, Public Health Statement," p. 49 (2001); on abnormalities in male babies, Schardein and Macina, *Human Developmental Toxicants,* p. 287.

28. Rogers and Kavlock, "Developmental Toxicology," pp. 353–356.

29. Kim N. Dietrich and David Bellinger, "Assessment of Neurobehavioral Development in Studies of the Effects of Prenatal Exposure to Toxicants," in *Prenatal Exposure to Toxicants: Developmental Consequences,* ed. H. L. Needleman and D. C. Bellinger (Baltimore: Johns Hopkins University Press, 1994), pp. 57–85, at p. 57.

30. James L. Schardein, *Chemically Induced Birth Defects,* 3rd ed., rev. and expanded (New York: Marcel Dekker, 2000), p. 5; Rogers and Kavlock, "Developmental Toxicology," p. 364; Ana Soto, "Does Breast Cancer Begin in the Womb?" (presentation at International Conference of Fetal Programming and Developmental Toxicity, Torshavn, Faroe Islands, May 20–24, 2007).

31. On variety of factors of molecules, Hood, "Principles of Developmental Toxicology Revisited," pp. 8–9; on size of drug molecules, Schardein, *Chemically Induced Birth Defects,* p. 5.

32. On substances transferred via breast milk, Philippe Grandjean et al., "The Faroes Statement: Human Health Effects of Developmental Exposure to Chemicals

in Our Environment," *Basic and Clinical Pharmacology and Toxicology* 102 (2008): pp. 73–75, at p. 74; on molecular factors affecting which compounds will enter breast milk, Birger G. J. Heinzow, "Endocrine Disruptors in Human Breast Milk and the Health-Related Issues of Breastfeeding," in *Endocrine-Disrupting Chemicals in Food,* ed. I. Shaw (Cambridge: Woodhead Publishing, 2009), pp. 322–355, at pp. 324–325 and 322–347.

33. Philippe Grandjean, Esben Budtz-Jorgensen, Dana B. Barr, Larry L. Needham, Pal Weihe, and Birger Heinzow, "Elimination Half-Lives of Polychlorinated Biphenyl Congeners in Children," *Environmental Science and Technology,* 42(18) (2008): pp 6991–6996, p. 6991.

34. Quotation on the general point, Hood, "Principles of Developmental Toxicology Revisited," p. 7; Claudio, et al., "Testing Methods for Developmental Neurotoxicity of Environmental Chemicals," p. 2.

35. Kenneth S. Korach, preface to *Reproductive and Developmental Toxicology,* ed. Kenneth S. Korach (New York: Marcel Decker, 1998), pp. v-vi, at p. v.

36. Quotation on exposures as causal factors in childhood diseases, Mark D. Miller et al., "Differences between Children and Adults: Implications for Risk Assessment at California EPA," *International Journal of Toxicology* 21 (2002): pp. 403–418, at p. 412.

37. On results found from backtracking chromosomal damage, C. M. McHale and M. T. Smith, "Prenatal Origin of Chromosomal Translocations in Acute Childhood Leukemia: Implications and Future Directions," *American Journal of Hematology* 75 (2004): pp. 25425–25427, and M. F. Greaves and J. Wiemels, "Origins of Chromosome Translocations in Childhood Leukaemia," *Nature Reviews Cancer* 3 (2003): pp. 639–649, all as cited by Martyn Smith, Professor of Toxicology, School of Public Health, University of California, Berkeley, personal email communication (May 7, 2009). On damage to DNA in stem cells, Miller et al., "Differences between Children and Adults," p. 412.

38. Richard M. Sharpe, "Toxicity of Spermatogenicesis and Its Detection, in *Reproductive and Developmental Toxicology,* ed. Kenneth S. Korach (New York: Marcel Decker, 1998), pp. 625–634, p. 626.

39. Grace K. Lemaster and Sherry G. Selevan, "Toxic Exposures and Reproduction: A View of Epidemiology and Surveillance," in *Reproductive Toxicology and Infertility,* ed. Anthony R. Scialli and Michael J. Zinaman (New York: McGraw-Hill, 1993), pp. 307–321, at p. 311.

40. Matthew D. Anway, Stephen S. Rekow, and Michael K. Skinner, "Transgenerational Epigenetic Programming of the Embryonic Testis Transcriptome," *Genomics* 91 (2008): pp. 30–40, quotation at p. 30; Kwan Hee Kim, "Transgenerational Changes after Embryonic Exposure to Plasticizer Phthalates" (presentation at PPTOX II: Role of Environmental Stressors in the Developmental Origins of Disease, Miami, Florida, December 7–10, 2009); Smita Salian Mehta, "Perinatal Exposures of Rats to Bisphenol A Affects Fertility of Male Offspring" (presentation at PPTOX II: Role of Environmental Stressors in the Developmental Origins of Disease, Miami, Florida, December 7–10, 2009).

41. Lemaster and Selevan, "Toxic Exposures and Reproduction," p. 312; on bisphenol A exposures in mice, P. A. Hunt et al., "Bisphenol A Exposure Causes

Meiotic Aneuploidy in the Female Mouse," *Current Biology* 13 (2003): pp. 546–553, and M. Susiarjo et al., "Bisphenol A Exposure in Utero Disrupts Early Oogenesis in the Mouse," *PLoS Genetics* 3 (2007): p. e5, both cited in Sarah Vogel, "Battles over Bisphenol A," www.defendingscience.org (accessed July 25, 2009).

42. Rogers and Kavlock, "Developmental Toxicology," p. 356.

43. Ibid., quotation at p. 357, and p. 357.

44. Ibid., quotations at p. 358 (emphasis in original), and p. 358. See also Larry L. Needham, Antonia M. Calafat, and Dana B. Barr, "Assessing Developmental Toxicant Exposures via Biomonitoring," *Basic and Clinical Pharmacology and Toxicology: Prenatal Programming and Toxicity* 102 (February 2008): pp. 100–108, at p. 101, for a figure illustrating periods of organ susceptibility.

45. Rogers and Kavlock, "Developmental Toxicology," p. 358.

46. Ibid., p. 358–359.

47. See Grandjean et al., "The Faroes Statement," pp. 2167–2168; J. Dobbing, "Vulnerable Periods in Developing Brain," in *Applied Neurochemistry*, ed. A. N. Davidson and J. Dobbing (Edinburgh: Blackwell, 1968): pp. 287–316; Patricia M. Rodier, "Developing Brain as a Target of Toxicity," *Environmental Health Perspectives* 103, no. 6 (1995): 73–76; D. Rice and S. Barone Jr., "Critical Periods of Vulnerability for the Developing Nervous System: Evidence from Humans and Animal Models," *Environmental Health Perspectives* 108, no. 53 (2000): pp. 511–533; on the uniquely sensitive nature of the brain, Prasada Rao S. Kodavanti, "Neurotoxicity of Persistent Organic Pollutants: Possible Mode(s) of Action and Further Considerations," *Dose-Response* 3 (2005): pp. 273–305.

48. Claudio et al., "Testing Methods for Developmental Neurotoxicity of Environmental Chemicals," pp. 3, 7; Miller et al., "Differences between Children and Adults," p. 412.

49. Miller et al., "Differences between Children and Adults," p. 410.

50. Philippe Grandjean and Philip Landrigan, "Developmental Neurotoxicity of Industrial Chemicals," *Lancet* 368 (December 16, 2006): pp. 2167–2178. See also Grandjean et al., "The Faroes Statement," pp. 2167–2168; Dobbing, "Vulnerable Periods in Developing Brain," pp. 287–316; Rodier, "Developing Brain as a Target of Toxicity," pp. 73–76; Rice and Barone, "Critical Periods of Vulnerability for the Developing Nervous System" pp. 511–533.

51. Rodier, "Developing Brain as a Target of Toxicity," quotations at p. 73–74; on the toxic effects on the nervous system, Claudio et al., "Testing Methods for Developmental Neurotoxicity of Environmental Chemicals," p. 9.

52. Miller et al., "Differences between Children and Adults," quotation at p. 411; see Rodney R. Dietert and Michael S. Piepenbrink, "Perinatal Immunotoxicity: Why Adult Exposure Assessment Fails to Predict Risk," *Environmental Health Perspectives* 114, no. 4 (2006): p. 477; R. R. Dietert et al., "Workshop to Identify Critical Windows of Exposure for Children's Health: Immune and Respiratory Systems Work Groups Summary," *Environmental Health Perspectives* 108 (Suppl. 3) (2000): pp. 483–490; Miller et al., "Differences between Children and Adults," p. 411.

53. Dietert and Piepenbrink, "Perinatal Immunotoxicity," p. 480.

54. Ibid.

55. Ibid.

56. Robert W. Luebke, David H. Chen, Rodney Dietert, Yung Yang, Marquea King, and Michael I. Luster, "The Comparative Immunotoxicity of Five Selected Compounds Following Developmental or Adult Exposure," *Journal of Toxicology and Environmental Health,* part B, 9 (2006): pp. 1–26, at p. 5.

57. Luebke et al., "The Comparative Immunotoxicity of Five Selected Compounds," quotation on DES outcomes at p. 5; Jason P. Hogaboam, Amanda J. Moore, and B. Paige Lawrence, "The Aryl Hydrocarbon Receptor Affects Distinct Tissue Compartments during Ontogeny of the Immune System, *Toxicological Sciences* 102 (2008): pp. 160–170, quotation on "functional alterations" at p. 160, on viral infection, at p. 480; Luebke et al., "The Comparative Immunotoxicity of Five Selected Compounds," quotation on lead sensitivity, p. 11.

58. Luebke et al., "The Comparative Immunotoxicity of Five Selected Compounds," p. 5; Rodney R. Dietert, Jamie C. DeWitt, Dori R. Germolec, and Judith T. Zelikoff, "Breaking Patterns of Environmentally Influenced Disease for Health Risk Reduction: Immune Perspectives," *Environmental Health Perspectives,* 118 (2010): pp. 1091–1099.

59. Prue Talbot, personal communication, December 10, 2009.

60. Radhika Kajekar, "Environmental Factors and Developmental Outcomes in the Lung," *Pharmacology and Therapeutics* 114 (2007): pp. 129–145, at p. 129.

61. Ibid.

62. U.S. Environmental Protection Agency, "Smog—Who Does It Hurt? What You Need to Know about Air Pollution and Your Health," www.airnow.gov/index.cfm?action=smog.page1#1 (accessed November 23, 2009). See also Frederica P. Perera, "Children Are Likely to Suffer from Fossil Fuel Addiction," *Environmental Health Perspectives,* 116, no. 8 (2008): pp. 987–990.

63. Korach, preface to *Reproductive and Developmental Toxicology,* pp. v-vi, quotations at p. v.

64. Barbara J. Davis and Jerrold J. Heindel, "Ovarian Toxicants: Multiple Mechanisms of Action," in *Reproductive and Developmental Toxicology,* ed. Kenneth S. Korach (New York: Marcel Dekker, 1998), pp. 373–395, at p. 374; Newbold, "DES Model: Cancer in Animals and Humans."

65. On the blood-brain barrier, see Grandjean and Landrigan, "Developmental Neurotoxicity," p. 2168; Douglas C. Anthony, Thomas J. Montine, William M. Valentine, and Doyle G. Graham, "Toxic Responses of the Nervous System," in *Casarett and Doull's Toxicology,* 6th ed., ed. Curtis D. Klaassen (New York: McGraw-Hill, 2001), pp. 535–563, at p. 536 (emphasis added).

66. Rodier, "Developing Brain as a Target of Toxicity," p. 75; Brian S. Schwartz and Walter F. Stewart, "Lead and Cognitive Function in Adults: A Questions and Answers Approach to a Review of the Evidence for Cause, Treatment, and Prevention," *International Review of Psychiatry* 19, no. 6 (2007): pp. 671–692, quotation at p. 672.

67. *American Heritage Dictionary* (Boston: American Heritage Publishing and Houghton Mifflin, 1969), p. 438; on enzymes, Miller et al., "Differences between Children and Adults," p. 406; on metabolic routes in the embryo, Rogers and

Kavlock, "Developmental Toxicology," p. 364; on vulnerability lasting until the age of seven, Karen Huen, Kim Harley, Jordan Brooks, Alan Hubbard, Asa Bradman, Brenda Eskenazi, and Nina Holland, "Developmental Changes in PON1 Enzyme Activity in Young Children and Effects of PON1 Polymorphisms," *Environmental Health Perspectives,* 117, no. 10 (2009): pp. 1632–1638, at 1638.

68. Miller et al., "Differences between Children and Adults," p. 406.

69. Ibid., p. 407.

70. Ibid., p. 410.

71. On exposure to polycyclic aromatic hydrocarbons, ibid., pp. 409–410; on adducts on DNA, R. Julian Preston and George R. Hoffman, "Genetic Toxicology," *Casarett and Doull's Toxicology,* 6th ed., ed. Curtis D. Klaassen (New York: McGraw-Hill, 2001), pp. 321–350, at p. 327; on the formation and industrial use of PAH, Centers for Disease Control, Agency for Toxic Substances and Disease Registry, "Polycyclic Aromatic Hydrocarbons: Fact Sheet," www.atsdr.cdc.gov (accessed August 4, 2009); on sidestream and secondhand smoke and PAH-DNA adducts, Miller et al., "Differences between Children and Adults," pp. 409–410, and F. P. Perera, W. Jedrychowski, V. Rauh, and R. M. Whyatt, "Molecular Epidemiologic Research on the Effects of Environmental Pollutants on the Fetus," *Environmental Health Perspectives* 107, no. S-3 (1999): pp. 451–460; on the effects of adducts on head circumference and other genetic damage, Perera, Jedrychowski, Rauh, and Whyatt. "Molecular Epidemiologic Research," p. 451.

72. Marla Cone and Emily Elert, "Urban Air Pollutants May Damage IQs Before Baby's First Breath, Scientists Say," *Environmental Health News,* July 26, 2010, located at www.environmentalhealthnews.org/ehs/news/air-pollutants-and-childrens-iqs (accessed July 26, 2010).

73. Brenda Eskenazi, Lisa G. Rosas, Amy R. Marks, Asa Bradman, Kim Harley, Nina Holland, Caroline Johnson, et al., "Pesticide Toxicity and the Developing Brain," *Basic and Clinical Pharmacology and Toxicology* 102 (2008): pp. 228–236, quotations at p. 235; Clement E. Furlong, Nina Holland, Rebecca, J. Richter, Asa Bradman, Alan Ho, and Brenda Eskenazi, "PON1 Status of Farmworker Mothers and Children as a Predictor of Organophosphate Sensitivity," *Pharmacogenetics and Genomics* 16 (2006): pp. 183–190, quotation at p. 189; on potential effects, Donald J. Ecobichon, "Toxic Effects of Pesticides," in *Casarett and Doull's Toxicology,* 6th ed., ed. Curtis D. Klaassen (New York: McGraw-Hill, 2001), pp. 763–810, at p. 773–778.

74. Grandjean et al., "The Faroes Statement," p. 74 (emphasis added).

75. On methylmercury concentrations in the fetal brain, Honda, Hylander and Sakamoto, "Recent Advances in Evaluation of Health Effects on Mercury," p. 173; Philippe Grandjean, Esben Budtz-Jorgensen, Dana B. Barr, Larry L. Needham, Pal Weihe, and Birger G. J. Heinzow, "Elimination Half-Lives of Polychlorinated Biphenyl Congeners in Children," *Environmental Science and Technology* 42, no. 18 (2008): pp. 6991–6996, at p. 6991.

76. David Bellinger and Herbert L. Needleman, "The Neurotoxicity of Prenatal Exposure to Lead: Kinetics, Mechanism and Expressions," in *Prenatal Exposure to Toxicants,* ed. Herbert L. Needleman and David Bellinger (Baltimore: Johns Hopkins University Press, 1994), pp. 89–111, at pp. 91–92.

77. On number of contaminants found in newborns, Peter Fimrite, "Study: Chemicals, Pollutants Found in Newborns," *San Francisco Chronicle*, December 3, 2009, located at www.sfgate.com (accessed December 3, 2009), and Environmental Working Group, "Pollution in People: Cord Blood Contaminants in Minority Newborns," December 2, 2009, www.ewg.org/minoritycordblood/fullreport (accessed December 20, 2009); on researchers urging on balance that women nurse, Heinzow, "Endocrine Disruptors in Human Milk and the Health-Related Issues of Breastfeeding," pp. 344–345.

78. On children's breathing and fluid intake, Miller et al., "Differences between Children and Adults," p. 406; on children's consumption of fruit, Sherry G. Selevan, Carole A. Kimmel, and Pauline Mendola, "Identifying Critical Windows of Exposure for Children's Health, *Environmental Health Perspectives* 108 (2000): pp. 451–455, at p. 454; on augmented absorption rates and ability to detoxify, see Grandjean and Landrigan, "Developmental Neurotoxicity," p. 2168; G. Ginsburg et al., "Incorporating Pharmacokinetic Differences between Children and Adults in Assessing Children's Risks to Environmental Toxicants," *Toxicology and Applied Pharmacology* 198 (2004): pp. 164–183; see, generally, National Research Council, *Pesticides in the Diets of Infants and Children* (Washington, D.C.: National Academy Press, 1993).

79. See Miller et al., "Differences between Children and Adults," p. 405.

80. L. G. Costa and G. Giordano, "Developmental Neurotoxicity of Polybrominated Diphenyl Ether (PBDE) Flame Retardants," *NeuroToxicology* 28 (2007): pp. 1047–1067, at p. 1050.

81. Jerrold J. Heindel, "Animal Models for Probing the Developmental Basis of Disease and Dysfunction Paradigm," *Basic and Clinical Pharmacology and Toxicology* 102 (2008): pp. 76–81, at p. 77.

82. Miller et al., "Differences between Children and Adults, p. 412; on long latency periods, Kenneth J. Rothman, *Modern Epidemiology* (New York: Little, Brown, 1986): pp. 14–15, 58–60; on other diseases conceived as one-, two-, or three-hit processes, Heindel, "Animal Models for Probing the Developmental Basis of Disease and Dysfunction Paradigm," p. 77.

83. Rice and Barone, "Critical Periods of Vulnerability for the Developing Nervous System," p. 525; on the baby boomer cohort, Dana B. Hancock, Eden R Martin, Gregory M. Mayhew, Jeffrey M. Stajich, Rita Jewett, Mark A Stacy, Burton L Scott, et al., "Pesticide Exposure and Risk of Parkinson's Disease: A Family-Based Case-Control Study," *BMC Neurology* 8, no. 6 (2008): pp. 1–12.

84. Claudio et al., "Testing Methods for Developmental Neurotoxicity of Environmental Chemicals," p. 2.

85. Centers for Disease Control and Prevention, DES Update: Consumers, "Recent DES Research," located at www.cdc.gov/DES/consumers/research/recent.html, citing L. Titus-Ernstoff, E. E. Hatch, R. N. Hoover, J. R. Palmer, E. R. Greenberg, et al., "Long-term Cancer Risk in Women Given Diethylstilbestrol (DES) During Pregnancy." *British Journal of Cancer*, 84 (2001): 125–133.

86. On DES causing vaginal cancer and increasing breast cancer in DES daughters, Andreas Kortenkamp, "Breast Cancer and Exposure to Hormonally Active

Chemicals: An Appraisal of the Scientific Evidence," background paper published by the Health and Environment Alliance and CHEM Trust, www.chemtrust.org.uk (accessed August 19, 2008); on effects of thalidomide, Landrigan et al., "Children's Health and the Environment," p. 259.

87. Landrigan et al., "Children's Health and the Environment," p. 259.

88. Ibid.

89. Miller et al., "Differences between Children and Adults," p. 412.

90. Luebke et al., "The Comparative Immunotoxicity of Five Selected Compounds," pp. 1, 22; on prenatal or nursing exposures to dioxin, Hogaboam, Moore, and Lawrence, "The Aryl Hydrocarbon Receptor," pp. 160–170. An accessible synopsis of this research is Michael D. Laiosa and Wendy Hessler, "Dioxin Alters Ability to Fight Infection, Mouse Study Finds," *Environmental Health News,* March 18, 2009, www.environmentalhealthnews.org (accessed March 20, 2009).

91. Miller et al., "Differences between Children and Adults," p. 412; Barbara A. Cohn, Mary S. Wolff, Piera M. Cirillo, and Robert I. Sholtz, "DDT and Breast Cancer in Young Women: New Data on the Significance of Age at Exposure," *Environmental Health Perspectives* 115, no. 10 (October 2007): pp. 1406–1414, at p. 1406.

92. Cohn et al., "DDT and Breast Cancer in Young Women," quotations and discussion of experimental studies at p. 1411.

93. National Research Council, *Toxicological Effects of Methylmercury,* as cited in Landrigan et al., "Children's Health and the Environment," pp. 261–262); Coleen A. Boyle, P. Decouflé, and M. Yeargin-Allsopp, "Prevalence and Health Impact of Developmental Disabilities in US Children," *Pediatrics* 93 (1994): pp. 399–403, at p. 399; on the possible overestimation of some effects, anonymous Harvard University Press referee, personal communication, October 2009.

94. Robert A. Goyer and Thomas W. Clarkson, "Toxic Effects of Metals," in *Casarett and Doull's Toxicology,* 6th ed., ed. Curtis D. Klaassen (New York: Mc-Graw-Hill, 2001). pp. 811–867, at p. 827; Donald T. Wigle and Bruce P. Lanphear, "Human Health Risks from Low-Level Environmental Exposures: No Apparent Safety Thresholds," in *PLoS Medicine* 2 (2005): pp. 1–3, www.plosmedicine.org (accessed August 5, 2009).

95. Jack Lewis, "Lead Poisoning: A Historical Perspective," *EPA Journal,* May 1985, www.epa.gov/history/topics/perspect/lead.htm (accessed June 11, 2008).

96. Ibid.

97. David Rosner, Gerald Markowitz, and Bruce Lanphear, "J. Lockhart Gibson and the Discovery of the Impact of Lead Pigments on Children's Health: A Review of a Century of Knowledge," *Public Health Reports* 120 (2005): pp. 296–300, at p. 297.

98. Lewis, "Lead Poisoning: A Historical Perspective."

99. Grandjean and Landrigan, "Developmental Neurotoxicity," p. 2169.

100. Mireya Navarro, "Lead Poisoning Is in Decline, but Proves a Stubborn Nemesis," *New York Times,* April 21, 2010, located at www.nytimes.com/2010/04/22/nyregion/22lead.html?hp (accessed May 2, 2010).

101. Collegium Ramazzini, "Collegium Ramazzini Statement: Call for Worldwide Reduction to Lead Exposure," www.collegiumramazzini.org/download/14_FourteenthCRStatement(2008).pdf (accessed July 17, 2009), pp. 1–10, at p. 2.

102. On target organs, Goyer and Clarkson, "Toxic Effects of Metals," p. 827; Ellen K. Silbergeld and Virginia M. Weaver, "Exposures to Metals: Are We Protecting the Workers?" *Occupational and Environmental Medicine* 64 (March 2007): pp. 141–142, at p. 141; on adult blood pressure, Robert A. Goyer, "Lead Toxicity: From Overt to Subclinical to Subtle Health Effects," *Environmental Health Perspectives* 86 (1990): pp. 177–181, at p. 178; on heart attacks and strokes, Ana Navas-Acien, Eliseo Guallar, Ellen K. Silbergeld, and Stephen J. Rothenberg, "Lead Exposure and Cardiovascular Disease—A Systematic Review," *Environmental Health Perspectives*, 115, no. 3 (2007): pp. 472–482.

103. Silbergeld and Weaver, "Exposures to Metals: Are We Protecting the Workers?," p. 141; U.S. Environmental Protection Agency, Region 10, "Health Effects of Lead," yosemite.epa.gov/R10, (accessed July 17, 2009).

104. U.S. Department of Health and Human Services, Public Health Service, Agency for Toxic Substances and Disease Registry, "Public Health Statement: Lead," www.atsdr.cdc.gov (accessed December 31, 2009): pp. 1–13, at p. 7; Collegium Ramazzini, "Collegium Ramazzini Statement: Call for Worldwide Reduction to Lead Exposure," www.collegiumramazzini.org/download/14_Fourteenth CRStatement(2008).pdf (accessed July 17, 2009), pp. 1–10, at p. 2; on fetal exposures, U.S. Department of Health and Human Services, "Public Health Statement: Lead," p. 7.

105. On problems when levels remained above EPA "safe" levels, Bellinger and Needleman, "The Neurotoxicity of Prenatal Exposure to Lead," p. 90; on even minor increases in prenatal increases in blood lead levels, John Paul Wright, Kim N. Dietrich, M. Douglas Ris, Richard W. Hornung, Stephanie D. Wessel, Bruce P. Lanphear, Mona Ho, et al., "Association of Prenatal and Childhood Blood Lead Concentrations with Criminal Arrests in Early Adulthood," *PLoS Medicine, 5*, no. 5 (2008): e115, www.ncbi.nlm.nih.gov/pmc/articles/PMC2689664/pdf/pmed.0050101.pdf (accessed December 31, 2009), and Thomas H. Maugh II and Marla Cone, "Lead Exposure in Children Linked to Violent Crime: A Study Finds That Even Low Levels Can Permanently Damage the Brain: The Research Also Shows That Exposure Is a Continuing Problem Despite Efforts to Minimize It," *Los Angeles Times,* May 28, 2008, www.latimes.com; Kim M. Cecil, Christopher J. Brubaker, Caleb M. Adler, Kim N. Dietrich, Mekibib Altaye, John C. Egelhoff, Stephanie Wessel, et al., "Decreased Brain Volume in Adults with Childhood Lead Exposure," *PLoS Medicine* 5, no. 5 (2008): pp. 741–450, www.ncbi.nlm.nih.gov/pmc/articles/PMC2689675/pdf/pmed.0050112.pdf (accessed December 31, 2009); David C. Bellinger, "Neurological and Behavioral Consequences of Childhood Lead Exposure, *PLoS Medicine* 5, no. 5 (2008): pp. 690–692, www.ncbi.nlm.nih.gov/pmc/articles/PMC2689677/pdf/pmed.0050115.pdf (accessed December 12, 2009).

106. Schwartz and Stewart, "Lead And Cognitive Function in Adults," p. 672; Edwin van Wijngaarden, James R. Campbell, and Deborah A. Cory-Slechta, "Bone

Lead Levels Are Associated with Measures of Memory Impairment in Older Adults," *NeuroToxicology* 30 (2009) pp. 572–580.

107. On there being no threshold for lead poisoning, Richard L. Canfield et al., "Low-Level Lead Exposure and Executive Functioning in Young Children," *Child Neuropsychology* 9 (2003): pp. 35–53, and David Bellinger and Herbert L. Needleman, "Intellectual Impairment and Blood Lead Levels," *New England Journal of Medicine* 349, no. 5 (2003): pp. 500–502; on effects of increments above zero, Bruce P. Lanphear, "Origins and Evolution of Children's Environmental Health," in "Essays on the Future of Environmental Health Research: A Tribute to Kenneth Olden," special issue, *Environmental Health Perspectives* (August 2005): p. 28, and Collegium Ramazzini, "Collegium Ramazzini Statement: Call for Worldwide Reduction to Lead Exposure."

108. Van Wijngaarden, Campbell, and Cory-Slechta, "Bone Lead Levels Are Associated with Measures of Memory Impairment in Older Adults," pp. 679, 688, 671.

109. Peggy O'Farrel, "Lead Exposure, Brain Damage Linked," *Cincinnati Enquirer,* December 1, 2009, www.enquirer.com (accessed December 5, 2009).

110. Rogers and Kavlock, "Developmental Toxicology," pp. 351–386, p. 374.

111. On mechanisms that have been conserved through evolution, U.S. Department of Health and Human Services, Centers for Disease Control and Prevention, introduction to PCBs and other dioxin-like substances, *Third National Report on Human Exposure to Environmental Chemicals,* 2005, p. 135, www.cdc.gov/exposurereport/report.htm (accessed July 15, 2008), and Ted Simon, Janice K. Britt, and Robert C. James, "Development of a Neurotoxic Equivalence Scheme of Relative Potency for Assessing the Risk of PCB Mixtures," *Regulatory Toxicology and Pharmacology* 48 (2007): pp. 149–150; on mechanisms activating enzymes that increase toxicity, see Elaine M. Faustman and Gilbert S. Omenn, "Risk Assessment," in *Casarett and Doull's Toxicology,* 6th ed., ed. Curtis D. Klaassen (New York: McGraw-Hill, 2001), pp. 83–104, at p. 86, and Andrew Parkinson, "Biotransformation of Xenobiotics," in *Casarett and Doull's Toxicology,* 6th ed., ed. Curtis D. Klaassen (New York: McGraw-Hill, 2001), pp. 133–224, at p. 193.

112. Kodavanti, "Neurotoxicity of Persistent Organic Pollutants," p. 274; on taking dioxin-like PCBs' additivity into account, Sherif A. Kafafi, Hussein Y. Afeefy, Ali H. Ali, Hakim K. Said, and Abdel G. Kafafi, "Binding of Polychlorinated Biphenyls to the Aryl Hydrocarbon Receptor," *Environmental Health Perspectives* 101, no. 5 (1993): pp. 422–428.

113. On impact of exposures to comparatively high doses of dioxins, Linda Birnbaum, "Developmental Effects of Dioxins," in *Reproductive and Developmental Toxicology,* ed. Kenneth S. Korach (New York: Marcel Deeker, 1998), p. 103; on PCB exposures and male deficits, Grandjean and Landrigan, "Developmental Neurotoxicity," p. 2172; on children's susceptibility to PCBs prenatally, Joseph L. Jacobson and Sandra W. Jacobson, "The Effects of Perinatal Exposure to Polychlorinated Biphenyls and Related Contaminants," *Prenatal Exposure to Toxicants,* ed. Herbert L. Needleman and David Bellinger. (Baltimore: Johns Hopkins University Press, 1994), pp. 130–147, at p. 144; on background levels of dixoins, Birnbaum, "Developmental Effects of Dioxins," p. 105.

114. Sharon K. Sagiv, Sally W. Thurston, David C. Bellinger, Paige E. Tolbert, Larisa M. Altshul and Susan A. Korrick, "Prenatal Organochlorine Exposure and Behaviors Associated with Attention Deficit Hyperactivity Disorder in School-Aged Children," *American Journal of Epidemiology,* 171 (2010): 593–601.

115. Birnbaum, "Developmental Effects of Dioxins," p. 106.

116. Francesca Lyman, "Are Boys An Endangered Species?," MSN Health and Fitness, health.msn.com/pregnancy/articlepage.aspx?cp-documentid=100171768 (accessed December 31, 2009).

117. Irva Hertz-Picciotto, Todd A. Jusko, Eric J. Willman, Rebecca J. Baker, Jean A. Keller, Stuart W. Teplin and M. Judith Charles, "A Cohort Study of *In Utero* Polychlorinated Biphenyl (PCB) Exposures in Relation to Secondary Sex Ratio," *Environmental Health* 7(2008): 37–44.

118. Poul Harremoës et al., *Late Lessons from Early Warnings: The Precautionary Principle, 1896–2000* (Copenhagen: European Environment Agency, 2001), p. 68; on other adverse health effects of non-dioxin-like PCBs, Kodavanti, "Neurotoxicity of Persistent Organic Pollutants," pp. 274, 283, and 297–298, and Simon, Britt, James, "Development of a Neurotoxic Equivalence Scheme," pp. 149–150; on the better-known toxicity effects traceable to non-dioxin-like variants, Kodavanti, "Neurotoxicity of Persistent Organic Pollutants," p. 275.

119. Masanori Kuratsune, Takesumi Yoshimura, and Junichi Matsuzaka, "Epidemiologic Study on Yusho, a Poisoning Caused by Ingestion of Rice Oil Contaminated with a Commercial Brand of Polychlorinated Biphenyls," *Environmental Health Perspectives* 1 (1972): pp. 119–128.

120. Paul W. Stewart, Edward Lonky, Jacqueline Reihman, James Pagano, Brooks B. Gump, and Thomas Darvill, "The relationship between prenatal PCB exposure and intelligence (IQ) in 9-year-old children," *Environmental Health Perspectives,* 116(10) (2008): pp. 1416-22. On memory effects, Kodavanti cites the study: Kodavanti, "Neurotoxicity of Persistent Organic Pollutants," p. 277.

121. Kodavanti, "Neurotoxicity of Persistent Organic Pollutants," p. 277; on Faroe Islands, Philippe Grandjean et al., "Neurobehavioral Deficits Associated with PCB in 7-year-old Children Prenatally Exposed to Seafood Neurotoxicants," *Neurotoxicology and Teratology* 23 (2001): pp. 305–317.

122. Kodavanti, "Neurotoxicity of Persistent Organic Pollutants," pp. 277–279.

123. Kim G. Harley, Amy R. Marks, Jonathan Chevrier, Asa Bradman, Andreas Sjödin, Brenda Eskenazi, "PBDE Concentrations in Women's Serum and Fecundability," *Environmental Health Perspectives,* 118 (2010): 699–704.

124. "Dust Harbors New Fire Retardants Associated with Hormone, Sperm Changes," Synopsis by Jonathan Chevrier, Ph.D. and Wendy Hessler,located at www.environmentalhealthnews.org/ehs/newscience/op-fire-retardants-in-dust-linked-to-hormone-sperm-changes/, last visited May 29, 2010 (summarizing the article by J.D. Meeker and HM Stapleton, "House Dust Concentrations of Organophosphate Flame Retardants in Relation to Hormone Levels and Semen Quality Parameters," *Environmental Health Perspectives* [2009], located at ehp03 .niehs.nih.gov/article/info%3Adoi%2F10.1289%2Fehp.0901332, [accessed May 29, 2010]).

125. Julie B. Herbstman, Andreas Sjödin, Matthew Kurzon, Sally A. Lederman,

Richard S. Jones, Virginia Rauh, Larry L. Needham, Deliang Tang, Megan Niedzwiecki, Richard Y. Wang, and Frederica Perera, "Prenatal Exposure to PBDEs and Neurodevelopment, *Environmental Health Perspectives*, 118(5) (2010): 712–719.

126. Kim Hooper and Thomas A. McDonald, "The PBDEs: An Emerging Environmental Challenge and Another Reason for Breast-Milk Monitoring Programs," *Environmental Health Perspectives,* 108, no. 5 (2000): pp. 387–392, p. 388; Rogers and Kavlock, "Developmental Toxicology," p. 374; Lucio G. Costa and Gennaro Giordano, "Developmental Neurotoxicity of Polybrominated Dipheny Ether (PBDE) Flame Retardants," pp. 1047–1048.

127. Kodavanti, "Neurotoxicity of Persistent Organic Pollutants," on substantial research showing PCBs' adverse developmental effects, p. 276; on rats, primates, and mice, pp. 276, 278; on neurotoxicity in humans, p. 275; on behavior changes and learning deficits from adult and perinatal exposure, p. 277.

128. On potential application of PCB data to PBDEs, Kodavanti, "Neurotoxicity of Persistent Organic Pollutants," pp. 297–298; Arnold Schecter et al., "Polybrominated Diphenyl Ether Flame Retardants in the U.S. Population: Current Levels, Temporal Trends, and Comparison with Dioxins, Dibenzofurans, and Polychlorinated Biphenyls," *Journal of Occupational and Environmental Medicine* 47 (2005): pp. 199–200. See, generally, L. Birnbaum and D. Staskal, "Brominated Flame Retardants: Cause for Concern?" *Environmental Health Perspectives* 112, no. 1 (2003): pp. 9–17; S. Hallgren and P. Darnerud, "Effects of Polybrominated Diphenyl Ethers (PBDEs), Polychlorinated Biphenyls (PCBs) and Chlorinated Paraffins (CPs) on Thyroid Hormone Levels and Enzyme Activities in Rats," *Organohalogen Comendium* 35 (1998): pp. 391–394.

129. Costa and Giordano, "Developmental Neurotoxicity of Polybrominated Diphenyl Ether (PBDE) Flame Retardants," on the need for reliance on animal data for PBDEs, p. 1061, above n96; on concentration in infants and adults, p.1051; on concentrations in the United States and Europe, pp. 1049–1050; Melissa Rose, Debora H. Bennett, Ake Bergman, Britta Fangstrom, Issac N. Pessah, and Irva Hertz-Picciotto, "PBDEs in 2-5 Year-Old Children from California and Associations with Diet and Indoor Environment," *Environmental Science and Technology,* 44 (2010): 2648–2653.

130. Costa and Giordano, "Developmental Neurotoxicity of Polybrominated Diphenyl Ether (PBDE) Flame Retardants," p. 1061; Andreas Sjodin, Lee-Yang Wong, Richard S. Jones, Annie Park, Yalin Zhang, Carolyn Hodge, Emily Dipietro, et al., "Serum Concentrations of Polybrominated Diphenyl Ethers (PBDEs) and Polybrominated Biphenyl (PBB) in the United States Population: 2003–2004," *Environmental Science and Technology* 42 (2008): pp. 1377–1384, at p. 1378.

131. Kodavanti, "Neurotoxicity of Persistent Organic Pollutants," at p. 298; see, generally, P. Eriksson et al., "Brominated Flame Retardants: A Novel Class of Developmental Neurotoxicants in Our Environment?" *Environmental Health Perspectives* 109, no. 9 (2001): pp. 903–908.

132. R. I. Canfield, C. R. Henderson, D. A. Cory-Slechta, C. Cox, T. A. Jusko, et. al, "Intellectual Impairment in Children with Blood Lead Concentrations below

10 Micrograms per Deciliter," *New England Journal of Medicine* 348 (2003): pp. 1517–1526, as cited in Bruce P. Lanphear, Charles V. Vorhees, and David C. Bellinger, "Protecting Children from Environmental Toxins," *PLoS Medicine* 2 (2005): pp. 199–204, at p. 201, www.plosmedicine.org (accessed July 18, 2009).

133. Costa and Giordano, "Developmental Neurotoxicity of Polybrominated Diphenyl Ether (PBDE) Flame Retardants," quotation at p. 1062; on bodies being contaminated with other developmental neurotoxicants, p. 1062.

134. Per Eriksson, Henrik Viberg, Niclas Johnsson, and Anders Fredriksson, "Effects of Perfluorinated Compounds and Brominated Flame Retardants on Brain Development and Behavior in a Rodent Model" (presentation at PPTOX II: Role of Environmental Stressors in the Developmental Origins of Disease, Miami, Florida, December 7–10, 2009).

135. Herbstman, et al., "Prenatal Exposure to PBDEs and Neurodevelopment," at pp. 712, 718.

136. On rates of breast cancer in the United States, American Cancer Society, "Breast Cancer Facts and Figures 2007–2008," www.cancer.org, p. 11 (accessed August 3, 2008); on rates of breast cancer in the United States compared with those in Japan, Statistics from the International Agency for Research on Cancer (2004), as cited by Women's Health Resource, "Breast Cancer: Statistics on Incidence, Survival, and Screening," p. 1, www.imaginis.com/breasthealth/statistics.asp#1 (accessed August 3, 200*); on incidence rates not being due to improved screening, Kortenkamp, "Breast Cancer and Exposure to Hormonally Active Chemicals," p. 4.

137. On percentage of breast cancer cases traceable to inherited genetic predispositions, Kortenkamp, "Breast Cancer and Exposure to Hormonally Active Chemicals," p. 4; on tumor-suppressor genes, Zoltan Gregus and Curtis D. Klaassen, "Mechanisms of Toxicity," *Casarett and Doull's Toxicology,* 6th ed., ed. Curtis D. Klaassen (New York: McGraw-Hill, 2001), pp. 35–81, pp. 75–76.

138. Kortenkamp, "Breast Cancer and Exposure to Hormonally Active Chemicals," quotation at p. 4; on study of breast cancer in twins, p. 5.

139. National Institutes of Health, National Cancer Institute, "Cell Biology and Cancer," science-education.nih.gov (accessed August 3, 2008).

140. Kortenkamp, "Breast Cancer and Exposure to Hormonally Active Chemicals," p. 5, and U.S. National Institutes of Health, the National Cancer Institute, "Pregnancy and Breast Cancer Risk," www.cancer.gov (accessed August 5, 2008).

141. Kortenkamp, "Breast Cancer and Exposure to Hormonally Active Chemicals," p. 5.

142. Helen A. Weiss et al., "Prenatal and Perinatal Risk Factors for Breast Cancer in Young Women," *Epidemiology* 89 (1992): pp. 181–187, at 181.

143. Kortenkamp, "Breast Cancer and Exposure to Hormonally Active Chemicals," p. 6.

144. Ibid., p. 7, and A. S. Robbins and C. A. Clarke, "Regional Changes in Hormone Theory and Breast Cancer Incidence in 2003 in California from 2001 to 2004," *Journal of Clinical Oncology* 25 (2007): pp. 3437–3439.

145. Kortenkamp, "Breast Cancer and Exposure to Hormonally Active Chemicals," p. 6.

146. On BPA contamination, National Institute of Environmental Health Sciences, National Institutes of Health, U.S. Department of Health and Human Services, "NTP Brief on Bisphenol A," April 14, 2008, p. 4, cerhr.niehs.nih.gov/chemicals/bisphenol/BPADraftBriefVF_04_14_08.pdf (accessed August 7, 2008); on BPA seeming less potent than natural estrogen and DES, Dolores Ibarreta and Shanna H. Swan, "The DES Story: Long-Term Consequences of Prenatal Exposure," in Harremoës et al., *Late Lessons from Early Warnings*, pp. 84-92, at p. 84; Frederick S. vom Saal et al., "Chapel Hill Bisphenol A Expert Panel Consensus Statement: Integration of Mechanisms, Effects in Animals and Potential Impact to Human Health at Current Exposure Levels," *Reproductive Toxicology* 24 (2007): pp. 131–138, at p. 136.

147. Vom Saal et al., "Chapel Hill Bisphenol A Expert Panel Consensus Statement," p. 136.

148. Ibid.

149. National Institute of Environmental Health Sciences, "NTP Brief on Bisphenol A," p. 37; National Research Council, *Hormonally Active Agents in the Environment* (Washington, D.C.: National Academy Press, 1999), pp. 114–115.

150. Andreas Kortenkamp, "Low-Level Exposure to Multiple Chemicals: Reason for Human Health Concerns?" *Environmental Health Perspectives* 115, S-1 (2007): pp. 106–114, quotation at p. 113; Andreas Kortenkamp, "Ten Years of Mixing Cocktails: A Review of Combination Effects of Endocrine-Disrupting Chemicals," *Environmental Health Perspectives* 115, S-1 (2007): pp. 98–105, quotation at p. 104 (emphasis added).

151. Kortenkamp, "Ten Years of Mixing Cocktails," p. 104.

152. E. Diamanti-Kandarakis et al.."Endocrine-Disrupting Chemicals: An Endocrine Society Scientific Statement," *Endocrine Reviews* 30 (2009): pp. 293–342, at p. 300.

153. On associations between exposures to phthalates and early stages of penile dysgenesis syndrome, Shanna H. Swan, Katharina M. Main, Fan Liu, Sara L. Stewart, Robin L. Kruse, Antonia M. Calafat, Catherine S. Mao, et al., and the Study for Future Families Research Team, "Decrease in Anogenital Distance among Male Infants with Prenatal Phthalate Exposure," *Environmental Health Perspectives* 113, no. 8 (2005): pp. 1056–1061; on warning flags about developmental effects from multiple endocrine-disrupting chemicals, Ulla Hass, Martin Scholze, Sofie Christiansen, Majken Dalgaard, Anne Marie Vinggaard, Marta Axelstad, Stine Broeng Metzdorff, et al., "Combined Exposure to Anti-Androgens Exacerbates Disruption of Sexual Differentiation in the Rat," *Environmental Health Perspectives* 115, S-1 (2007): pp. 122–128.

154. T. J. Woodruff, L. Zeise, D. A. Axelrad, K. Z. Guyton, S. Janssen, M. Miller, G. G. Miller, et al., "Meeting Report: Moving Upstream—Evaluating Adverse Upstream End Points for Improved Risk Assessment and Decision-Making," *Environmental Health Perspectives* 16, no. 11 (2008): pp. 1568–1575, at p. 1570.

155. Ibid.

156. Ibid., at p. 1570 (emphasis added).

157. Woodruff et al., "Moving Upstream," pp. 1571–1572; Hass et al., "Combined Exposure to Anti-Androgens Exacerbates Disruption of Sexual Differentiation in the Rat," p. 122; Sofie Christiansen, Martin Scholze, Majken Dalgaard, Anne Marie Vinggaard, Marta Axelstad, Andreas Kortenkamp, and Ulla Hass, "Synergistic Disruption of External Male Sex Organ Development by a Mixture of Four Antiandrogens," *Environmental Health Perspectives* 117 (2009): pp. 1839-1846.

158. Committee on the Health Risks of Phthalates, National Research Council, *Phthalates and Cumulative Risk Assessment The Task Ahead* (Washington, D.C.: National Academy Press, 2008), p. 4 (emphasis added).

159. Heindel, "Animal Models for Probing the Developmental Basis of Disease and Dysfunction Paradigm," p. 76.

160. Ibid., both quotations at p. 76.

161. Caroline McMillen and Jeffrey S. Robinson, "Developmental Origins of the Metabolic Syndrome: Prediction, Plasticity, and Programming," *Physiological Review* 85 (2005): pp. 571–633, at p. 609.

162. Ibid.

163. James A. Armitage, Imran Y. Khan, Paul D. Taylor, Peter W. Nathanielsz, and Lucilla Poston, "Developmental Programming of the Metabolic Syndrome by Maternal Nutritional Imbalance: How Strong Is the Evidence from Experimental Models in Mammals?" *Journal of Physiology* 561 (2004): pp. 355–377, at pp. 356–357; C. Nicholas Hales and David J. P. Barker, "The Thrifty Phenotype Hypothesis," *British Medical Bulletin* 60 (2001): pp. 5–20, at p. 10; Caroline McMillen, Severence M. MacLaughlin, Beverly S. Muhlhausler, Sheridan Gentilli, Jaime L. Duffield, and Janna L. Morrison, "Developmental Origins of Adult Health and Disease: The Role of Periconceptional and Foetal Nutrition," *Basic and Clinical Pharmacology and Toxicology* 102 (2008): pp. 82–89, at p. 82.

164. On sheep as animal models of metabolic syndrome, McMillen et al., "Developmental Origins of Adult Health and Disease," pp. 83–84; quotation at p. 88.

165. Brian K. Barlow, Deborah A. Cory-Slechta, Eric K. Richfield, and Mona Thiruchelvam, "The Gestational Environment and Parkinson's Disease: Evidence for Neurodevelopmental Origins of a Neurogenerative Disorder," *Reproductive Toxicology* 23 (2007): pp. 457–470, at p. 458.

166. Heindel, "Animal Models for Probing the Developmental Basis of Disease and Dysfunction Paradigm," p. 80.

167. Ibid., pp. 77, 80.

168. On DES and uterine carcinogens, ibid., p. 78; on endocrine-mimicking substances and on effects of atrazine, ibid., p. 78.

169. Ibid., p. 79.

170. Ibid., p. 79.

171. Ibid., p. 78.

172. Swan, Main, Liu, Stewart, Kruse, Calafat, Mao, et al., and the Study for Future Families Research Team, "Decrease in Anogenital Distance among Male Infants with Prenatal Phthalate Exposure," *Environmental Health Perspectives* 113, no. 8 (2005): pp. 1056–1061.

173. On effects of methoxychlor or vinclozolin, Heindel, "Animal Models for

Probing the Developmental Basis of Disease and Dysfunction Paradigm," p. 79; M. D. Anway, A. S. Cupp, M. Uzumcu, and M. Skinner, "Epigenetic Transgenerational Actions of Endocrine Disruptors on Male Fertility," *Science* 308 (2005): pp. 1466–1469; M. D. Anway, C. Leathers; M. Skinner, "Endocrine Disruptor Vinclozolin Induced Epigenetic Transgenerational Adult-Onset Disease," *Endocrinology* 55 (2006): pp. 5515–5523. On similar results demonstrated for other compounds, Kim, "Transgenerational Changes after Embryonic Exposure to Plasticizer Phthalates"; Mehta, "Perinatal Exposures of Rats to Bisphenol A Affects Fertility of Male Offspring."

174. Anway, Leathers et al., "Endocrine Disruptor Vinclozolin Induced Epigenetic Transgenerational Adult-Onset Disease," p. 5518.

175. Brian K. Barlow et al., "The Gestational Environment and Parkinson's Disease," p. 458.

176. D. B. Calne et. al, "Positron Emission Tomography After MPTP: Observations Relating to the Cause of Parkinson's Disease," *Nature* 317 (1985): pp. 246–248.

177. On effects of parquat and maneb, Heindel, "Animal Models for Probing the Developmental Basis of Disease and Dysfunction Paradigm," p. 79; on paraquat's chemical structure, David A. Eastmond, Chair, Department of Cell, Molecular, and Developmental Biology, University of California, Riverside, personal communication, July 2009.

178. On the general hypothesis, D. B. Calne, A. Eisen, E. McGeer, and P. Spencer, "Hypothesis: Alzheimer's Disease, Parkinson's Disease, and Motoneurone Disease: Abiotropic Interaction between Ageing and Environment?" *Lancet* 328 (1986): pp. 1067–1070, at p. 1067; on the plausible mechanism, Barlow et al. "The Gestational Environment and Parkinson's Disease." p. 459.

179. Dana B. Hancock et al., "Pesticide Exposure and Risk of Parkinson's Disease: A Family-Based Case-Control Study"; Sadie Costello, Myles Cockburn, Jeff Bronstein, Xinbo Zhang, and Beate Ritz, "Parkinson's Disease and Residential Exposure to Maneb and Paraquat from Agricultural Applications in the Central Valley of California," *American Journal of Epidemiology* 169 (2009): pp. 919–926, quotation at p. 919.

180. Don M. Gash, Kathryn Rutland, Naomi L. Hudson, et al., "Trichloroethylene: Parkinsonism and Complex 1 Mitochondrial Neurotoxicity," *Annals of Neurology* 63 (2008): pp. 184–192, at p. 191.

181. Thomas H. Maugh II, "Industrial Solvent Linked to Increased Risk of Parkinson's Disease," *Los Angeles Times*, February 27, 2010, located at http://latimesblogs.latimes.com/booster_shots/2010/02/industrial-solvents-sharply-increase-risk-of-parkinsons-disease.html (accessed May 27, 2010).

182. On both the two hundred human neurotoxicants and numerous other compounds, Grandjean and Landrigan, "Developmental Neurotoxicity," pp. 2167–2178.

183. Hooper and McDonald, "The PBDEs: An Emerging Environmental Challenge," pp. 387–388; see also Costa and Giordano, "Developmental Neurotoxicity of Polybrominated Diphenyl Ether (PBDE) Flame Retardants," p. 1047; Koda-

vanti, "Neurotoxicity of Persistent Organic Pollutants," pp. 273–305; Schecter et al., "Polybrominated Diphenyl Ether Flame Retardants in the U.S. Population," p. 200.

184. John F. Lucas, *Introduction to Abstract Mathematics* (Lanham, Md.: Roman and Littlefield, 1990), p. 75–76.

185. Schardein, *Chemically Induced Birth Defects*, p. 5.

5. RECKLESS NATION:
HOW EXISTING LAWS FAIL TO PROTECT CHILDREN

1. Josée Doucet, Brett Tague, Douglas L. Arnold, Gerard M. Cooke, Stephen Hayward, and Cynthia G. Goodyear, "Persistent Organic Pollutant Residues in Human Fetal Liver and Placenta from Greater Montreal, Quebec: A Longitudinal Study from 1998–2006," *Environmental Health Perspectives Online* (2008) www .ehponline.org/members/2008/0800205/0800205.pdf (accessed June 15, 2009); Kim Harley, "Flame Retarding Chemicals Pollute Fetal Tissue," *Environmental Health News,* January 16, 2009, www.environmentalhealthnews.org/ehs/newscience/ pbdes-pollute-fetal-tissue (accessed January 17, 2009).

2. Julie B. Herbstman, Andreas Sjödin, Matthew Kurzon, Sally A. Lederman, Richard S. Jones, Virginia Rauh, Larry L. Needham, Deliang Tang, Megan Niedzwiecki, Richard Y. Wang, and Frederica Perera, "Prenatal Exposure to PBDEs and Neurodevelopment, *Environmental Health Perspectives*, 118(5) (2010): 712–719.

3. Peter Fimrite, "Study: Chemicals, Pollutants Found in Newborns," *San Francisco Chronicle,* December 3, 2009, located at www.sfgate.com (accessed December 3, 2009).

4. U.S. Congress, Office of Technology Assessment, *Identifying and Regulating Carcinogens* (Washington, D.C.: Government Printing Office, 1987), pp. 199–220.

5. Peter Barton Hutt, "Philosophy of Regulation Under the Food, Drug and Cosmetic Act," *Food and Drug Law Journal* 50 (1995): pp. 101–109, at p. 101; Wendy Wagner, "Using Competition-Based Regulatory to Bridge the Toxics Data Gap," *Indiana Law Journal* 83 (2008): pp. 629–659, at p. 631; and Richard Merrill, "FDA Regulatory Requirements as Tort Standards," *Journal of Law and Policy* 12 (2004): pp. 549–558, at p. 552.

6. Richard Merrill, "FDA Regulatory Requirements as Tort Standards," *Journal of Law and Policy* 12 (2005): 549–558, at p. 552.

7. U.S. Congress, *Identifying and Regulating Carcinogens,* p. 109.

8. Ibid. p. 136.

9. Ibid., p. 200.

10. Quoted phrase on air pollutants, Clean Air Act, 42 U.S.C. § 7412(a)(2) (1995); on the list that the EPA is required to create and revise, U.S. Congress, *Identifying and Regulating Carcinogens,* p. 209; for the list of toxicants, see the U.S. Environmental Protection Agency, Air and Radiation Program, www.epa.gov/ ttn/atw/188polls.html (accessed June 20, 2009).

11. Quotation on residual risk standards, Clean Air Act, 42 U.S.C. § 7412(f)(2)(A) (1995) (emphasis added); on residual risk standard for carcinogens, ibid.

12. On "hypothetical and actual maximally exposed individuals," Clean Air Act, 42 U.S.C. § 7412(o)(2) (1995); on how risk assessments for children should be conducted, National Research Council, Committee on Risk Assessment of Hazardous Air Pollutants, *Science and Judgment in Risk Assessment* (Washington, D.C.: National Academy Press, 1994), pp. 210–211.

13. Wendy E. Wagner, "The Triumph of Technology-Based Standards," *University of Illinois Law Review* 2000, no. 1 (2000): pp. 83–113, at p. 93.

14. Ibid., pp. 94–103.

15. Safe Drinking Water and Enforcement Act of 1986, California Health and Safety Code § 25249.5 (1989).

16. California Code of Regulations, Title 22, § 2601(a) (2007).

17. Carl F. Cranor, "Information Generation and Use under Proposition 65: Model Provisions for Other Postmarket Laws?" *Indiana Law Journal* 83 (2008): pp 609–627, at pp. 620–624.

18. Ibid., pp. 620–624.

19. State of California, Environmental Protection agency, Office of Environmental Health Hazard Assessment, Safe Drinking Water and Toxic Enforcement Act of 1986, "Chemicals Known to the State to Cause Cancer or Reproductive toxicity," June 11, 2010, located at oehha.ca.gov/prop65/prop65_list/files/P65single061110.pdf (accessed June 17, 2010)

20. Edward Weil, Deputy Attorney General, State of California (presentation to the Northern California Society for Risk Analysis, Berkeley, California, October 2005).

21. U.S. Congress, *Identifying and Regulating Carcinogens*, p. 127; Cranor, "Information Generation and Use under Proposition 65."

22. Bruce A. Ackerman and Richard B. Stewart, "Comment: Reforming Environmental Law," *Stanford Law Review* 37 (1985): pp 1333–1365.

23. *Industrial Union Dept., AFL-CIO v. American Petroleum Institute*, 448 U.S. 607 (1980).

24. U.S. Congress, *Identifying and Regulating Carcinogens*, p. 79.

25. *American Federation of Labor and Congress or Industrial Organizations v. Occupational Safety and Health Administration* 965 F.2d 962, 980 and 987 (1992).

26. U.S. Congress, *Identifying and Regulating Carcinogens*, pp. 127–128, at 200–201, 215 (emphasis added).

27. On the Delaney Clause, ibid. p. 200; on the OSH Act, ibid., p. 78.

28. Ibid., pp. 206–207.

29. See Consumer Product Safety Act, 15 U.S.C. §§ 2051–2084 (2006); Federal Hazardous Substances Act, 15 U.S.C. §§ 1261–1278 (2008).

30. Reference to the language (defining "unreasonable risk") at § 2057. See also, Rachael Rawlins, "Teething on Toxins: In Search of Regulatory Solutions for Toys and Cosmetics," *Fordham Environmental Law Review*, 20 (2009): 1–50, at p. 23.

31. On voluntary standards, 15 U.S.C. §2056(b)(1); on preapproval to release information about risks from a product, 15 U.S.C. § 2055(3). I owe these points to

Dr. Jennifer Sass, Senior Scientist, National Resources Defense Counsel, June 9, 2010.

32. U.S. Congress, *Identifying and Regulating Carcinogens*, pp. 103–104.

33. *Aqua Slide 'n' Dive v. the Consumer Product Safety Commission*, 569 F.2d 831, 838–840 (5th Cir. 1978); *Gulf South Insulation v. Consumer Products Safety Commission*, 701 F.2d 1137 (5th Cir. 1983); on particular procedural and substantive rules, U.S. Congress, *Identifying and Regulating Carcinogens*, pp. 208–209.

34. U.S. Environmental Protection Agency, Endocrine Disruptor Screening Program, www.epa.gov/endo (accessed December 26, 2009).

35. On benzene decision, *Industrial Union Dept.*, 448 U.S.; on "chilling effect," Carl F. Cranor, "The Regulatory Context for Environmental and Workplace Health Protections: Recent Developments," in *The Blackwell Guide to Business Ethics*, ed. Norman Bowie (Malden, Mass.: Blackwell Publishers 2002), pp. 77–101, and Thomas O. McGarity and Sidney A. Shapiro, *Workers at Risk: The Failed Promise of the Occupational Safety and Health Administration* (New York: Praeger, 1993), pp. 257–258.

36. *Industrial Union Department, AFL-CIO v. Hodgson*, 499 F.2d 467, 472 (1974).

37. Carl F. Cranor, *Regulating Toxic Substances: A Philosophy of Science and the Law* (New York: Oxford University Press 1993), p. 27; U.S. Congress, *Identifying and Regulating Carcinogens*, p. 17; interview with risks assessors (who reported such long timelines for testing), at the U.S. EPA Air Toxics Program in Washington, D.C., at the U.S. EPA main office, April 1986.

38. National Research Council, Committee on Toxicity Testing and Assessment of Environmental Agents, et al., *Toxicity Testing for Assessment of Environmental Agents: Interim Report* (Washington, D.C.: National Academies Press 2006), pp. 134–138.

39. On other committees recognizing legitimacy of animal evidence, V. J. Cogliano, R. A. Baan, K. Straif, Y. Grosse, M. Secretan, F. El Ghissassi, and P. Kleihues, "The Science and Practice of Carcinogen Identification and Evaluation," *Environmental Health Perspectives* 112, no. 13 (2004): pp. 1269–1274, at p. 1272; National Toxicology Program, "Introduction," *Eleventh Report on Carcinogens*, ntp.niehs.nih .gov/index.cfm?objectid=32BA9724-F1F6–975E-7FCE50709CB4C932 (accessed July 5, 2008); Institute of Medicine and National Research Council, Committee on the Framework for Evaluating the Safety of Dietary Supplements, *Dietary Supplements: A Framework for Evaluating Safety* (Washington, D.C.: National Academy Press, 2005), pp. 255–256, 262. On animal and other kinds of evidence legitimately providing evidence of disease risks to humans, Institute of Medicine, Committee on Evaluation of the Presumptive Disability Decision-Making Process for Veterans, *Improving the Presumptive Disability Decision-Making Process for Veterans*, ed. J. M. Samet and C. C. Bodurow (Washington, D.C.: National Academies Press 2008), pp. 175–195; National Research Council, Committee on Toxicity Testing and Assessment of Environmental Agents, et al., *Toxicity Testing in the Twenty-first Century: A Vision and a Strategy* (Washington, D.C.: National Academies Press, 2007), pp. 21–38.

40. On gaps to be closed by TSCA, John S. Applegate, "Synthesizing TSCA and REACH: Practical Principles for Chemical Regulation Reform," *Ecology Law Quarterly* 35 (2009): 101–149, at pp. 104–105; U.S. Senate Report, no. 94–698, Mar. 16, 1976, on Public Law 94–469, Toxic Substances Control Act, at section 2, p. 2. Toxic Substances Control Act, 15 U.S.C. §§ 2601–2692 (2006).

41. U.S. Council on Environmental Quality, *Toxic Substances,* Washington, D.C., 1971, p. 21 (emphasis added).

42. Applegate, "Synthesizing TSCA and REACH," p. 107.

43. Toxic Substances Control Act, 15 U.S.C. §§ 2601–2692 (2006).

44. U.S. Congress, *Identifying and Regulating Carcinogens,* pp. 216–217.

45. Applegate, "Synthesizing TSCA and REACH," quotation at p. 115, p. 127.

46. U.S. Environmental Protection Agency, "Appendix: The Toxic Substances Control Act: History and Implementation," p. 111, located at www.epa.gov/oppt/newchems/pubs/chem-pmn/appendix.pdf, accessed on 4/22/10.

47. U.S. Congress, *The Information Content of Premanufacture Notices* (Washington, D.C.: Government Printing Office, 1983), as cited in Applegate, "Synthesizing TSCA and REACH," p. 127 (emphasis added).

48. Larry D. Claxton, Gisela de A. Umbuzeiro, David M. DeMarini, "The *Salmonella* Mutagenicity Assay: The Stethoscope of Genetic Toxicology for the 21st Century," forthcoming in *Environmental Health Perspectives,* online 2 August 2010 located at doi: 10.1289/ehp.1002336 (accessed August 3, 2010). See also, Joseph H. Guth, Richard A. Denison, Jennifer Sass, "Require Comprehensive Safety Data for All Chemicals," *New Solutions* 17(3) (2007): 233–258, at pp. 237–238, (on veil of confidential business informration, p. 241).

49. Larry D. Claxton, Gisela de A. Umbuzeiro, David M. DeMarini, "The *Salmonella* Mutagenicity Assay," p. 14–15 of draft.

50. Amber R. Wise, Jackie Schwartz, Tracey J. Woodruff, *Policy Recommendations for Addressing Potential Health Risks from Nanomaterials in California* (Draft report for the California Science Advisory Panel on nanomaterials, June 14, 2010), p. 27.

51. Peter Wick, Antoine Malek, Pius Manser, Danielle Meili, Xenia Maeder-Althaus, Liliane Diener, et al., "Barrier Capacity of Human Placenta for Nanosized Materials, *Environmental Health Perspectives* 118 (3) (2010): 432–436; Benedicte Trouiller, Ramune Reliene, Aya Westbrook, Parrisa Solaimani, and Robert H. Schiestl, "Titanium Dioxide Nanoparticles Induce DNA Damage and Genetic Instability in vivo in Mice," *Cancer Research*, 69(22) (2009): 8784-8789.

52. U.S. Congress, *House Report,* no. 94–1341, 1976, p. 15.

53. Applegate, "Synthesizing TSCA and REACH," p. 108.

54. Applegate, "Synthesizing TSCA and REACH," p. 117 (Citing U.S. Congress, House Report, No. 94-1341, at 55–56 (1976); U.S. Congress, House Report, No. 94-1679, at 96 (1976); see also U.S. Congress, Senate Report No. 94-698, at 28 (1976).

55. Applegate, "Synthesizing TSCA and REACH," p. 117.

56. Applegate, "Synthesizing TSCA and REACH," p. 118.

57. U.S. Congress, *Chemical Regulations: Options Exist to Improve EPA's Ability to Assess Health Risks and Manage Its Chemicals Review Program, GAO-05,* 1.

58. Ronald Melnick, Senior Toxicologist and Director of Special Programs, Environmental Toxicology Program, National Institute of Environmental Health Services, personal communication explaining how few substances are tested for toxicity each year on animals, October 25, 2002.

59. Based on estimates from the National Research Council, *Toxicity Testing: Strategies to Determine Needs and Priorities* (Washington, D.C.: National Academy Press 1984), p. 11.

60. Merrill, "FDA Regulatory Requirements as Tort Standards," p. 553.

61. The U.S. Food and Drug Administration, The Center for Drug Evaluation and Research, "The CDER Handbook," located at www.fda.gov/downloads/AboutFDA/CentersOffices/CDER/UCM198415.pdf (accessed June 15, 2008), at pp. 5–6.

62. Ibid., p. 7 (emphasis added).

63. Ibid., pp. 8–9.

64. Ibid., p. 47.

65. U.S. Department of Health and Human Services, Food and Drug Administration, *Sandoz Pharmaceuticals Corp.; Bromocriptine Mesylate (Parlodel) for the Prevention of Physiological Lactation; Opportunity for a Hearing on a Proposal to Withdraw Approval of the Indication,* August 1994, pp. 43347–43351, at p. 43348.

66. Ibid., pp. 43348.

67. Ibid.

68. Ibid., p. 43351.

69. U.S. Environmental Protection Agency, "Pesticides," www.epa.gov/pesticides/about/aboutus.htm (emphasis added) (accessed June 21, 2009).

70. Thomas O. McGarity, "Politics by Other Means: Law, Science and Policy in EPA's Implementation of the Food Quality Protection Act," *Administrative Law Review,* 53 (2001): 103-222, at. p. 137.

71. Ibid.

72. U.S. Environmental Protection Agency, "Framework for Addressing Key Science Issues Presented by the Food Quality Protection Act (FQPA) as Developed through the Tolerance Reassessment Advisory Committee (TRAC)," *Federal Register* 63 (October 1998): pp. 58038–58045, www.epa.gov/fedrgstr/EPA-PEST/1998/October/Day-29/p29013.htm (accessed June 21, 2009) (emphasis added).

73. U.S. Environmental Protection Agency, "Pesticides," www.epa.gov/pesticides.custhelp.com/cgi-bin/pesticides.cfg/php/enduser/std_alp.php (accessed June 22, 2009).

74. U.S. Environmental Protection Agency, Office of Pesticide Programs; "General Principles for Performing Aggregate Exposure and Risk Assessments," November 28, 2001, www.epa.gov/pesticides/trac/science/aggregate.pdf (emphasis in original) (accessed June 21, 2009).

75. Ibid. (emphasis in orginal).

76. U.S. Environmental Protection Agency, Office of Pesticide Programs, "Guidance for Identifying Pesticide Chemicals and Other Substances That Have a Common Mechanism of Toxicity," January 29, 1999, www.epa.gov/fedrgstr/EPA-PEST/1999/February/Day-05/6055.pdf (accessed October 9, 2008).

77. Donald J. Ecobichon, "Toxic Effects of Pesticides," in *Casarett and Doull's Toxicology*, 6th ed., ed. Curtis Klaassen (New York: Pergamon Press, 2001): pp. 763–810, at p. 775.

78. Ibid., pp. 774–775.

79. Brenda Eskenazi, Lisa G. Rosas, Amy R. Marks, Asa Bradman, Kim Harley, Nina Holland, Caroline Johnson, et al., "Pesticide Toxicity and the Developing Brain," *Basic and Clinical Pharmacology and Toxicology* 102 (2008): 228–236; Clement E. Furlong, Nina Holland, Rebecca, J. Richter, Asa Bradman, Alan Ho, and Brenda Eskenazi, "PON1 Status of Farmworker Mothers and Children as a Predictor of Organophosphate Sensitivity," *Pharmacogenetics and Genomics* 16 (2006): 183–190.

80. Rowan Hooper, "Top 11 Compounds in US Drinking Water," *New Scientist,* January 12, 2009, www.newscientist.com (accessed February 12, 2009); Tanya Tillett, "Meeting Report: Summit Focuses on Pharmaceuticals in Drinking Water," *Environmental Health News* 117, no. 1 (2009): A16.

81. Philippe Grandjean and Philip Landrigan, "Developmental Neurotoxicity of Industrial Chemicals," *Lancet* 368 (December 16, 2006): pp. 2167–2178, at pp. 2172–2173.

82. McGarity, "Politics by Other Means," pp. 142–144.

83. Ibid., pp. 193–200.

84. Keith Schneider, "Faking It," *The Amicus Journal*, 1983, reprinted and available at planetwaves.net/contents/faking_it.html (accessed June 17, 2010).

85. Tony Honoré, "The Morality of Tort Law—Questions and Answers," *Philosophical Foundations of Tort Law,* ed. David G. Owen (Oxford: Clarendon Press, 1995), pp. 73–95, at p. 79. See also W. Page Keeton et al., *Prosser and Keeton on the Law of Torts,* 5th ed., ed. W. Page Keeton (St. Paul, Minn.: West Publishing, 1984), pp. 5–6.

86. Honoré, "The Morality of Tort Law," p. 79.

87. C. Eads and P. Reuter, *Designing Safer Products: Corporate Responses to Product Liability Law and Regulation* (Santa Monica, Calif.: Rand Corporation, 1983); Charles Nesson, "The Evidence or the Event? On Judicial Proof and the Acceptability of Verdicts," *Harvard Law Review* 98 (1985): pp. 1357–1392; Michael J. Saks, "Do We Really Know Anything about the Behavior of the Tort Litigation System—and Why Not?" *Pennsylvania Law Review* 140 (1992): pp. 1147–1289.

88. Not all jurisdictions insist on this rigid distinction; see *Donaldson v. Central Illinois Public Service,* 199 Ill. 2d 63 (2002).

89. On innovative legal theory, *Sindell v. Abbott Laboratories,* 26 Cal. 3d 588, 607 P. 2d 924 (1980); on resulting liability line of cases, James L. Schardein and Orest T. Macina, *Human Developmental Toxicants: Aspects of Toxicology and Chemistry* (Boca Raton, Fla.: Taylor and Francis, 2007), p. 286.

90. Carl F. Cranor, *Toxic Torts: Science, Law and the Possibility of Justice* (New York: Cambridge University Press, 2006), p. 224; *In re Agent Orange Prod. Liab. Litig.,* 611 F.Supp. 1223, 1231 (E.D.N.Y. 1985); *Lynch v. Merrell-National Lab., Div. of Richardson-Merrell Inc.,* 830 F.2d 1190 (1st Cir. 1987); *Brock v. Merrell*

Dow Pharm. Inc., 874 F.2d 307, 312 (5th Cir. 1989), *cert. denied*, 494 U.S. 1046 (1990); *Richardson v. Richardson-Merrell Inc.*, 857 F.2d 823, 825, 831n59 (D.C. Cir. 1988), *cert. denied*, 493 U.S. 882 (1989); *Renaud v. Martin Marietta Corp.*, 749 F. Supp. 1545 (D. Colo. 1990); *Carroll v. Litton Sys. Inc.*, No. B-C88–253, 1990 WL 312969, at *47 (W.D.N.C. Oct. 29, 1990); *Thomas v. Hoffman-La-Roche Inc.*, 731 F. Supp. 224 (N.D. Miss. 1989), *aff'd on other grounds*, 949 F.2d 806 (5th Cir. 1992); *Chambers v. Exxon Corp.*, 81 F. Supp. 2d 661 (M.D. La. 2000).

91. On especially strict demands for human evidence, *E. I. du Pont de Nemours and Company Inc., Petitioners v. C. R. Robinson*, 923 S.W.2d 549 (1995), and *Merrill-Dow Pharmaceuticals Inc., v. Havner*, 953 S.W.2d 706 (1997); on bringing tort law closer to its potential, *Kumho Tire v. Carmichael*, 526 U.S. 137, 152 (1999), and Cranor, *Toxic Torts*, chap. 7.

92. Margaret A. Berger, "Eliminating General Causation: Notes towards a New Theory of Justice and Toxic Torts," *Columbia Law Review* 97 (1997): pp. 2117–2152, at pp. 2135, 2117, 2135 (emphasis added). See also Barry I. Castleman, "Regulations Affecting Use of Carcinogens," in *Cancer Causing Chemicals*, ed. N. Irving Sax (New York: Van Nostrand Reinhold, 1981), pp. 78-98, at p. 78; David E. Lilienfeld, "The Silence: The Asbestos Industry and Early Occupational Cancer Research—A Case Study," *American Journal of Public Health* 81 (1991): pp. 791–798, at p. 791; David Michaels, "Waiting for the Body Count: Corporate Decision Making and Bladder Cancer in the U.S. Dye Industry," *Medical Anthropology Quarterly* 2 (1988): pp. 215–227; Donald R. Mattison and John E. Craighead, "Reproductive System," in *Pathology of Environmental and Occupational Disease,* ed. John E. Craighead (St. Louis: Mosby, 1995), pp. 559–572.

93. Carl F. Cranor and David A. Eastmond, "Scientific Ignorance and Reliable Patterns of Evidence in Toxic Tort Causation," *Law and Contemporary Problems*, 64, (2010): 5-48, at p. 5.

94. Thomas O. McGarity and Wendy E. Wagner, *Bending Science: How Special Interests Corrupt Public Health Research (Cambridge, MA: Harvard University Press, 2008),* at pp. 60-96.

95. David Michaels, *Doubt Is Their Product: How Industry's Assault on Science Threatens Your Health* (New York: Oxford University Press, 2008), pp. ix–xii., 3–11.

96. Brown and Williamson Tobacco Company, "Smoking and Health Proposal," Brown and Williamson document no. 680561778–1786, 1969, legacy.library.ucsf.edu/tid/nvs40f00 (accessed June 30, 2008).

97. K.S. Santone, and G. Powis, (1991). "Mechanism of and Tests for Injuries," In W. J. Hayes, Jr. and E. R. Laws, Jr. (eds.), *Handbook of Pesticide Toxicology* (New York: Harcourt Brace Jovanovich, 1991), pp. 169–214, at p. 169.

98. Cranor, *Toxic Torts*, pp. 200–204; see also Poul Harremoës et al., eds., *Late Lessons from Early Warnings: The Precautionary Principle, 1896–2000* (Copenhagen: European Environment Agency, 2001),: pp. 38–51; David Brodeur, *Outrageous Misconduct: The Asbestos Industry on Trial* (New York: Pantheon Books, 1985), pp. 113, 327, 348; Gerald Markowitz and David Rosner, *Deceit and*

Denial: The Deadly Politics of Industrial Pollution (Berkeley: University of California Press, 2002), p. 303; Michaels, "Waiting for the Body Count," pp. 215, 218–221.

99. Cranor, "The Regulatory Context for Environmental and Workplace Health Protections," pp. 77–78.

100. Michaels, *Doubt Is Their Product*, pp. 60–78.

101. Ibid., p. 46.

102. Dina Cappiello, "Oil Industry Funding Study to Contradict Cancer Claims," *Houston Chronicle*, April 29, 2005, A1; Susanna Rankin Bohme, John Zorabedian, and David S. Egilman, "Maximizing Profit and Endangering Health: Corporate Strategies to Avoid Litigation and Regulation," *International Journal of Occupational and Environmental Health* 11 (2005): pp. 338–348. I also possesses internal American Petroleum Institute documents indicating what research into benzene exposures would show before the studies were conducted. These documents, labeled "API.Shanhai.Studies.Internal.Documents," were provided from discovery documents in a toxic tort case.

103. On studies designed to blunt regulation or frustrate tort suits, Bohme, Zorabedian, and Egilman, "Maximizing Profit and Endangering Health," p. 340; quotation of appellate judge describing abuses by pharmaceutical company, *Blum v. Merrell Dow Pharmaceuticals*, 33 Phila. Co. Rptr. 193–258, 247 (1996); on tougher disclosure policies for scientific papers, Natasha Singer and Duff Wilson, "Medical Editors Push for Ghostwriting Crackdown," *New York Times*, September 18, 2009, www.nytimes.com (accessed October 10, 2009).

104. John C. Bailar III, "How to Distort the Scientific Record without Actually Lying: Truth, and the Arts of Science," *European Journal of Oncology* 11 (2006): pp. 217–224.

105. Alex Berenson, "Follow-Up Study on Vioxx Safety Is Disputed," *New York Times*, May 13, 2006; Richard A. Oppel, Jr., Environmental Tests "Falsified," U.S. Says, *New York Times*, Sept. 22, 2000, at A14; Melody Petersen, "Settlement Is Approved in Diet Drug Case," *New York Times*, Aug. 29, 2000, at C2; David Willman, The Rise and Fall of the Killer Drug Rezulin; People Were Dying as Specialists Waged War Against Their FDA Superiors, *Los Angeles Times*, June 4, 2000, at A1; David Willman, Risk Was Known as FDA Ok'd Fatal Drug, *Los Angeles Times*, Mar. 11, 2001, at A1.

106. National Research Council et al., *Toxicity Testing: Strategies*, p. ix.

107. Ibid., p. 47.

108. John C. Bailar III and Eula Bingham, Collegium Ramazzini, personal communication about possible update to the NRC report, 2002; U.S. Congress, *Screening and Testing Chemicals in Commerce* (1995), p. 1.

109. U.S. Environmental Protection Agency, Office of Pollution Prevention and Toxics, "Chemical Hazard Data Availability Study: What Do We Really Know about the Safety of High Production Volume Chemicals?" www.epa.gov/HPV/pubs/general/hazchem.pdf (accessed January 15, 2009).

110. "EPA, EDF, CMA Agree on Testing Program Targeting 2,800 Chemicals," *Environmental Health Letter* 37 (October 1998): p. 193.

111. Richard Dennison, "High Hopes, Low Marks: A Final Report Card on the

High Production Volume Chemical Challenge," *Environmental Defense*, July 2007, p. 3; www.environmentaldefense.org/documents/6653_HighHopesLowMarks.pdf (accessed June 25, 2009).

112. Harremoës et al., *Late Lessons from Early Warnings*, pp. 11, 168–169.

113. Philip J. Landrigan, Carole A. Kimmel, Adolfo Corea, and Brenda Eskenazi, "Children's Health and the Environment: Public Health Issues and Challenges for Risk Assessment," *Environmental Health Perspectives*, 112, no. 2 (2004): 257–265, at p. 259.

114. Marla Cone, "Salmon in Near-Shore Pacific Contaminating Killer Whales," *Environmental Health News*, January 20, 2009, www.environmentalhealthnews .org (accessed January 20, 2009).

115. James Huff and David P. Rall, "Relevance to Humans of Carcinogenesis Results from Laboratory Animal Toxicology Studies," in *Maxcy-Rosenau-Last, Public Health and Preventive Medicine*, 13th ed., ed. John M. Last and Robert B. Wallace (Norwalk, Conn.: Appleton and Lange, 1992), pp. 433, 439.

116. Hutt, "Philosophy of Regulation Under the Food, Drug and Cosmetic Act," p. 101; Wendy Wagner, "Using Competition-Based Regulatory to Bridge the Toxics Data Gap," p. 631; and Merrill, "FDA Regulatory Requirements as Tort Standards," p. 552.

117. Merrill, "FDA Regulatory Requirements as Tort Standards," p. 553. Similar principles apply to foods and dietary supplements also administered by the FDA but not to therapeutic drugs and new medical devices (ibid., pp. 555–556).

118. Cranor, "The Regulatory Context for Environmental and Workplace Health Protections," pp. 89–98.

119. Wagner, "Using Competition-Based Regulatory to Bridge the Toxics Data Gap," pp. 631–633.

120. Ibid., pp. 635–636.

121. Cranor, *Toxic Torts*, pp. 205–279.

122. Bryan A. Garner, *Black's Law Dictionary, eighth edition,* (St. Paul, Minnesota: Thompson West, 1999), p. 293.

123. Keeton et al., *Prosser and Keeton on Torts*, p. 617.

124. Ibid., pp. 619–620.

125. Ibid.

126. In particular, a plaintiff must show that 1) the "defendant acted with the intent of interfering with the use and enjoyment of the land by those entitled to that use; 2) there was some [actual] interference with the use and enjoyment of the land of the kind intended"; 3) the interference was substantial and 4) constituted an "unreasonable" interference with it (ibid., p. 622).

127. William H. Rodgers, Jr., *Environmental Law, Second Edition (St. Paul, Minn., 1994)*, pp. 248, 252, 125.

128. Ibid., p. 113.

129. Ibid., pp. 184–189.

130. On kinds of nuisances, Keeton et al., *Prosser and Keeton on Torts*, pp. 619–620.

6. A MORE PRUDENT
APPROACH TO REDUCE TOXIC INVASIONS

1. National Research Council, Committee on Toxicity Testing and Assessment of Environmental Agents, et al., *Toxicity Testing for Assessment of Environmental Agents: Interim Report* (Washington, D.C.: National Academy Press, 2008), p. 190.

2. See, generally, European Community, *Registration, Evaluation, Authorisation, and Restriction of Chemicals (REACH)*, establishing a European Chemicals Agency, December 18, 2006, no. 1907/2006 (U.K.) (hereafter, REACH).

3. U.S. Food and Drug Administration, Center for Drug Evaluation and Research, "The New Drug Development Process: Steps from Test Tube to New Drug Application Review," www.fda.gov/CDER/HANDBOOK/DEVELOP.HTM (accessed July 29, 2009).

4. Ibid.

5. Ibid., p. 7 (emphasis added).

6. Ibid.

7. On special protections for infants and children, U.S. Environmental Protection Agency, "Framework for Addressing Key Science Issues Presented by the Food Quality Protection Act (FQPA) as Developed through the Tolerance Reassessment Advisory Committee (TRAC)," *Federal Register* 63 (October, 1998): pp. 58038–58045, www.epa.gov/fedrgstr/EPA-PEST/1998/October/Day-29/p29013 .htm (accessed June 30, 2009); on sensitivity of developing bodies, U.S. Environmental Protection Agency, "Pesticides," www.epa.gov/pesticides.custhelp.com/ cgi-bin/pesticides.cfg/php/enduser/std_alp.php (accessed June 30, 2009); on recognition of multiple pathways for single chemicals, U.S. Environmental Protection Agency, Office of Pesticide Programs, "General Principles for Performing Aggregate Exposure and Risk Assessments," November 28, 2001, www.epa.gov/pesticides/ trac/science/aggregate.pdf; on cumulative risk assessments, U.S. Environmental Protection Agency, Office of Pesticide Programs, "Guidance for Identifying Pesticide Chemicals and Other Substances That Have a Common Mechanisms of Toxicity," January 29, 1999, www.epa.gov/fedrgstr/EPA-PEST/1999/February/Day-05/ 6055.pdf (accessed October 9, 2008); on organophosphate and carbamate insecticides, Donald J. Ecobichon, "Toxic Effects of Pesticides," in *Casarett and Doull's Toxicology*, 6th ed., ed. Curtis Klaassen (New York: Pergamon Press, 2001), pp. 763–810, at p. 775.

8. Tom L. Beauchamp and LeRoy Walters, "Research Involving Human and Animal Subjects," in *Contemporary Issues in Bioethics*, 6th ed., ed. Tom L. Beauchamp and LeRoy Walters (Belmont, Calif.: Wadsworth Publishing, 2003), pp. 345–354.

9. On abuses by Nazis, Japanese, U.S. government, ibid., pp. 345–354; on Nuremberg Code, the Declaration of Helsinki, and Belmont Report, National Commission for the Protection of Human Subjects of Biomedical and Behavioral Research, Office of Human Subjects Research, National Institutes of Health, "The Belmont Report: Ethical Principles and Guidelines for the Protection of Human Subjects of Research," April 18, 1979, www.ohsr.od.nih.gov/guidelines/belmont

.html (hereafter, Belmont Report). Abbreviations NC and DH refer to specific provisions of the Nuremberg Code and the Declaration of Helsinki, e.g., NC.6, DH.7.

10. Beauchamp and Walters, "Research Involving Human and Animal Subjects (citing the Nuremberg Code)," p. 354.

11. Ibid., p. 357.

12. *Mink v. University of Chicago,* 460 F. Supp. 713, 715 (1978).

13. W. Page Keeton et al., *Prosser and Keeton on the Law of Torts,* 5th ed., ed. W. Page Keeton (St. Paul, Minn. West Publishing, 1984), pp. 39 (emphasis added) and 40 (emphasis added).

14. William Prosser, *Handbook of the Law of Torts,* 4th ed. (St. Paul, Minn.: West Publishing, 1971, section 9, at 35, as cited in *Mink,* 460 F. Supp. at 717–718 (emphasis added).

15. Prosser, *Handbook of the Law of Torts,* p. 35.

16. On the historical background of torts, Keeton et. al., *Prosser and Keeton on Torts,* pp. 28–30.

17. Ibid., quotations at pp. 67–68.

18. Ibid., quotations at pp. 75–76.

19. On vindication of the legal right, ibid., p. 76; on aspects of trespass, ibid., pp. 70–71.

20. See *Martin v. Reynolds Metals Co.,* 342 P.2d 790 (Or. 1959), on holding a manufacturing operation that caused fluoride gases and particulates to become airborne and settle upon owner's land liable for direct trespass, and *Borland v. Sanders Lead Company Inc.,* 369 So.2d 523 (Ala. 1979), on characterizing dangerous accumulation of lead particulates and sulfoxide deposits as trespass.

21. See *Sullivan v. Dunham,* 55 N.E. 923, 926–927 (N.Y. 1900), extending trespass liability to personal injury caused by debris from blast on one's own property.

22. Arthur Ripstein, "Beyond the Harm Principle," *Philosophy and Public Affairs* 34 (2006): pp. 215–245, at p. 218.

23. Ibid., pp. 227, 234.

24. See Raymond Neutra, "Risk Assessment, *Science* 237 (1987): p. 235.

25. Ripstein, "Beyond the Harm Principle," at pp. 229–245, esp. pp. 231, 233–236, 241–242.

26. Ibid., pp. 228, 233–236, 241–242.

27. My colleague Paul Hoffman suggested this point.

28. Caroline M. Tanner, "Occupational and Environmental Causes of Parkinsonism," in *Occupational Medicine: State of the Art Reviews* 7 (1992): pp. 503–513, at p. 506; J. William Langston and Phillip A. Ballard, "Parkinson's Disease in a Chemist Working with 1-methyl-4-phenyl-1,2,3,6-tetrahydropyridine," *New England Journal of Medicine* 309 (1983): pp. 979–980.

29. T. J. Woodruff, L. Zeise, D. A. Axelrad, K. Z. Guyton, S. Janssen, M. Miller, G. G. Miller, et al., "Meeting Report: Moving Upstream—Evaluating Adverse Upstream End Points for Improved Risk Assessment and Decision-Making," *Environmental Health Perspectives* 16, no. 11 (2008): pp. 1568–1575.

30. Belmont Report, p. 8.

31. Alicia J. Fraser, Thomas F. Webster, and Michael D. McClean, "Diet Contributes Significantly to the Body Burden of PBDEs in the General U.S. Population," *Environmental Health Perspectives*, 117, no. 10 (2009): pp. 1520–1525.

32. On adverse reproductive and developmental effects, Moira Welsh, "Call for Ban on Chemicals in Cosmetics, Cleaners: Children Exposed to Nail Polish, Perfumes or Household Cleaners Are at Risk, Say Health Groups," *Toronto Star,* November 8, 2008, www.thestar.com (accessed November 9, 2008); on carcinogens, Breast Cancer Fund, *State of the Evidence 2008: The Connection between Breast Cancer and the Environment,* ed. by Janet Gray, www.breastcancerfund.org (accessed November 19, 2008).

33. BNA Inc., *Chemical Regulation Reporter* 31 (2007): 1–6, at p. 1.

34. On definition of injunction, *Black's Law Dictionary,* ed. Bryan A. Garner (St. Paul, Minn.: Thompson-West, 2004): p. 800; on requirements of many common injunctions, Adam Liptak, "Supreme Court Rules for Navy in Sonar Case," *New York Times,* November 12, 2008 (the ruling rejected the preliminary injunction because the court was skeptical that sonar caused "irreparable damage"); on serving social interest, Keeton et al., *Prosser and Keeton on Torts,* pp. 630–632.

35. John Murphy, "Rethinking Injunctions in Tort Law," *Oxford Journal of Legal Studies,* 27 (2007): pp. 509-535, p. 535 (emphasis added).

36. On professions needing licenses, American Bar Association, Division for Public Education, "When You Need a Lawyer: What Are Professional Requirements for Becoming a Lawyer?" www.abanet.org/publiced/practical/lawyer_requirements.html (accessed July 10, 2009); on licenses permitting otherwise illegal behavior, *Black's Law Dictionary,* p. 938.

37. National Research Council et al., *Toxicity Testing for Assessment of Environmental Agents,* pp. 12–13, quotation at p. 187. See also National Research Council, Committee on Toxicity Testing and Assessment of Environmental Agents, et al., *Toxicity Testing in the Twenty-first Century: A Vision and a Strategy* (Washington, D.C.: National Academies Press, 2007), quotation at p. 19.

38. National Research Council et al., *Toxicity Testing for Assessment of Environmental Agents,* p. 190.

39. "Kids Safe Chemical Act of 2008," U.S. Senate Bill S.3040, 110th Congress, 2d Session, section 504 (a)(1)(A) and (B) pp. 13–14.

40. U.S. Department of Health and Human Services, National Institutes of Health, National Institute of Environmental Health Sciences, National Toxicology Program, *Draft NTP Brief on Bisphenol A,* Washington, D.C., April 14, 2008, p. 32 (emphasis added).

41. See, generally, REACH.

42. Ibid.

43. REACH, O. J. (L 396) 1, 20 (EC), art. 1(1)(3), and art. 1, ¶ 4.

44. REACH, quotation at art. 1, ¶ 12; quotation at art. 1, ¶ 18–19; on number and types of testing and cumulative tests, art. 1, ¶ 28–29.

45. Ibid., art. 1, ¶ 19–21.

46. See REACH, at art. L 396/339, L 396/352-53, L 396/3643-66.

47. Ibid., art. I, ¶ 14.

48. National Research Council et al., *Toxicity Testing for Assessment of Environmental Agents,* p. 181.

49. Ibid., p. 186.

50. Ibid.

51. U.S. Environmental Protection Agency, Office of Pesticide Programs, "Available Information on Assessing Exposure from Pesticides in Food: A User's Guide," July 28, 2003, at p. 3, located at www.hc-sc.gc.ca/cps-spc/alt_formats/pacrb-dgapcr/pdf/pubs/pest/pol-guide/spn/spn2003-03-eng.pdf (accessed June 25, 2010).

52. National Research Council et al., *Toxicity Testing for Assessment of Environmental Agents,* p. 188.

53. Ibid., pp. 13, 188.

54. M. J. Brunner, T. M. Sullivan, A. W. Singer, M. J. Ryan, J. D. Toft II, R. S. Menton, S. W. Graves, et al. "An Assessment of the Chronic Toxicity and Oncogenicity of Aroclor-1242, Aroclor-1254 and Aroclor-1260 Administered in Diet to Rats" (Batelle Study no. SC920192, Chronic Toxicity and Oncogenicity Report, Columbus, Ohio, 1996); T. C. Hornshaw, R. J. Aulerich, and H. E. Johnson. "Feeding Great Lakes Fish to Mink: Effects on Mink and Accumulation and Elimination of PCBs by Mink," *Journal of Toxicology and Environmental Health* 11 (1983): pp. 933–946; R. J. Aulerich, R. K. Ringer, and J. Safronoff, "Assessment of Primary and Secondary Toxicity of Aroclor 1254 to Mink," *Archives of Environmental Contamination Toxicology* 15 (1986): 393–399.

55. David Eastmond, Chair, Environmental Toxicology Program, University of California, Riverside, personal communication, July 28, 2008.

56. David L. Eaton and Curtis D. Klaassen, "Principles of Toxicology," in *Casarett and Doull's Toxicology,* 6th ed., ed. Curtis Klaassen (New York: Pergamon Press, 2001): pp. 11–34, at p. 11.

57. Ibid., p. 13.

58. National Research Council et al., *Toxicity Testing for Assessment of Environmental Agents,* quotation at p. 188; National Academy of Sciences, "Executive Summary," in *Science and Decisions: Advancing Risk Assessment* (Washington, D.C.: National Academy Press, 2008), pp. 1–32, at p. 7, www.nap.edu/catalog/12209.html (accessed October 9, 2009).

59. T. J. Woodruff, et al., "Meeting Report: Moving Upstream," p. 1570; on compounds causing greater effects than additive doses would predict, Sofie Christiansen, Martin Scholze, Majken Dalgaard, Anne Marie Vinggaard, Marta Axelstad, Andreas Kortenkamp, and Ulla Hass, "Synergistic Disruption of External Male Sex Organ Development by a Mixture of Four Antiandrogens," *Environmental Health Perspectives* 117, 12 (2009): pp. 1839-1846, p. 1839; Andreas Kortenkamp, "Breast Cancer and Exposure to Hormonally Active Chemicals: An Assessment of the Scientific Evidence," www.chemtrust.org.uk (accessed August 19, 2008).

60. Ulla Hass, Martin Scholze, Sofie Christiansen, Majken Dalgaard, Anne Marie Vinggaard, Marta Axelstad, Stine Broeng Metzdorff, et al., "Combined Exposure to Anti-Androgens Exacerbates Disruption of Sexual Differentiation in the Rat," *Environmental Health Perspectives* 115, S-1 (2007): 122–128, p. 127.

61. Sheldon Krimsky, "Plastics in Our Diet," *Scientific American Earth 3.0,* 18, no. 4 (2008): pp. 30–31, at p. 31.

62. National Academy of Sciences, Committee on Improving Risk Analysis Approaches Used by the U.S. EPA, National Research Council, "Science and Decisions: Advancing Risk Assessment" (free executive summary), pp. 8, www.nap.edu/catalog/12209.html (Accessed July 27, 2009).

63. This is known as the Kerry-Moran Bill because it is sponsored by Senator John Kerry and Congressman Jim Moran. Endocrine Society, "The Endocrine Society Endorses Moran/Kerry Bill Aimed at Protecting Public from Exposure to Harmful Chemicals," www.newswise.com/articles/view/559294/?sc=rsmn (accessed December 21, 2009).

64. Ibid.

65. Kim Hooper and Thomas A. McDonald, "The PBDEs: An Emerging Environmental Challenge and Another Reason for Breast-Milk Monitoring Programs," *Environmental Health Perspectives* 108, no. 5 (2000): pp. 387–392, at p. 388.

66. Donald T. Wigle and Bruce P. Lanphear, "Human Health Risks from Low-Level Environmental Exposures: No Apparent Safety Thresholds," *PLoS Medicine* 2 (2005): pp. 1–3, www.plosmedicine.org (accessed August 5, 2009).

7. WHAT KIND OF WORLD DO WE WANT TO CREATE?

1. U.S. Department of Health and Human Services, U. S. Food and Drug Administration, Center for Food Safety and Applied Nutrition, Office of Cosmetics and Colors, "Phthalates and Cosmetic Products," www.foodsafety.gov/~dms/cos-phth.html (accessed July 20, 2009).

2. S. A. Roach and S. M. Rappaport, "But They Are Not Thresholds: A Critical Analysis of the Documentation of Threshold Limit Values," *American Journal of Industrial Medicine* 17 (1990): pp. 727–753, at p. 731.

3. Ibid., p. 741.

4. Robert A. Goyer and Thomas W. Clarkson, "Toxic Effects of Metals," in *Casarett and Doull's Toxicology,* 6th ed., ed. Curtis D. Klaassen (New York: McGraw-Hill, 2001), pp. 811–867, at pp. 823–825.

5. Renee Schoof, "Coal Ash Is Damaging Water, Health in 34 States, Groups Say," *McClatchy Newspapers,* May 7, 2009, www.mcclatchydc.com/251/story/67753.html (accessed May 11, 2009).

6. U.S. Environmental Protection Agency, "Mercury," www.epa.gov/hg/control_emissions/index.htm (accessed April 29, 2009).

7. Ibid.

8. State of Washington, Department of Ecology, "Dirt Alert: Tacoma Smelter Plume," www.ecy.wa.gov/programs/tcp/sites/tacoma_smelter/ts_hp.htm (accessed May 4, 2009).

9. I owe this point to Mary Lyndon, Professor of Law, St. John's University Law School, personal communication, May 7, 2009.

10. National Research Council, Steering Committee on Identification of Toxic and Potentially Toxic Chemicals for Consideration by the National Toxicology

Program, *Toxicity Testing: Strategies to Determine Needs and Priorities* (Washington, D.C.: National Academy Press, 1984), p. 363.

11. Carl F. Cranor, *Regulating Toxic Substances: A Philosophy of Science and the Law* (1993; repr., New York: Oxford University Press, 1997), pp. 8, 71–78, 142–146; Carl F. Cranor "The Social Benefits of Expedited Risk Assessment," *Risk Analysis* 15 (1995): pp. 353–358.

12. National Research Council, *Toxicity Testing,* pp. 367–370; Lester Lave and Gil Omenn, "Cost Effectiveness of Short-Term Tests for Carcinogenicity," *Nature* 324 (1986): pp. 29–34; Toby Page, "A Framework for Unreasonable Risk in the Toxic Substances Control Act," in *Annals of the New York Academy of Science: Management of Assessed Risk for Carcinogens,* ed. W. J. Nichols (New York: New York Academy of Sciences, 1981), pp. 145–166; and Cranor, "Social Benefits of Expedited Risk Assessment," pp. 354–355.

13. Clean Air Act (1970, amended 1974, 1977, 1990), 42 U.S.C. 7408 (a)(2); on attention focused on children, the elderly, preexisting conditions, Discussion of Provisions, Public Law 101–549:1990, and American Thoracic Society, Committee of the Environmental and Occupational Health Assembly, "Health Effects of Outdoor Air Pollution," *American Journal of Respiratory Critical Care Medicine* 53, no. 1 (1996): pp. 3–50.

14. Safe Drinking Water Act, 42 U.S.C. Sec. 300g-1(b)(a)(b) (emphasis added) and Sec. 300g-1(b)(4).

15. Occupational Safety and Health Act: 1970, 29 U.S.C. 655(b)(5) (emphasis added).

16. Safe Chemicals Act of 2010, 111th Congress, 2d Session, introduced by Frank R. Lautenberg. Located at lautenberg.senate.gov/assets/SCA2010.pdf (accessed July 7, 2010).

17. Ibid, at section 6(b)(1)(B)(i) for burden of proof and section 4(2)(23) for definition of safety standard.

18. For the requirement that aggregate and cumulative exposures must present a reasonable certainty of no harm, ibid. at section 4(2)(23).; for definition of adverse effect, Ibid., at 4(2)(14); ibid., at section 4(a)(1-3); ibid., at section 8(a)(4)(A-B).

19. Ibid., at sections (6(a)(1) and 6(a)(1-2); ibid., at section 4(a)(1-3); ibid., at section 8(a)(4)(A-B).

20. Joseph H. Guth, Richard A. Denison, Jennifer Sass, "Require Comprehensive Safety Data for All Chemicals," *New Solutions,* 17(3) (2007): 233–258, p. 242.

21. Massachusetts Department of Environmental Protection, Toxic Use Reduction Act, described at www.mass.gov/dep/toxics/tura/turaover.htm (accessed May 15, 2009).

22. Toxics Use Reduction Institute, "Five Chemicals Study: Alternatives Assessment Study: Executive Summary," updated August 1, 2008, www.turi.org/library/turi_publications/five_chemicals_study (accessed May 5, 2009).

23. Ibid.

24. John Rawls, *A Theory of Justice* (Cambridge, Mass: Belknap Press of Harvard University Press, 1971), p. 73, 83–89, 298–303.

25. Norman Daniels, "Health Care Needs and Distributive Justice," *Philoso-*

phy and Public Affairs, 10 (1981): pp. 146–179; Norman Daniels, *Just Health Care* (New York: Cambridge University Press, 1985).

26. Eunice Kennedy Shriver National Institute of Child Health and Human Development "Institute Mission and Accomplishment Highlights," www.nichd.nih .gov/about/overview/mission/index.cfm (accessed December 21, 2009).

27. Philip Shabecoff and Alice Shabecoff, *Poisoned Profits: The Toxic Assault on Our Children* (New York: Random House, 2008), pp. 263–264; Environmental Working Group, www.ewg.org/health (accessed December 12, 2009).

28. Shabecoff and Shabecoff, *Poisoned Profits,* pp. 263–264.

29. Ibid.

30. Ibid., pp. 271, 272–273.

31. Heather Hamlin and Wendy Hessler, "Sleeping with the Enemy: Indoor Airborne Contaminants," www.environmentalhealthnews.org/ehs/newscience/ indoor-measure-of-airborne-contaminants (accessed April 28, 2009); Robert W. Gale, Walter L. Cranor, David A. Alvarez, James, N. Huckins, Jimmie D. Petty, and Gary L. Robertson, "Semivolatile Organic Compounds in Residential Air along the Arizona-Mexico Border," *Environmental Science and Technology* 43, no. 9 (2009): pp. 3054–3060, pubs.acs.org/doi/pdfplus/10.1021/es803482u (accessed on May 28, 2009).

32. Gale et. al, "Semivolatile Organic Compounds in Residential Air," pp. 3056, 3057, 3054, 3056.

33. U.S. Environmental Protection Agency, Office of Pesticide Programs, "General Principles for Performing Aggregate Exposure and Risk Assessments," November 28, 2001, www.epa.gov/pesticides/trac/science/aggregate.pdf (accessed August 15, 2009).

34. Martin Mittelstaedt, "Fragile Labrador Ecosystem Overcomes a Toxic Past," *Globe and Mail,* December 21, 2009, www.theglobeandmail.com/news/ national/fragile-labrador-ecosystem-overcomes-a-toxic-past/article1408308 (accessed December 21, 2009); on blood lead levels, U.S. Department of Health and Human Services, Centers for Disease Control and Prevention, National Center for Environmental Health, *Fourth National Report on Human Exposure to Environmental Chemicals,* December 2009, p. 4, www.cdc.gov/exposurereport (accessed December 25, 2009).

35. Max Neiman, *Defending Government: Why Big Government Works* (Upper Saddle River, N.J.: Prentice Hall, 2004), p. 192.

36. Neva R. Goodman and Tom Tietenberg, "Externality," *The Encyclopedia of the Earth,* www.eoearth.org/article/Externality, November 20, 2008 (accessed April 12, 2009).

37. On symbolic or concrete externalities, psychological and subjective externalities, quotation at Tom Tietenberg, *Environmental Economics and Policy* (Boston: Pearson Education, 2004), p. 73, Neiman, *Defending Government,* p. 192.

38. Marc A. Franklin, *Injuries and Remedies: Cases and materials on Tort Law and Alternatives* (Mineola, NY: The Foundation Press, Inc, 1979), p. 762.

39. Ibid., p. 61.

40. James L. Schardein, *Chemically Induced Birth Defects,* 3rd rev. and expanded ed. (New York: Marcel Dekker, 2000): pp. 12–20.

41. On economic costs of diminished productivity before lead was reduced in gasoline, Philippe Grandjean and Philip Landrigan, "Developmental Neurotoxicity of Industrial Chemicals," *Lancet* 368 (December 16, 2006): pp. 2167–2178, at p. 2174. On costs of lost productivity today, ibid; see also, Philip J. Landrigan, Clyde B. Schechter, Jeffrey M. Lipton, Marianne C. Fahs, and Joel Schwartz, "Environmental Pollutants and Disease in American Children: Estimates of Morbidity, Mortality, and Costs for Lead Poisoning, Asthma, Cancer, and Developmental Disabilities," *Environmental Health Perspectives* 110, no. 7 (2002): pp. 721–728.

42. Landrigan et al. "Environmental Pollutants and Disease in American Children," p. 727; Leonardo Trasande, Philip J. Landrigan, and Clyde Schechter, "Public Health and Economic Consequences of Methyl Mercury Toxicity to the Developing Brain," *Environmental Health Perspectives* 113, no. 5 (2005): pp. 590–596, p. 590.

43. Landrigan et al., "Environmental Pollutants and Disease in American Children," pp. 726–727.

44. Jamie Lincoln Kitman, "The Secret History of Lead," *Nation*, March 20, 2000, www.thenation.com/doc/20000320/kitman (accessed May 27, 2009).

45. On company postmarket surveillance of a product, U.S. Food and Drug Administration, "Report to Congress: Reports on Postmarketing Studies," April 4, 2002, quotation at p. 5, www.fda.gov/cber/fdama/pstmrktfdama130.htm (accessed May 26, 2009); on warning labels required by FDA, Food, Drug, and Cosmetic Act, 21 U.S.C. § 201.80(e); on 2007 amendments requiring company updating of labels, *Wyeth v. Levine*, 129 S. Ct. 1187, 1196 (2009).

46. Carl F. Cranor, *Toxic Torts: Science, Law and the Possibility of Justice* (New York: Cambridge University Press, 2006).

47. Robert D. Morris, Anne-Marie Audex, Italo F. Angelillo, Thomas C. Chalmers, and Frederick Mosteller, "Chlorination, Chlorination By-Products, and Cancer: A Meta-analysis," *American Journal of Public Health* 82 (1992): pp. 953–963, at p. 957.

48. Frank Ackerman, "The Unbearable Lightness of Regulatory Costs," *Fordham Urban Law Journal* 33 (2006): pp. 1071–1096, pp. 1076–1077.

49. Ibid., pp. 1077–1078.

50. European Commission, "REACH in Brief at 14," September 15, 2004, ecb.jrc.it/documents/reach/overview/reach_in_brief-2004_09_15.pdf (accessed July 10, 2010).

51. Ackerman, "Unbearable Lightness of Regulatory Costs," p. 1077.

52. Ibid.

53. Sally Deneen, "Frustrated local mom testifies to Senate: Why can't I protect my baby from chemicals?" Seattle Globe Reporter, Friday, February 05, 2010, located at seattlepostglobe.org/2010/02/04/infuriated-local-mom-testifies-to-senate-why-cant-i-protect-my-body-from-chemicals, (accessed 4/29/10).

Some people might be inclined to think that single-authored books are solo enterprises. While there may be rare circumstances in which this is true, the research and comments of many scientists, philosophers and legal scholars have contributed to this project.

An invitation from Philippe Grandjean to the Faeroe Islands conference on Prenatal Programming and Toxicity, Torshavn, Faeroe Islands, May 2007, triggered the project, where I presented some early thoughts on the legal implications of the developing science. The research from this event, a later follow-up conference and the literature suggest substantial paradigm shifts for science and the law. Input from Ana Soto, Fred vom Saal, Jerry Heindel, Beate Ritz, Philip Landrigan, Bruce Lanphear, Deborah Cory-Schlechta, Larry Needham, Caroline McMillan, Morando Soffritti, Birger Hienzow, Brenda Eskenazi, and David Gee along with numerous papers at the conference supported and further inspired the project, for which I am grateful. The published proceedings were an early and continual guide to the issues. Beyond that event Philippe has been continuously supportive of this project for which I owe him special gratitude.

An invitation to give a keynote address to the Cain Conference at the Chemical Heritage Foundation (March 2007) provided a scientific venue to introduce some ideas arising out of this research. David Caudill and Jody Roberts have been valuable interlocutors on some of these issues since then. This event was followed by a James Martin Special Lecture at Oxford University (October 2007), where Matthew Liao, Julian Savalescu, and other attendees provided insightful comments. Also in the fall of 2007 participation in a confrence on low dose exposures at the Villanova Law School invited feedback from legal scholars and practitioners.

The keynote address to the Twentieth Anniversary celebration of the

University of California, Humanities Initiative, at the Medical Humanities Consortium of the University of California, San Francisco (April 2008), provided an occasion to present some of the main ideas to a medical audience. Brian Dolan, Paul Blanc, Tracey Woodruff, Max Neiman and Lauren Zeise all provided feedback and support for the project.

Jonathan Wolff's Ethics of Risk conference, University College London (September 2008), resulted in interesting discussions from him and his British colleagues that muted some normative paths I had been emphasizing. This event was quickly followed by John Deigh's invitation to the Philosophy and Law Colloquium at the University of Texas Law School. Wendy Wagner was a supportive but searching commentator while John Deigh, Jane Stapleton, Larry Sager and colloquium participants provided other input.

Feedback from Arlene Blum's Green Science Policy Symposium, The Fire Retardant Dilemma (May 2008), brought additional insight into some of the issues concerning the brominated fire retardants that are likely to plague our country and many others for decades. Arlene's dedicated efforts are resulting in some reduction of these persistent products around the world.

The developmental origins of disease raise problems of causation that I broached at the University of Aberdeen Law School conference on causation (June 2009). Ultimately the law likely will be forced to address some of these issues. Richard Goldberg and Joe Sanders were especially valuable and enjoyable interlocutors.

Service on California's Nanotechnology Science Advisory Panel reinforced the idea that the legal problems concerning industrial chemicals are not restricted to those products. Sourya Boudia's and Nathalie Jas's invitation for a keynote address to a conference on carcinogens, mutagens and reproductive toxicants at the University of Strasbourg in March 2010 introduced me to a different group of researchers. Yet many of the topics expressed similar concerns about toxicants in our midst. Discussions with Sourya, Nathalie, Nancy Langston, Barbara Allen, Jody Roberts and Sheldon Krimsky enriched the project.

Beyond specific contributions from participants at particular colloquia numerous others enhanced the book. My colleague Coleen McNamara read early drafts of a law review article that became the seeds of the book. A graduate assistant, Megs Gendreau, provided both editorial and footnote assistance early and late in the life of this book project. My recently deceased colleague Paul Hoffman, despite having quite different philosophical expertise, took a substantial interest in the topic of the book and offered trenchant comments that I incorporated into the book. I miss him and his acute mind.

A discussion of the central legal chapter of the book at the Southern California Law and Philosophy Seminar with Steve Munzer, Sharon Lloyd, Chris Naticchia, Marshall Cohen, Craig Ihara, Zac Cogley, Aaron James, Paul Hoffman, and John Davis was particularly valuable in sharpening some of its main ideas. Mary Lyndon, Tom McGarity, Wendy Wagner, and John Applegate also provided valuable legal insights and ongoing support for the project.

Tracey Woodruff and Melanie Marty read drafts of the health effects chapter. Lauren Zeise of the California Environmental Protection Agency invariably provides insights into public health policy and toxicology. And, I could always call on my science colleagues David Eastmond, Prue Talbot, and Maggie Curras-Collazo to clarify toxicological and developmental issues. No doubt I have forgotten some from whom I have learned, but I hope they will forgive any inadvertent omissions.

Elizabeth Knoll of Harvard University Press was a most supportive editor from the beginning, who embraced the project, and then read early and later drafts, leading to improvements every time. Her advice was unerringly good. Tonnya Norwood and Martha Carlson-Bradley carefully copy edited the book, producing substantial improvements in its style and consistency. I appreciate early publicity efforts on behalf of the book from Rose Ann Miller of Harvard Press and look forward to their continuation.

Finally, I greatly appreciate the patience and love of my wife Crystal, and my children, Chris and Taylor, who might not have had as much recreation time with me as they should have during the duration of this project simply because it was so compelling. Now we are off on a vacation!